W9-CRY-876

New Trends in Bio-inorganic Chemistry

Edited by

R.J.P. WILLIAMS
Inorganic Chemistry Laboratory
University of Oxford
England

J.R.R.F. DA SILVA
Centro de Estudos de Química Estrutural
Instituto Superior Tecnico
Portugal

1978

ACADEMIC PRESS

London · New York · San Francisco

A Subsidiary of Harcourt Brace Jovanovich, Publishers

6308-3966

CHEMISTRY

ACADEMIC PRESS INC. (LONDON) LTD.
24/28 Oval Road,
London NW1

United States Edition published by
ACADEMIC PRESS INC.
111 Fifth Avenue
New York, New York 10003

Copyright © 1978 by
ACADEMIC PRESS INC. (LONDON) LTD.

All Rights Reserved

No part of this book may be reproduced in any form by photostat, microfilm, or any other
means, without written permission from the publishers

Library of Congress Catalog Card Number: 78–18034
ISBN: 0–12–755050–X

Printed in Great Britain by
Whitstable Litho Ltd., Whitstable, Kent.

QP531
N47
CHEM

CONTRIBUTORS

HILL, H.A.O. *Inorganic Chemistry Laboratory, South Parks Road, Oxford*

VALLEE, B.L. *Biophysics Laboratory, Peter Bent Brigham Hospital, Hungtington Ave., Boston, Mass. 02115, U.S.A.*

BIRCHALL, J.D. *Imperial Chemical Industries Ltd., P.O. Box No. 8, The Heath, Runcorn, Cheshire.*

MALMSTROM, B.G. *Dept. of Biochemistry, University of Göteborg and Chalmers Institute of Technology, Fack S-402 20 Göteborg 5, Sweden.*

BIRCH, N.J. *Dept. of Biochemistry, University of Leeds, 9 Hyde Terrace, Leeds LS2 9LS.*

GILLARD, R.D. *Dept. of Chemistry, University College, P.O. Box 78, Cardiff CF1 1XL, Wales.*

LECHENE, C. *Biotechnology Resource in Electron Probe Microanalysis, Harvard Medical School, 45 Shattuk Street, Boston, Mass. 02115, U.S.A.*

PAUTARD, F.G.E. *MRC Mineral Metabolism Unit, The General Infirmary, Great George Street, Leeds LS1 3EX.*

WILLIAMS, R.J.P. *Inorganic Chemistry Laboratory, South Parks Road, Oxford, England.*

DA SILVA, J.J.R.F. *Centro de Quimica Estrutural da Universidade de Lisboa, I.S.T., Lisboa 1, Portugal.*

MOURA, J.J.G. *Centro de Quimica Estrutural da Universidade de Lisboa, I.S.T., Lisboa 1, Portugal.*

XAVIER, A.V. *Centro de Quimica Estrutural da Universidade de Lisboa, I.S.T., Lisboa 1, Portugal.*

CONTENTS

List of contributors   v
Preface   viii

1.   INTRODUCTION   1
     R.J.P. Williams

2.   ZINC BIOCHEMISTRY AND PHYSIOLOGY AND
     THEIR DERANGEMENTS   11
     B.L. Vallee

3.   SOME ASPECTS OF STRUCTURE AND FUNCTION
     IN COPPER CONTAINING OXIDASES   59
     Bo. G. Malmström

4.   MOLYBDENUM IN PROTEINS   79
     J.J.G. Moura and A.V. Xavier

5.   HIGH REDOX POTENTIAL CHEMICALS IN
     BIOLOGICAL SYSTEMS   121
     R.J.P. Williams and J.J.R.F. Da Silva

6.   THE SUPEROXIDE ION AND THE TOXICITY OF
     MOLECULAR OXYGEN   173
     H.A.O. Hill

7.   SILICON IN THE BIOSPHERE   209
     J.D. Birchall

8.   A SHORT NOTE ON SELENIUM BIOCHEMISTRY   253
     R.J.P. Williams

9.   PHOSPHORUS AND BONE   261
     F.G.E. Pautard

10. HEAVY METALS IN MEDICINE   355
     R.D. Gillard

11. LITHIUM IN MEDICINE   389
     N.J. Birch

12. ELEMENT DETERMINATION IN BIOLOGICAL MATERIALS
     USING ELECTRON PROBE MICROANALYSIS   437
     C. Lechene

13. INTERACTION OF THE CHEMICAL ELEMENTS WITH
     BIOLOGICAL SYSTEMS   449
     J.J.R.F. Da Silva

Subject Index   485

PREFACE

This volume is part of a series 'Frontiers of Knowledge'
written to mark the bicentenary celebration of the foundation in 1779 of the Academia Das Ciencias de Lisboa. The
authors wish to express their hopes for the continuing
health and well-being of the Academy. They are sure that
they are joined in this wish by all scientists from all
over the world.

PUBLISHER'S NOTE

This work has been put together as part of
the 200th anniversary celebrations of the Portugese
Academy of Sciences. A special limited edition is to
be published by the Academy in their series 'Frontiers
of Knowledge'. We are grateful to the Academy for their
agreement to the publication of this general edition
and congratulate them on their anniversary.

1 INTRODUCTION

R.J.P. Williams

Inorganic Chemistry Laboratory, Oxford University, South Parks Road, Oxford, OX1 3QR, England

The disciplines of inorganic chemistry have now been applied in some detail to biological systems for about 30 years. Such has been the success of these studies that some have felt it necessary to call this area of study by a name, Bio-inorganic Chemistry. The advantages of such a title can be lost if the extent of the role of inorganic elements in biology is hidden, by those who stray from inorganic chemistry into the study of biological systems, through an adherence to the whims and fashions of inorganic chemistry itself especially as it has been studied in the last thirty years. Biology embraces a totality of chemistry restricted only by the chemicals, temperatures, and pressures etc. available to it. It is a major aim of this book therefore to show the width of the subject to the bio-inorganic chemist. One way in which this is done here is to point to the work of those early leaders in the field, who came from biological sciences, and were unprejudiced by inorganic thinking and who examined a few particular areas of bio-inorganic chemistry. We shall then turn to areas which should have been included in all texts on bio-inorganic chemistry but in fact are sadly neglected. Today the well-populated areas of the subject are mainly the three topics of metal ion solution chemistry, i.e. of metals such as Sodium,

Potassium, Magnesium and Calcium; of transition metals
plus Zinc; and of poisonous metals. The major missing
areas are the discussion of the roles of the non-metals,
of the solid state, and of analysis.

  In two early chapters Vallee and Malmström, two of the
first biochemists in this field, describe their approaches to the examination of the way in which zinc and
copper respectively act in biology. We asked Prof. Vallee
specifically to show how he had gone about the analysis
for zinc and to look introspectively at the key leads in
his work. The reason for this request was that zinc is
not an easy trace element to study since like silicon,
phosphorus and selenium and unlike iron or copper it does
not reveal its presence by any simple spectroscopic property. Thus the unravelling of zinc biochemistry seems to
us to show how to unravel the chemistry of those elements
which have no reporter property, both metals and non-
metals, starting from the long-forgotten art of analysis
— truly a lost part of inorganic as well as bio-inorganic
chemistry. Prof. Vallee goes on to show the way in which
zinc is involved in the well-being of the whole animal
thereby stressing a link with medicine which is another
theme of the book. Prof. Malmström's role in the examination of copper biochemistry is quite different for detection and even analysis of the element in biology is not
so much the problem as an understanding of the variety of
types of copper, their valences and their biological
bondings. This is the second stage in the attempted understanding of the role of an element in biological systems
at the molecular level. The third stage is to discover
how binding to biological macromolecules modulates the
behaviour of an element giving it functional value.
Malmström discusses this approach too. A similar study to
that of copper is clearly required in work on molybdenum
— which relatively is only beginning, see the chapter by
Xavier and Moura. In both cases the elements are involved

in multi-electron transfer processes which are hard to understand. A greatly expanded parallel effort will be required in work on selenium and perhaps even on the halogens to bring the non-metal redox chemistry to the level at which the redox active metals have been described. The fourth stage in the examination of the role of an element returns to the appreciation of total function. As an example Da Silva and Williams attempt an appraisal of the value of the high redox states (mainly of metals) generated in biology by the evolution of an oxygen atmosphere, and Hill gives a parallel account of a special non-metal compound, superoxide.

A quite different biochemistry of metals is in the area of pollution and medicine. Elements in themselves are not good or evil and we recognise a general concentration dependence of the action of all the elements. This is described in more detail by Da Silva. There is a dose/effect response curve, Fig. 1, different for each element and

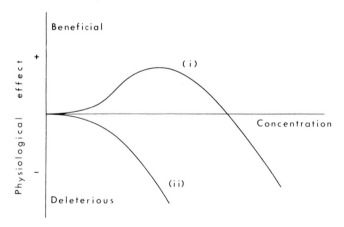

Fig. 1. The physiological effect of an element can be beneficial but it is only so over a restricted concentration range. A few elements have no known functional value, curve (ii).

even in its biological impact on each individual in a population. We see in the chapter on lithium drugs (Birch) and in that on the development of heavy metals in drugs,

Pt, Au, etc. (Gillard), the revival of the lost art of
inorganic medicine which had its beginnings many thousands
of years ago.

Heavy metals can also be used in the probe of biological samples, for example in X-ray crystallography (Pt,
Hg) and in electron microscopy (Os, Pb). These are not
areas where the inorganic chemist has been particularly
helpful to date though surely he alone can put his finger
on the correct reagents and conditions.

A further part of the book describes some *non-metal*
biochemistry of increasing biological interest. While
oxygen and the halogens are covered in some measure in
chapters on metals, the roles of silicon (described by
Birchall) and of selenium (summarised by Williams) are
much more specialised. In the future we look forward to
parallel accounts of the inorganic biochemistry of elements such as iodine, bromine, fluorine and boron.

While the underlying molecular principles take pride
of place throughout the book it must be a part of our
effort to appreciate the distribution in biological space
of all the elements *in vivo*. Microprobe analysis using
the electron microscope is one procedure and another will
be the use of nuclear magnetic resonance especially in an
increasing variety of field gradients. We include a chapter by Lechene on the use of the electron microscope but
refer only to a symposium on the potentials of NMR *in
vivo* as this field is only at its beginning (Dwek *et al.*,
1977).

Much as the solid state has been rather neglected of
late in inorganic chemistry so has it been neglected in
biology. The problem lies with methods. How does one
study the amorphous solid state? We hope that an approach
is clearly shown in the chapter on bone written by
Pautard, a biologist with a deep appreciation of this
problem. This chapter also looks at some phosphorus biochemistry.

There are deliberate omissions in this volume. The
fields of sodium, potassium, magnesium, calcium, iron and
cobalt biochemistry are so well advanced that many separate volumes are available describing their bio-inorganic
chemistries. By way of contrast the roles of elements
such as manganese, chromium, nickel, vanadium, aluminium
and tin are so poorly developed that we deemed it better
to use copper and zinc as illustrations of how these subjects should be developed than to give rather vague sur-
veys — all that is possible at present. We have left out
of the book also the use of substitutional inorganic
chemistry, inorganic chemical modifications backed by
physical methods, since there is a vast literature on the
use of the elements in this way, Table I.

Introduction to the Major Elements in Biology

The important metal ions of biological systems can be
divided into groups which follow the Periodic Table, Table
II. This Table is a table of chemical properties and it
is not surprising that the chemical classification becomes in biology a functional division. The functions are
tightly defined in biology of course for evolution has
had to operate under prescribed abundance/availability
conditions. The first divisions in Table II are the metals
which occur largely as free ions, Na^+ and K^+, and which
act to provide an osmotic balance and an electrolyte current. Like the second group, Mg^{2+} and Ca^{2+}, they may have
little strictly catalytic significance in enzyme action
although they can assist or prevent binding. The second
group of metal ions, Mg^{2+} and Ca^{2+}, are regulators of
binding since they can be pulsed in and out of biological chambers. The ions then act through binding re-
arrangements and they can therefore control the protein
triggers of biological activity. Well-known examples are
found in hormone-release, muscle contraction etc. They
also neutralise charge especially of phosphate, e.g. in

TABLE I

Examples of the use of extrinsic probes in NMR studies

	Probe	System
Binding of paramagnetic cations	Lanthanides	lysozyme
	Transition metals	troponin-C kinases lysozyme
Binding of paramagnetic anions	$[Cr(CN)_6]^{3-}$	lysozyme cytochrome *c* phospholipase A2
	$[Fe(CN)_6]^{3-}$	cytochrome *c* phospholipase A2
	$[Ln(ATP)]^{-}$	phosphoglycerate kinase
Isomorphous replacement of metals in metalloenzymes	$Fe^2 \rightarrow Co^{2+}$	cytochrome *c*
	$Fe^2 \rightarrow Fe^{3+}$	cytochrome *c* haemoglobin
	$Cu^{2+} \rightarrow Cu^{+}$	azurin plastocyanin
	$Cu^{2+} \rightarrow Hg^{2+}$	superoxide dismutase
	$Ca^{2+} \rightarrow Cd^{2+}$	troponin-C
Whole Organs	Mn^{2+}	adrenal medulla

See R.J.P. Williams, 'Chemistry in Britain' (1977).

TABLE II

The Division of the Metals in Biology

	Group IA	Group IIA	Group IIB	Transition Metals
Examples	Na^{+}, K^{+}	Mg^{2+}, Ca^{2+}	$Zn^{2+}(Ni^{2+})$	Mn, Fe, Co, Cu, Mo
Exchange Rates	Fast	Medium	Do not exchange	Do not exchange
Function	Osmotic Control Electrolytes Ion Currents	Triggers and Conformational Control (Structures)	Acid Catalysts	Redox Catalysts

the binding of ATP, and form insoluble precipitates e.g.
bone. The third group of metals is usually represented
by the element zinc (but should now include nickel) which
is a firmly bound metal with two functions. It assists
acid/base catalysis and it can cross-link proteins. It
is the first trace element. In the fourth group are the
transition elements, largely Mn, Fe, Co, Cu, Mo, which
are firmly (almost covalently) bound to fixed sites and
which act in the vast majority of cases by change of
oxidation state. The last group of metals, not in the
Table, are those which do not occur *in vivo* but can be
introduced deliberately as poisons or for medicinal purposes, e.g. Cd, Hg, Au, Pt, Li, Bi. Their bio-chemistry
is little understood.

This is the background against which we shall look at
the occurrence and properties of the non-metal elements
in biology. Table III is a very simplified over-view.
There is a striking general parallel in the division of
functional responsibilities with those illustrated for
the metals but there is additional diversity within the
pattern — as is well known in the non-metal chemistry.
The first group of non-metals is the halogens which are
to be compared with the Group IA metals. The over-riding
functional role of the only really abundant halogen,
chlorine, is in osmotic control and electrolytic currents
where it acts as the chloride ion. The less common anion,
fluoride, forms insoluble salts with calcium and is incorporated into bone. The other halogens are trace elements, bromine and iodine, and find use in covalent
attachment to organic molecules e.g. in the thyroid hormones. Thus they are not used in oxidation state -1 but
in the oxidation state zero. In fact there is a very different chemistry of all these elements in their higher
oxidation states where they are vigorously reactive e.g.
Cl^+, ClO^-, I^+. The use to which biology puts these states
is extremely interesting, see chapter V, for they become

TABLE III

The Division of the Non-Metals in Biology

	Atomic Anions	Stable Oxy-Anions	Unstable States	Metabolites and Polymer Formation	Redox Catalysts
Examples	Cl^-	$PO_4{}^{3-}$, $(SiO_3{}^{2-})_n$	ClO^-, I^+	Reduced States of C,N,O and S	Se
Exchange Rages	Fast	Medium	?	Slow	No exchange
Function	Osmotic Control Electrolytes	Triggers and Conformational Control (Structures)	Protective Devices	Basic units of Biochemistry	Redox Catalyst

protective agents and biological killers — e.g. hypochlor-
ites, very much as man uses bleach, iodine and chlorate.

Passing to a second group of non-metals and seeking
for a parallel with the biological chemistry of magnesium
and calcium we come to the elements P and Si in the form
of their highest oxidation states, phosphates and silicates which are virtually the only forms in which they
occur in biology. A major role of these elements is as
structure formers and regulators in both solid structures
and in association with nucleotides, proteins, and polysaccharides. They then act as regulators of bio-polymer
structure and the functional correspondence between the
doubly charged M^{2+} (cations) and these XO_4^{n-} (anions) is
close. One chapter in this volume is devoted to the cooperative interactions of calcium and phosphate in bone
formation, another is devoted to the role of silicon.
Both these subjects belong in part to the *solid state*
function of inorganic elements in biology. Of course
phosphate has many other roles in biology associated with
condensation reactions as in DNA and RNA but these are
described *in extenso* in conventional volumes of biochemistry — usually bio-organic chemistry.

The non-metals which are involved as the attacking
groups of the catalysts of biology, enzymes, are in the
group of elements, C, N, O, S, Se. We have not thought it
to be necessary to re-describe in this volume the roles
of the conventional bulk non-metals of biology, C, N and
O. Sulphur is not quite so common an element but its
biological chemistry was for a long time usually included
within organic biochemistry. Its functions are extensively
covered elsewhere. The chemistry of selenium is then quite
distinct. It is a trace element, it is retained by proteins (no exchange); it has variable valence. It can then
be compared with the transition metal ions in its function, Table III. This function is analysed in chapter
VIII. Selenium like iron is much associated with protec-

tive devices.

Within this book there is also an area where metals and non-metals overlap since they must occur together in solid salts. The understanding of material such as bone and teeth has attracted few inorganic chemists to date. We hope that now there has been a return of interest to solid-state inorganic chemistry there will be a start by inorganic chemists on the road to an understanding of these remarkable materials made largely from calcium salts. The parallels with shells and silicate deposits in biology (Chapter VII) are left for the reader to explore.

As is true for the metals so it is the case for the non-metals that there are elements which biology uses but slightly if at all. Here the possibility arises of new medicine based on the chemistry of such elements as Ge, As, Sb and Te for example. (Of course we do not imply that the elements found in biology can not be used in medicines).

We hope that this book will widen the coverage of bio-inorganic chemistry and will act as a stimulant for those who are just entering the field to look at areas but poorly covered by the present generation of inorganic chemists. We hope too that the inorganic chemist entering this field will take with him the approaches and techniques shown to be so successful by biologists. The understanding of the roles of inorganic elements in biology demands an appreciation of *functional* significance.

Reference

Dwek, R., Campbell, I.D., Richards, R.E. and Williams, R.J.P. (Eds.) (1977). 'NMR in Biology', Academic Press, London.

2 ZINC BIOCHEMISTRY AND PHYSIOLOGY AND THEIR DERANGEMENTS

Bert L. Vallee

Biophysics Research Laboratory, Department of Biological Chemistry, Harvard Medical School and Division of Medical Biology, Peter Bent Brigham Hospital, Boston, Massachusetts

Knowledge regarding the biological occurrence and function of zinc, the development of methods that permit its detection and those aspects of biochemistry which have allowed an appreciation of the manner in which it participates in generating biological specificity have grown exponentially in the last few years. The chemical basis of enzyme specificity, the properties of metalloenzymes that have served to channel this information and the interaction of the metal in cellular growth and development, as that relates to the development of the whole organism, are of particular interest.

In addition, a vast amount of information now exists on the physiological, nutritional and medical roles of zinc, but much of it is not readily accessible to interested biochemists who might also find it profitable. Such practical consequences to human welfare as are now emerging are not less important than the basic knowledge now accumulated and awareness of the interrelationships should materially accelerate progress. Hence, these features of the field will receive more attention than is conventional in biochemically oriented treatments of the subject. This synopsis will indicate those avenues which have proven profitable to the understanding of the role of zinc, regardless of the discipline from which they have emerged.

The importance of zinc in biochemical and physiological processes is now well established at all levels of
cellular complexity, and its role in bacteria, fungi,
higher plants, animals and man have been studied inten-
sively. Zinc had long been referred to as a trace element.
In the past, this collective designation has generally
implied that the concentration in which an element occurs
is vanishingly low and, hence, its quantitative detection
is either not possible or presents major problems. Beyond
this, however, it has also been perceived to denote great
uncertainty — or even negativity — regarding any biological significance of its occurrence. Indeed, zinc was
thought to be present in biological matter in very small
amounts, an assumption which has proven incorrect. In the
past, its occurrence at best provided fertile conjectures
and inductive hypotheses regarding its function. At one
time or another its deficiency, imbalance or intoxication
was invoked as a cause for multitudes of human disease
processes of unknown cause. As a consequence, such hypotheses, involving juxtaposition and correlation of *two*
unknowns, have mostly generated confusion. Since it has
been difficult to *disprove* such conjectures, they have
often been accepted widely, albeit without adequate documentation. It is appreciated easily that ultimate dis-
illusionment on finding a given set of claims invalid
readily gives a dubious connotation to a subject — rather
than to the hypothesis. As in the present case the result
is retardation of progress by default. Such 'guess work'
has been prevalent partly because valid information has
proven so hard to obtain. No doubt, the analytical chemistry of zinc and the incapacity to perform precise quantitative measurements of the metal were among the most
troublesome blocks to definitive assessment of its biological role.

Analytical Chemistry of Zinc

As the problems regarding the analytical determination of zinc in biological matter have been solved, the element has been found to be present in much larger amounts than had been assumed. Simultaneously, knowledge of its very critical role in complex biological systems and processes has grown exponentially. Of primary interest in this regard are the mechanisms whereby zinc serves catalytic, structural, conformational and regulatory functions in proteins or other macromolecules which are of principal biological consequence. The essentiality of zinc to human nutrition and its possible roles in disease are not less important and portend its possible determination as a diagnostic index of illness. In fact, appreciation and understanding of such correlations have increased almost exactly in parallel with the ability to measure accurately very low concentrations of zinc in readily accessible body fluids, particularly those of man.

Analytical Methods

Pertinent analytical techniques are now well established; yet others are still in a period of very rapid development, both in regard to methodology and application; understandably, they are still suffering growing pains. Hence, a summary of the present state would seem indicated and should prove helpful to scientists contemplating studies in this area. Biochemists, nutritionists and physicians are continually offered a bewildering array of instrumentation with ever-increasing analytical capabilities, by and large developed by analytical chemists whose primary work is performed with relatively simple inorganic matrices. Users in biology can seldom be fully aware of the limitations of such equipment when employed for biological samples which are generally of rather complex composition and nature. Hence, there is

little mutual awareness of the requisite knowledge and
appreciation of the problems inherent to the specialities
of the analyst and instrument manufacturer on the one hand
and the user on the other.

In the following brief summary, spectrophotometric
and fluorometric, X-ray and other nuclear methods, electrochemical, mass-spectrometric approaches and electron
probes are not discussed, largely owing to their relatively specialized capacities for routine zinc analyses.
However, this omission should not be taken to be a judge-
ment regarding their suitability either for specialized
purposes or, indeed, for other metals (see for instance
Morrison, 1965; Winefordner, 1976).

While microchemical methods led the analytical revolu-
tion on zinc determination some 30 years ago, they have
proven too cumbersome for broadscale studies. The methods
most widely employed now for examination of zinc in biological matter are based on atomic spectroscopy, i.e.,
the interaction of analyte atoms with electromagnetic
radiations. Broadly, these procedures fall in three categories: atomic emission, atomic absorption (Vallee, 1969)
and atomic fluorescent spectroscopy (Veillon, 1976). Over
the last two decades all of these have been developed into
highly sensitive, specific and rapid means of chemical
analysis and also are applicable to elements other than
zinc, of course. Instrumentally, all three techniques and
their variants share many similar features, but each has
unique advantages, dependent in most cases on the application, type, number and amounts of samples, the sensitivity
of detection required and information desired. These instrumental categories, their virtues and limitations,
limited to those arrangements — whether developed com-
mercially or not — that have been shown to be applicable
to analysis of biological materials, especially enzymes,
have been evaluated (Veillon and Vallee, 1978). Only high-
lights will be indicated here and those approaches which

may be potentially useful but are not as yet reduced to practice will be neglected.

All instruments now employed incorporate a source, an element for the dispersion of radiation and a detector. However, the choice of each of these components, but particularly their combinations, are critical.

Atomic Emission and Absorption Spectrometry

Instruments which measure the emission of electromagnetic radiation fall into several classes, based largely on the type of source used to atomize the analyte atoms to produce the desired emission — or to be absorbed by the sample, as in atomic absorption — and the receiver employed to quantitate the radiation emitted or absorbed, respectively. Instruments are designed either to perform single element or multi-element analyses. Instrumental arrangements for multi-element determinations allow *simultaneous* measurements of different elements at numerous wavelengths on a single sample. Alternately, multi-element determinations can be performed by scanning a single sample successively and exceedingly rapidly. We will not itemize the multi-element nature of available instruments, since our interest centres exclusively on the determination of zinc for which atomic absorption employing a long path length absorption cell has proven exceptionally effective (Fuwa and Vallee, 1963; Fuwa *et al.*, 1964). It should be emphasized, however, that the capacity of an instrument to perform simultaneous analyses of yet other elements can generate a considerable advantage in experiments which emphasize imbalances, intoxications, or interelement effects of zinc with other elements, as is often the case.

Various atomic excitation sources are employed to bring about atomization in both kinds of instruments. These include chemical flames, high current arcs, sparks, hollow cathode lamps, laser sources, electrodeless — plasma —

discharges produced by constricted arcs, radiation fields,
or microwave fields ('open plasma' being simply a highly
stylized, electrically conducting gas).

The recognition that zinc — and other metals are components of enzymes and other biochemical molecules or
biological substances has been aided by the resultant
emission spectrography or spectrometry which utilize the
fact that elements such as zinc, when excited by sufficient energy, emit radiation of a characteristic spectral distribution. Although excitation of zinc generates
many hundreds of spectral lines, generally the identification of 3 prominent ones is accepted to assure its
presence with virtual certainty. The failure to detect
its most sensitive line confirms the absence of the element. In this manner, the presence of zinc and its quantitative occurrence can be established easily, since the
quantity of light emitted is proportionate to the number
of atoms vaporized and excited. Limitation of space precludes a discussion of the mechanisms of excitation and
the relative advantages and disadvantages of various
sources. The interested reader is referred to some simple
and concise treatments of this and related subjects which,
further, offer references for more exhaustive study
(Schenk, 1975; Veillon and Vallee, 1978). Suffice it to
say here that flames, lasers, arcs or sparks, plasmas and
hollow cathodes excite atoms to higher electronic energy
states followed by emission of radiation on return to lower
energy states, or — when ionization takes place — by complete ejection of electrons from the atom. The emission of
radiation at very specific wavelengths reflects the various
transitions from higher to lower states and is diagnostic
of the element (see above).

Which electronic transitions will occur depends upon
electronic, magnetic and thermal influences to which a particular atom is subjected. Different forms of excitation
will, of course, result in the population and repopulation

of specific energy levels, producing varying effects. Hot
flames and sparks are the sources of excitation in most common use. Spectrographs (recording photographically) and
spectrometers (recording photoelectrically) differ most in
regard to their sources since gratings are the dispersion
devices employed most frequently. In addition to gratings,
echelles or videcon tubes, are used in dispersing elements.
Prisms — once the principal means of diffraction — are now
used infrequently, and when they are employed it is usually
in combination with one of the above.

The receivers encompass photographic plates (in spectrographs) now used less frequently due to the technical
inconvenience of processing them and quantitating line
intensities to correlate these with concentrations to be
determined. Principal among such problems is the tedium of
suitable calibration both as a function of wavelength and
intensity. However, the great advantage of photographic
plates remains the simultaneous detection and quantitation
of many different elements and their characteristic lines
while they constitute a permanent record. However, this
form of detection is quite insensitive as compared with
photoelectric devices now available.

Among the spectral approaches Emission Spectrography
and Atomic Absorption Spectrometry have been particularly
helpful in the determination of zinc both in the past and
at present (Fuwa and Vallee, 1963; Fuwa *et al.*, 1964;
Vallee, 1969). They are now established as 'standard'
procedures. The reader desirous of acquiring further details on the technical matters pertaining to these techniques will find abundant information and is referred to
Veillon and Vallee, 1978; Slavin, 1968, 1971; Schenk, 1975
(above); Morrison, 1965 (above); Christian and Feldman,
1970; Pinta, 1970; and Winefordner, 1976.

Microwave Excitation Emission Spectroscopy has recently
come into its own and, therefore, will be given particular
emphasis here since it has played an important role in

work to be discussed later. It will no doubt continue to
acquire increasing value for biological work owing to its
low limits of detection while demanding but miniscule
amounts of sample whose size and availability often limit
biological investigations.

Microwave Plasmas

The low-pressure, microwave-induced plasma, an emission
source developed recently, has been shown to be one of the
most sensitive yet designed for the determination of trace
elements in aqueous solutions (Kawaguchi and Vallee,
1975). Basically, the system operates as follows: a helium
flow at low pressure is contained in a quartz discharge
tube. The tube passes through a microwave resonant cavity
into which is fed about 50 W of microwave power at a frequency of 2450 MHz. A plasma discharge is initiated in the
tube and is sustained by the microwave field. Above the
plasma, in the helium carrier gas stream, is a small
tantalum filament onto which microvolume samples are
placed and dried. The filament is then pulse-heated by
capacitive discharge, vaporizing the sample which is then
swept into the plasma. The resultant atomic emission is
then measured using a spectrometer. Sample volumes of only
5 µl are used, detection limits (on an absolute basis) are
in the 1-50 *picogram* range, and little or no sample preparation is required.

A block diagram of the instrumental arrangement is
shown in Fig. 1. The instrument is very compact, elec-
tronically sophisticated, and highly automated in its
operation.

This analytical system was designed primarily for the
determination of *picogram* quantities of metals in metalloenzyme preparations. To verify its utility for this purpose, Kawaguchi and Vallee (1975) determined zinc stoichiometry in highly-purified, well-characterized preparations of several zinc metalloenzymes, and compared these

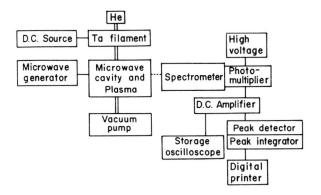

Fig. 1. Block diagram of microwave excitation spectrometer.

data to those obtained for the same material by atomic absorption spectrometry. The enzymes used as standard reference materials were bovine carboxypeptidase A, human carbonic anhydrase, horse liver alcohol dehydrogenase, and *E. coli* alkaline phosphatase. In addition to complete agreement between the two independent methods, it is interesting to note that the atomic absorption method required about 100 µg of enzyme per determination, while the microwave-induced emission measurements required only about 0.1 µg of enzyme. Subsequently, they were able to demonstrate that reverse transcriptase from avian myeloblastosis virus — available only in miniscule amounts — contained stoichiometric amounts of zinc, while Cu, Fe and Mn are absent (Auld *et al.*, 1974a,b). Following this, the reverse transcriptases from mammalian tumour viruses (murine and simian leukemic viruses) have been shown similarly to contain stoichiometric amounts of zinc, employing this instrumental system (Auld *et al.*, 1975) as have been RNA polymerases I and II from *E. gracilis*, the RNA polymerase I from yeast, and methionyl-tRNA transferase. The main features of the method we summarized in Table I and compared with those of atomic spectrometry.

Following these initial studies, the system has been utilized further in investigations of mercury-substituted

TABLE I

*Comparison of the Capabilities of Atomic Absorption
and Microwave Emission Spectrometry*

	Atomic Absorption	Microwave Emission
Detection Limit (g-atom)	10^{-10}	10^{-15}
Protein (µg)	100	0.1
Volume of Sample (ml)	2	0.005
Effective Range	10^{1}	10^{5}

The experimental data pertinent to the characteristics of micro-wave emission spectrometry are from Kawaguchi and Vallee, 1975.

(Atsuya *et al.*, 1977a) and arsenic-labelled metallo-enzymes (Atsuya *et al.*, 1977b).

This emission source is unique in that it operates at a low pressure in an atomic gas having rather unusual metastable energy levels. Efforts have been made to under-stand and explain the abnormal excitation observed in this plasma (Kawaguchi *et al.*, 1977; Atsuya *et al.*, 1977c). Extraordinary enhancement effects have been observed for several elements when samples are placed in a KCl matrix (Kawaguchi and Vallee, 1975). This has also been studied by Busch and Vickers (1973) in a similar source, and re-volves mainly around the incongruity of a relatively low gas (i.e., kinetic energy) 'temperature' and an extremely high effective excitation 'temperature' in the plasma dis-charge. The KCl matrix alters this, and apparently serves as a spectroscopic and matrix buffer. Table II compares the detection limits, and, hence, relative sensitivities of porous cup spark emission and atomic absorption spectro-metry with those of microwave excitation emission spectro-metry for a number of elements showing their relative capabilities.

TABLE II

Spectroscopic Detection Limits* for Elements

Element	Spark	Flame At. Abs.	Microwave
Zn	2,000	5.2	0.30
Cd	2,000	1	0.30
Hg	5,000	200	0.25
Pb	50	30	10
Mg		0.5	0.05
Mn	60	6	0.3
Cu	60	5	0.2
Fe		10	2
Co	2,000	10	7

*Detection Limits (ppb)

The Biological Role of Zinc

Clearly, these methods of analysis have markedly lowered the limits for detecting zinc— and other elements — with corresponding gains in understanding of their biological roles. However, when observing the quantitative zinc content of an organ, how can this knowledge be translated into function in an organism? The problem resembles that confronting an organic chemist trying to establish the structural and functional characteristics of an unknown compound contained in a mixture of others by means of elemental analysis. This is precisely the situation encountered by a biologically oriented spectroscopist or a biochemist attempting to solve problems concerning the function of, e.g., zinc, by analysis of tissues. In the biological material it may exist as a constituent of many different compounds or macromolecules serving quite different functions. Since many biological zinc-containing compounds and their functions have not been discerned as yet, the sum of their constituent zinc atoms will be measured and, hence, in the absence of other criteria, the biological interpretation of the analytical results is not only problematic, but virtually impossible.

The association of zinc with specific biological func-

tions is detected by isolating cellular or sub-cellular
components and individual molecules whose biological functions are known, measuring both activity and zinc.

The earliest indications that zinc plays a biological
role date to 1869 when Raulin (1869) found that *Asper-
gillus niger*, the common bread mould, does not grow in
its absence. Shortly thereafter, zinc was found in plants
and animal tissues, and studies regarding its importance
in plant growth began in the early part of the current
century. In the subsequent two decades there was very
little biological work that revealed any functional significance of zinc. Toxic effects of zinc, a result of
inhalation of its fumes during smelting of its ores, became of paramount interest to Public Health, however. Zinc
was recognized to be an industrial hazard which was eliminated by suitable ventilation of work spaces and removal
from processing plants. Thus, a toxic action of the element manifesting as 'Metal Fume Fever' was established
long before its essential biological functions were known.
Yet, suitable credit should be given to an exceptionally
alert team of investigators. While investigating the cause
and basis of Metal Fume Fever, Collier and Drinker (1926)
developed a chemical method — exceedingly imprecise and
insensitive by present standards — for zinc analysis in
organs. Nevertheless, they could recognize that zinc is
present in all organs of many species, and this presence
could not be accounted for by the then prevalent theories
on toxicity. They appreciated and pointed out that this
universal distribution implied important biological functions rather than incidental contamination, an epochal
insight considering the state of the art and knowledge at
that time. This past history of studies of zinc in biology and its implications are still timely. While it may
be relatively easy to define toxicological effects of a
metal, such observations do not allow the conclusion that
this same metal is not biologically essential also.

Most of the work leading to the present awareness that
zinc is biologically indispensable has been performed
largely during the last forty years. Neglecting the analytical problems already detailed, this realization was
the result of a number of approaches.

The nutritional evidence came first through the examination of its effects on animals using highly purified
and specially constituted diets to supply inadequate
amounts of zinc and through studies of a number of naturally occurring, endemic diseases of animals suspected to
be due to the deficiency of zinc.

Major progress in the isolation and characterization
of composition, structure and function of metalloenzymes
came next, albeit delayed by many years, but the cell
biological role of the metal remains that which was and
still is neglected most. Even the halting appreciation
of its nutritional importance is in large measure due to
the fact that the required deprivation has to be extreme.
It involved housing animals under conditions which very
strictly control the dietary zinc intake over long periods
to induce deficiency so that its clinical consequence
becomes apparent.

In this regard it is helpful to recall Gabriel Bertrand's realization, formulated in 1912, that the physiological and toxic effects of zinc and other metals should
be considered a continuum. Bertrand pointed out that the
physiological response to zinc would be expected to be
biphasic as shown in Fig. 2. Potentially, the biological
function of zinc and other metals can be assessed properly only against the background of a deficiency state.
Maximum functional response to zinc administration then
constitutes Phase I of this dose response curve. Essential functions may be hampered, however, if the relationship of zinc to all other metals present is not balanced
properly, resulting in imbalance, as indicated in the
middle of the figure. Concentration of zinc higher than

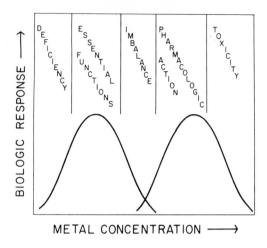

Fig. 2. Potential dose response to a metal.

those required to maintain essential functions may have
secondary or pharmacological effects, Phase II of the dose
response curve. At still higher concentrations zinc – like
every other element – is potentially toxic and – eventually – lethal, a pattern which pertains to virtually all
elements for which adequate information now exists. Current knowledge has established the position of zinc in
each portion of such a triphasic curve, and we are now
left with the resolution of specific mechanisms leading
to each of the responses.

Physiology and Metabolism

A 70 kg human has been estimated to contain between 2
and 3 gm of zinc, distributed in all tissues and varying
from 10 to 200 μg of zinc per gm wet weight. Most organs
contain 20 to 30 μg, while liver, voluntary muscle and
bone contain 60 to 180 μg of zinc per gm. The concentrations of zinc in the tissues of the eye, in particular of
the iris, retina and choroid, are somewhat higher. No
tissue is known to accumulate zinc preferentially, although its concentrations are remarkably high in the prostate (860 μg per gm in normal prostate), prostatic secre-

tions and spermatozoa. Neither carbonic anhydrase nor
alkaline phosphatase activities can account for this nor
is there any other functional explanation for the high
concentration of zinc in the male reproductive tract. In
most instances, differences in organ content of different
species are not remarkable. Normal human whole blood contains about 900 μg of zinc per 100 ml and serum zinc concentrations average 96 ± 20 μg/100 ml (Falchuk, 1977).
Normal erythrocytes contain 1.4 mg/100 ml of packed red
blood cells. Three per cent of all zinc in blood is found
in leukocytes, 3.2×10^{-2} μg zinc per million cells, about
25 times more than in a comparable number of erythrocytes.
There are no seasonal, diurnal, or sexual variations of
zinc concentrations in blood. In effect, all zinc in serum
is protein-bound and is distributed in multiple — at least
two and possibly more — fractions. Globulins bind zinc
most firmly; an α_2-zinc macroglobulin has been isolated
and accounts for 40–50% of all zinc in serum (Parisi and
Vallee, 1970). However, it does not seem to be a transport
protein. Most of the remainder seems to be bound loosely
to albumin or protein fractions electrophoretically
close to the albumins, although it is not clear whether or
not albumin itself binds zinc specifically or non-specifi-
cally. The existence of a specific transport protein remains the objective of persistent yet so far frustrated
research. Carbonic anhydrase seems to account for a large
percentage of zinc in erythrocytes and alkaline phosphatase for that in leukocytes. Serum zinc decreases during
pregnancy and with ACTH administration (Falchuk, 1977).

The distribution of ^{65}Zn injected into mice and dogs
indicates that the liver accumulates the largest fraction
of the total dose, and its turnover is most rapid in that
organ followed by the pancreas, kidney, and pituitary.
About 10 per cent of a dose of ^{65}Zn given is excreted in
the pancreatic juice within the first few days, while bone
and red blood cells accumulate it.

Zinc is widely distributed in a variety of foods, particularly from animal sources, and shellfish, in particular,
is rich in this element. The average human dietary intake
is about 10 to 15 mg per day of which about 5 mg are retained.

The sites and mechanisms of absorption of zinc from the
intestine are not known with certainty. Phytic acid, a
phosphorus storage compound of plant seeds which forms
insoluble zinc-phytate complexes, decreases the availability of zinc for absorption, which can be decreased
further by calcium through the formation of the more insoluble mixed zinc-calcium-phytate complex. Once absorbed,
zinc is excreted primarily in gastrointestinal and pancreatic secretions. Human urine contains about 0.5 mg per
day, an amount which is apparently independent of intake
and urinary volume. Studies in the rat indicate that, in
contrast to iron, the body stores of zinc are not readily
mobilized and, hence, there is an unusual dependence upon
a regular, exogenous supply of the element, particularly
during periods of growth.

Zinc deficient female rats produce progeny with multiple congenital malformations including clubbed feet,
adactyly, hydrocephalus, anecephaly, spina bifida and
cleft palate, among others. The decreased zinc content
of the mothers is an index of the percentage of teratological defects observed. (Hurley and Shrader, 1972). (See
also below.)

Chemical Features of Zinc Enzymes

Zinc, a IIB element with a completed d subshell and two
additional s electrons, chemically combines with the +2
oxidation states. There is no evidence that it is oxidized
or reduced in biological reactions. It generally forms
tetrahedral complex ions, but many octahedral complexes
are known.

The characteristics of zinc in e.g., simple halo-,

cyano-, and amino- complexes likely differ from those in metalloenzymes, the stereochemistry being determined largely by ligand size, electrostatic and covalent binding forces. The three dimensional structure of proteins, heterogeneity of ligands and the degree of vicinal polarity of the metal binding site may jointly generate atypical coordination properties (Vallee and Williams, 1968). Unusual bond lengths, distorted geometries, and/or an odd number of ligands can generate a metal binding site on the enzyme, which when occupied by a metal, can be kinetically more energetic than metal ions when free in solution where they are complexed to water or simple ligands. As a result, in enzymes zinc is thought to be poised for its intended catalytic function in the *entatic state* (Vallee and Williams, 1968). In this context the term entasis indicates the existence of a condition of tension or stress in a zinc — or other metal — enzyme prior to combining with substrate.

In effect, the entatic state is thought to originate in the genetic heritage of the cell. The primary structure of the enzyme protein dictates the relative spatial positions of those amino acid side chains destined to serve as ligands when the apoprotein combines with the metal ion. Evidence suggests that the metal ion is not incorporated into the growing, ribosome-bound polypeptide chain (Harris and Coleman, 1968), until the protein is fully formed. According to this view, the metal does not induce its own coordination site, but its interaction awaits the expression of the genetic message.

The lack of suitable physical-chemical probe properties of the diamagnetic zinc atom led to a search for means to replace it with paramagnetic metals, e.g., cobalt, which could signal information on the nature and environment of the active site (Vallee, 1973). The spectra of such cobalt substituted zinc enzymes (Lindskog and Nyman, 1964; Simpson and Vallee, 1968; Vallee and Latt,

1970; Holmquist *et al.*, 1975) differ significantly from those of model Co^{2+} complexes and are thought to reflect the entatic state of the cobalt (and zinc) ions in these enzymes (Vallee and Williams, 1968).

Identification of the metal binding ligands of metallo-enzymes has been difficult by means other than x-ray crystallography. Cysteinyl, histidyl, tyrosyl residues and the carboxyl groups of aspartic and glutamic acids have been implicated most commonly (Vallee and Wacker, 1970). Thus far, metal complex ions in which the mode of metal coordination is known quite precisely have not proven adequate to define metal binding in enzymes, invariably lacking the entatic environment seemingly characteristic of metalloenzymes (Vallee and Williams, 1968). Moreover, the complexes which are known most thoroughly and have been studied most extensively are bidentate, but present evidence indicates that in metallo-enzymes, metals are most likely coordinated to at least three ligands (Vallee and Wacker, 1970). With few excep-tions (Herskovitz *et al.*, 1972) multidentate complex ions, suitable for appropriate comparisons with metallo-enzymes, have not been studied. Yet, absorption (Latt and Vallee, 1971), magnetic circular dichroic (Kaden *et al.*, 1974) and electron paramagnetic resonance spectra (Kennedy *et al.*, 1972) combined with kinetic studies (Vallee and Latt, 1970) of cobalt-substituted and chemi-cally modified zinc metalloenzymes have enlarged under-standing both of the possible modes of interaction of zinc with the active sites of zinc metalloenzymes and their potential mechanisms of action.

The molecular details of metalloenzyme action have been elucidated greatly in the last few years (Riordan and Vallee, 1974). Crystal structures for bovine carboxy-peptidase A (Quiocho and Lipscomb, 1971), thermolysin (Colman *et al.*, 1972) and horse liver alcohol dehydro-genase (Branden *et al.*, 1975) are now available, and

chemical and kinetic studies have defined the role of
zinc in substrate binding and catalysis (Vallee and
Wacker, 1970; Riordan and Vallee, 1974). In fact, many of
the significant features elucidating the mode of action
of enzymes, in general, have been defined at the hands
of zinc metalloenzymes.

Biochemistry

Work performed in the past 20 years has shown that
zinc is essential to enzymatic reactions (Vallee and
Wacker, 1970). The functional role of iron in oxidative
processes was recognized 100 years ago and suggested an
essential role for that element and pointed the way to
the discovery of metalloenzymes in general. However, the
idea that other first transition and IIB groups of metals
might have significant roles remained unappreciated for
a long time and biochemical knowledge in this area was
almost nonexistent.

The search for an explanation of the physiologic role
of zinc emphasized its interaction with enzymes. Such
metalloenzymes are characterized by their stability constants, whose magnitudes have served as the operational
basis for the two extremes of metal-protein interactions,
i.e., the very stable metalloenzymes and the more labile
metal-enzyme complexes (Vallee, 1955), the latter exemplified by sodium, potassium, calcium and magnesium, the
most abundant cations of mammalian species.

The transition metals and zinc, because of their
electronic structures, tend to form stable complexes with
proteins with characteristic functions. The lack of visible colour of zinc proteins accounts for the long delay
in their recognition, while the red iron (Preyer, 1866)
and blue copper proteins (Harless, 1847) early called
attention to themselves. Carbonic anhydrase was the first
zinc enzyme to be isolated and purified (Keilin and Mann,
1940), bovine pancreatic carboxypeptidase A was the

second to be recognized in 1954 (Vallee and Neurath, 1955). Additional zinc metalloenzymes were discovered quite rapidly (Vallee, 1955, 1959). Zinc is now known to be an integral component of a large variety of proteins and enzymes and, hence, the total zinc content of tissues cannot serve as a guide to the multiple functions of the metal (see above). The number of functionally diverse types of zinc metalloenzymes which are now known document the importance of this metal in metabolism. More than 90 zinc-containing enzymes and proteins have been discovered and isolated from multiple sources encompassing all phyla, and they participate in a wide variety of metabolic processes including carbohydrate, lipid, protein and nucleic acid synthesis or degradation. Each of the six major categories of enzymes, designated by the IUB Commission on Enzyme Nomenclature according to function, contains at least one example of a zinc including enzyme, among them, several dehydrogenases, aldolases, peptidases, phosphatases and isomerases, a transphosphorylase and aspartate transcarbamylase. Thus, there are oxidoreductases, transferases, hydrolases, lyases, isomerases and ligases (Vallee and Wacker, 1976). Table III shows a representative list.

Detailed studies of the structure and function of many of these enzymes have become crucial to present understanding of enzymatic catalysis. Alcohol dehydrogenases, alkaline phosphatases, carboxypeptidases and carbonic anhydrases have also been obtained from human tissues and erythrocytes, among other sources. In these, as in other zinc metalloenzymes, the metal is indispensable for catalytic function and/or the structural integrity of the molecule.

Alcohol dehydrogenase, which is present in liver as well as other organs, oxidizes ethanol, other primary and secondary alcohols, sterols as well as vitamin A alcohol. Retinene reductase of the retina is apparently

TABLE III

Representative Zinc Metallo-enzymes and -proteins

Enzyme	Source
Alcohol Dehydrogenase	Horse and human liver, yeast
D-Lactate cytochrome c reductase	Yeast
Aspartate transcarbamylase	*E. coli*
RNA polymerase	*E. coli*, yeast, *E. gracilis*
DNA polymerase	*E. coli*, sea urchin
Reverse transcriptase [RNA dependent - DNA Polymerase]	Mammalian viruses
Alkaline phosphatase	Mammalian livers, kidneys, placenta and leukocytes, *E. coli*
Phospholipase C	*B. cereus*
α-D-mannosidase	Jack bean
AMP-aminohydrolase	Rabbit muscle
Leucine aminopetidase	Mammalian kidney, bovine lens, porcine kidney
Carboxypeptidase C	Orange peel
Carboxypeptidase A	Beef, human pancreas
Carboxypeptidase B	Beef, pig pancreas
Dipeptidase	Pig kidney, mouse ascites, tumour cells, *E. coli*
Aldolase	Yeast, *Aspergillus niger*
Carbonic anhydrase	Mammalian erythrocytes
δ-Aminolevulinate dehydratase	Beef and rat liver, erythrocytes
Phosphomannose isomerase	Yeast
Pyruvate carboxylase	Pig liver, yeast
Protein synthesis elongation factor	Rat liver
Methionyl t-RNA synthetase	*E. coli*
Enzyme Containing Zinc and Copper	
Superoxide dismutase	Govine liver, erythrocytes
Enzyme Containing Zinc and Cobalt	
Oxaloacetate Transcarboxylase	Propionibacterium shermannii
Enzyme Containing Zinc and Calcium	
Thermolysin	*B. thermolyticus*

identical or very similar to the liver enzyme. Horse and human liver alcohol dehydrogenases contain 4 atoms of zinc per molecule of protein, and zinc is essential not only to the catalytic function of the enzyme, but also to maintain the subunit structure (Drum *et al.*, 1967; Li, 1977). In addition to the above substrates, alcohol

dehydrogenase from *human* liver also oxidizes methanol
and ethylene glycol, serving as the primary mechanism of
detoxification of these and other similar compounds
(Blair and Vallee, 1966; Grant and Coombs, 1970).

Erythrocytes contain carbonic anhydrase in high concentrations. The enzyme catalyzes the reaction
$CO_2 + H_2O \leftrightharpoons H_2CO_3$ and many tissues exhibit this enzymatic
activity. In its absence the rate of carbon dioxide
elimination is insufficient to sustain life; hence, this
enzyme is as important to the elimination of carbon dioxide as is the many hemoglobins to the absorption and
transport of oxygen. Carbonic anhydrases from the ox,
monkey, human and other erythrocytes all contain 1 gm
atom of zinc per mole of enzyme. Under both normal and
pathologic conditions there is a close correlation between the zinc content and this enzymatic activity of red
blood cells. Substitution of cobalt for zinc results in
an enzymatically active carbonic anhydrase (Lindskog and
Malmstrom, 1962).

Both carboyxpeptidase A and B from pancreatic juice
contain 1 atom of zinc per molecule of protein. The single
zinc atom is indispensable for the catalytic activities
of the enzymes which hydrolyze peptide bonds to liberate
the amino acids from the carboxy-terminal ends of proteins
and peptides. The zinc in carboxypeptidase A can be replaced *in vitro* by other metals, e.g., Co, Mn, Ni, Cd,
Hg, and Pb, with consequent dramatic alterations in
catalytic activity and substrate specificity (Vallee *et
al.*, 1963). Both carboxypeptidase A and B are excreted
into the gastrointestinal tract and are essential to
proteolysis and digestive processes.

Four zinc atoms of alkaline phosphatase from *E. coli*
are essential for its activity. The enzyme consists of
two subunits; in each of the two subunits one zinc atom
is responsible for function and another for structure
stabilization and — in addition — two Mg atoms control

the activity of the holoenzyme (Bosron *et al.*, 1975). A
decrease in alkaline phosphatase activity has long been
noted in zinc-deficient experimental animals, a change
which may now be attributed to a failure of synthesis of
the active holoenzyme. The alkaline phosphatase of leukocytes may well be identical with the zinc-containing protein of human leukocytes which contains 0.3 per cent of
zinc per gm dry weight of protein, and is responsible for
80 per cent of all zinc found in human leukocytes (Hoch
and Vallee, 1952).

Substantial quantities of firmly bound zinc in RNA and
DNA have revealed additional important avenues of its
role in biology. Zinc appears to stabilize the secondary
and tertiary structure of RNA and to play an important
role in protein synthesis. To date we have little or no
information on the association of zinc with lipids or
carbohydrates, glyco- and lipoproteins, but their systematic analysis might turn out to be important.

There are unusual opportunities to study the manner in
which the specific function of zinc proteins and enzymes
is achieved through interaction of the metal with the
protein. By removal and restoration of zinc with concomitant loss and restitution of enzymatic activity, it
is possible to achieve the differential chemical labelling
of the sites of the protein to which the metal binds (see
above). This permits physico-chemical approaches to discern what contribution zinc may make to overall protein
structure and stability. Other metals can be substituted
successfully in a number of enzymes for the native zinc,
with corresponding characteristic alterations of physical
properties and/or catalytic function. These, as well as
other approaches, have assisted in the delineation of the
role of zinc in the function and structure of enzymes.
In addition to metal substitution, these include inhibition of chelating agents, kinetics of metal exchange,
absorption, circular dichroism and magnetic circular

dichroism spectroscopy, resonance spectroscopy, resonance
Raman spectroscopy, chemical modifications and, last but
not least, X-ray crystallography.

The discovery that DNA and RNA polymerases from both
procaryotic and eucaryotic microorganisms are zinc
metalloenzymes is one of the important recent advances
(Falchuk *et al.*, 1976). Although these enzymes have yet
to be identified similarly from mammalian sources, they
are likely to be zinc metalloenzymes also, since impaired
growth, protein synthesis and DNA and RNA metabolism are
such prominent features in zinc-deficient mammals.

The relationship of zinc to cellular growth has recently been investigated in *Euglena gracilis* and a number
of striking morphologic and chemical changes accompany
zinc deficiency-induced growth arrest (Falchuk *et al.*,
1975a,b): cell volume and size increase while osmophilic
granules and paramylon accumulate; cellular DNA content
doubles while the protein content decreases; peptides,
amino and other organic acids, nucleotides and poly-
phosphates accumulate (Figure 3). Further studies have
shown that zinc is present in subnuclear structures and
is trans-located to and from the nucleolus, the mitotic
spindle apparatus and chromosomes at each state of
mitosis. In zinc deficiency, the progression of the cell
through the various phases of mitosis (G_1, S, G_2) is impaired. These findings indicate that zinc is not only
essential for protein and nucleic acid synthesis, but
also is critical to the processes of cellular development, division and differentiation. Zinc has also been
shown to be essential for poly-ribosome formation (Weser
et al., 1969) and to be a component of protein synthesis
elongation factor I from rat liver, which catalyzes the
binding of amino acyl-tRNA to ribosomes (Katsiopoulos
and Mohr, 1975). It has now been shown that the DNA dependent RNA polymerases I and II of *E. gracilis* are
zinc metalloenzymes, and most recently its polymerase III

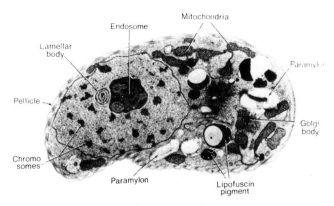

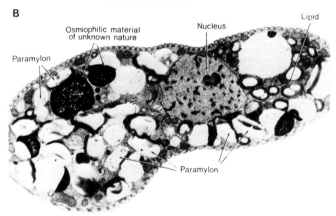

Fig. 3. Electron micrographs permitting comparison of the ultra-structure of *Euglena* grown 10 days with or without zinc, respectively. The zinc-deficient organism is significantly larger, its cytoplasm contains an abundance of paramylon and large masses of dense osmiophilic material presumably rich in lipid. The size difference is actually greater than it appears, since the micrograph of the zinc-deficient *Euglena* is shown at somewhat lower magnification so that the whole organism can be included. X 3825 and 2925, respectively.

has been shown to contain functional zinc also (K.H. Falchuk, unpublished observations). In view of the importance of these data, the work leading to the characterization of these enzymes will be presented in greater detail to exemplify the approach.

DNA Dependent RNA Polymerases I and II of *E. gracilis*: Zinc
Metalloenzymes

Decisive advances during the last few years both in
methods of isolation of eukaryotic polymerases (Cold
Spring Harbor Symp., 1973) together with the advent of
microwave excitation emission spectrometry for metal
analysis (see above) (Kawaguchi and Vallee, 1975) have
allowed the characterization of these metalloenzymes as
zinc enzymes (Falchuk *et al.*, 1976).

Figure 4 outlines the purification procedure for RNA
polymerases I and II from zinc sufficient *E. gracilis* and
their ionic strength dependent separation (Falchuk *et al.*,
1976).

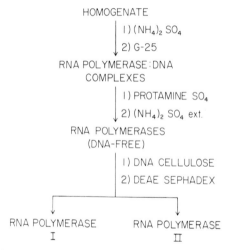

HOMOGENATE

1) $(NH_4)_2 SO_4$

2) G-25

RNA POLYMERASE:DNA
COMPLEXES

1) PROTAMINE SO_4

2) $(NH_4)_2 SO_4$ ext.

RNA POLYMERASES
(DNA-FREE)

1) DNA CELLULOSE

2) DEAE SEPHADEX

RNA POLYMERASE   RNA POLYMERASE
I   II

Fig. 4. Scheme for the isolation of the RNA polymerases I and II
from *Euglena gracilis*.

Both polymerases are entirely dependent on an exogenous DNA template for activity. RNA is the product of their
enzymatic reaction, as evidenced by an absolute substrate
requirement for ribonucleotide triphosphates and by digestion of the product by ribonuclease. As with other
polymerases, the *E. gracilis* enzymes are inactive in the
absence of Mg^{2+} and Mn^{2+}. Both these DNA dependent RNA

polymerases are homogeneous on polyacrylamide gels, and
their estimated molecular weights, determined on SDS
gels, lie between 650,000 and 700,000 for both polymerases. They are composed of multiple subunits of varied
but unknown molecular weights which remain to be determined to obtain the precise molecular weights of the
holoenzymes. These features are summarized in Table IV.

TABLE IV

Properties of E. gracilis *RNA Polymerases I and II*

Property	I	II
Template Dependence	DNA	DNA
Product	RNA	RNA
Activating Metals	Mg,Mn	Mg,Mn
M.W. (SDS-PAGE)	650,000	700,000

α-Amanitin serves to differentiate the type II RNA
polymerases, which are inhibited by this agent, from
those of type I, which are not (Falchuk *et al.*, 1977).

Both RNA polymerases I and II are inhibited by saturating amounts of chelating agents and 1,10-phenanthroline
and EDTA inhibit both their activities completely. Other
chelating agents such as 8-hydroxyquinoline and its
5-sulfonic acid, EDTA and α,α'-bipyridine, reduce the RNA
polymerase II activity from 70 to 50%. At saturating
amounts the non-chelating analogues of 1,10-phenanthro-
line, 1,7- or 4,7-phenanthroline, do not inhibit either
polymerase. Hence, the inhibition by the 1,10-isomer is
due to chelation of a functional metal atom. Both RNA
polymerases I and II are inhibited by 1,10-phenanthro-
line but with different pK_I's, 5.2 and 3.4, respectively,
further differentiating them (Falchuk *et al.*, 1976).

The inhibition of polymerase II by 1,10-phenanthroline
is instantaneous and completely reversible by dilution of
the assay mixture with buffer, suggesting the formation
of a mixed complex and confirming the presence of a metal

essential for function. Quantitative measurements of the
metal content have established the presence of zinc in
both RNA polymerase I and II by means of microwave excitation emission spectrometry. The *E. gracilis* RNA polymerases I and II contain 0.2 μg of Zn per mg of protein.
Keeping in mind the as yet provisional nature of the
molecular weights of these polymerases which form the
basis of the metal/protein ratio, the stoichiometry of
both is essentially the same, i.e., 2 g-atom of zinc per
mole. The sum of Cu, Fe and Mn is less than 0.2 g atom/
mole. Thus, both RNA polymerases I and II from zinc sufficient *E. gracilis* are zinc metalloenzymes and, as mentioned above, so is the RNA polymerase III. The eukaryotic DNA dependent RNA polymerase I of yeast is also a
zinc enzyme containing 2.4 g-atom of Zn per molecular
weight 650,000 (Auld *et al.*, 1976).

The demonstration that *E. gracilis* RNA polymerase I,
II and III are zinc metalloenzymes confirms the essentiality of this element for RNA synthesis, in conjunction with its involvement in other aspects of RNA metabolism (Wacker *et al.*, 1965; Prask and Plocke, 1971). The
isolation of the DNA dependent DNA polymerase of *E.
gracilis* and its inhibition by 1,10-phenanthroline
(McLennan and Keir, 1975a,b) further represent critical
steps in deciphering the series of events which may
jointly account for some of its requirements in the growth
of *E. gracilis* and of other eukaryotic organisms.

In a given zinc enzyme the metal can either be essential for activity, regulate it, stabilize its structure
or any combination of these (Stadtman *et al.*, 1968;
Rosenbusch and Weber, 1971; Anderson *et al.*, 1975).
While attempts to specify a particular enzymatic (or
other) step in zinc deficiency which might limit growth
would be conjectural at present, a number of alternatives
suggest themselves (Falchuk *et al.*, 1976).

Lack of zinc could preclude synthesis of any one or

all of these zinc enzymes or result in inactive apo-
enzyme(s). Further, the possibility must be considered
that functional metalloenzymes containing different metals
might be generated which could exhibit altered values of
K_{cat} or K_m — either for substrate or template. In this
regard, the accumulation of Fe, Cr, and Ni which occur
in response to zinc deficiency of *E. gracilis* (Falchuk
et al., 1975a; Wacker, 1962) could reflect a compensatory
mechanism, conceivably designed to overcome a metal-
dependent metabolic block. The simultaneous accumulation
of Mg and Mn, could similarly serve to regulate such
metal substituted RNA polymerases. Certainly, the requirements for both Mg and Mn of the corresponding native
Euglena enzymes are variable although they are all inhibited by high concentrations of Mn (Roeder and Rutter,
1969).

Whatever the specific mechanism, the zinc deficient
state might alter the relative proportions of existent
RNA polymerase activities or induce new variants. A number of observations would tend to support such conjectures. Thus, in sea urchin (Nemer and Infante, 1965;
Emerson and Humphreys, 1970; Roeder and Rutter, 1970),
liver (Blatti *et al.*, 1970) *Helianthus tuberosus* (Fraser,
1975), and amoebae (Yagura *et al.*, 1976), the activities
of RNA polymerase I and II vary as function of the state
of development. Thus, during normal growth, regulatory
mechanisms serve to synthesize or preferentially activate
different classes of RNA polymerases; these control processes might be called into play in zinc deficiency and
alter the classes of RNA synthesized.

Further, Pogo *et al.* (1967) have shown in experiments
with *E. coli* RNA polymerase that the relative preponderance of Mg and Mn can modify the nucleotide sequences of
the RNA which is the product of the polymerization reaction, a course of events which could obtain here. One
or all of these postulated processes could ultimately

lead to the synthesis of unusual RNA sequences and the
ensuing synthesis of altered or abnormal proteins (Ochoa
and Mazunder, 1974; Lucas-Lenard and Beres, 1974; Tate
and Caskey, 1974). Zinc deficient *E. gracilis* are known
to synthesize peptides and/or proteins whose amino acid
composition is unusual (Wacker *et al.*, 1965). Hence, it
is conceivable that there could be defects in protein
synthesis of zinc deficient *E. gracilis* either during
initiation, elongation or termination. The observation
that elongation factor I from rat liver contains functional zinc atoms (Kotsiopoulos and Mohr, 1975) is consistent with and could bear importantly on these considerations. At present, it is not possible to discriminate
between such alternatives. Clearly, critical experiments
to delineate between them are both indicated and feasible.

The present data provide evidence for a role of zinc
in transcription and translation of *E. gracilis* underlying
the essentiality of zinc in cell division. The demonstration that eukaryotic-RNA polymerase II and I (Auld *et al.*,
1976) are zinc enzymes supports the hypothesis that the
element is essential for RNA metabolism in all phyla
(Scrutton *et al.*, 1971; Slater *et al.*, 1971; Springgate
et al., 1973; Coleman, 1974; Auld *et al.*, 1974a,b,c;
Auld *et al.*, 1975). The data further suggest that extension of our studies on zinc deficient *E. gracilis* to
other organisms may assist in generalizing the emerging
biochemical role of zinc in growth, proliferation and
differentiation.

Zinc and the Biochemistry of Neoplastic Processes

The biological literature now documents profound
effects of zinc on normal growth and development, although
its possible role in abnormal growth has been examined in
but a cursory manner. The demonstration that a series of
reverse transcriptases are zinc metalloenzymes has un-
expectedly linked earlier observations of a role of zinc

in abnormal growth to contemporary enzymology.

That zinc was thought to be particularly abundant in
certain solid tumours (Cristol, 1922; Labbé, 1927), first
generated interest in its possible role in their formation. Quite the same, these and other similar early reports of abnormal concentrations and possibly altered
metabolism of this element in cancer would now seem to be
primarily of historical interest, considering the state
of the analytical chemistry then employed.

One facet of this early work, however, may be of
more lasting significance. The role of zinc in carcinogenesis was explored by injecting zinc chloride or sulfate into the cock testis generating teratomata which can
be transplanted (Michailowsky, 1928; Falin and Gromzewa,
1939); zinc acetate or zinc sterate proved ineffective
(Guthrie, 1964). These seasonal tumours develop only when
the injections are given between January and March, a
period of high sexual activity (Falin and Gromzewa,
1939; Guthrie, 1964). In this species spontaneous teratomata are rare.

The relationship of zinc to leukemia has also proven
to be of more enduring importance.

Studies of normal and leukemic leukocytes more than
25 years ago revealed a strikingly lowered zinc content
in leukemic, compared with normal white cells, generating
the postulate that alterations of zinc-dependent enzyme(s)
might be critical to the pathophysiology of leukemias
(Vallee *et al.*, 1949). This hypothesis has recently found
support in studies of the metal content of the RNA-
dependent DNA polymerases (reverse transcriptases) from
type C oncogenic viruses. Thus, the enzymes from avian
myeloblastosis, murine, simian, feline and RD-114 RNA
tumour viruses have now been shown to be zinc metalloenzymes. These viruses are associated with the induction
of leukemia in these species. Moreover, studies of CCRF-
CEM lymphoblasts, human lymphoid cells grown in culture

from the blood of leukemic patients, have shown that low
concentrations of the chelating agent, 1,10-phenanthro-
line, inhibit the growth of both normal and leukemic
CEM lymphoblasts. Interestingly, however, the chelating
agent is cytotoxic to the leukemic cells at concentrations which cause only growth arrest of normal cells
(Vallee, 1976). This unusual sensitivity may perhaps
relate to the known difference in the zinc content between
normal and leukemic leukocytes, and might lead to novel
therapeutic approaches. Figure 5 diagrams the many important metabolic pathways in which zinc is an essential
participant.

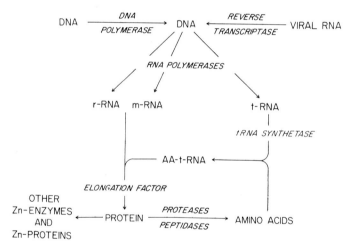

Fig. 5. The role of zinc in translation, transcription and formation
of zinc enzymes and proteins. Italics denote zinc-containing enzymes.

Other Biochemical Roles of Zinc

Beyond the critical role of zinc in the mechanism of
action of zinc metalloenzymes, the addition of zinc increases the activities of many other enzymes, much as
they do not contain zinc. This effect may not be specific,
since other metal ions also activate most of these systems. The zinc-activated metal-enzyme complexes are
numerous and, among others, include carnosinase, histi-

dine deaminase, enolase, dinucleotide pyrophosphatase, phosphoenolpyruvate carboxylase and pyridoxal kinase (Vallee and Coleman, 1964). Zinc has also been thought to stabilize cell membranes and to protect them against peroxidative damage.

A relationship between zinc and porphyrins has long been known and the excretion of zinc uroporphyrin and coproporphyrin complexes in urine has been described in human diseases such as porphyria. However, the significance of these compounds is uncertain since zinc porphyrins are known to form spontaneously, i.e. nonenzymatically. However, the enzymatic incorporation of zinc into porphyrins by an enzyme, zinc chelatase, has also been demonstrated. It is present in chromatophores of *Rhodopseudomonas spheroids* (Neuberger and Tait, 1964) and also found in mammalian organs. At present, however, biologically active zinc porphyrins are not known to exist.

A relationship of zinc to the action of insulin, glucagon, corticotropin and other hormones has been postulated. Despite much study and circumstantial evidence, a role for zinc in hormonal function has not been established with certainty. The crystallization of insulin at pH values near 6 is accomplished in the presence of zinc which associates with 6 molecules of insulin to form a hexamer of MW 36,000, Porcine proinsulin also interacts with zinc (Grant and Coombs, 1970). Yet, physiologically active insulin can be prepared in both amorphous and crystalline form, entirely free of zinc and other metals. Thus, the association of zinc with insulin *in vitro* has not been shown to be a compositional or structural feature essential for function. From the evidence available, it cannot be judged whether an insulin-zinc complex forms *in vivo* and is necessary for biological activity.

Nutritional Effects of Zinc

In the late 1920s and early 1930s, Bertrand in France and Elvehjem and their associates in the U.S.A., respectively, designed diets sufficiently low in zinc content to render rodents deficient. Since then zinc deficiency has been induced in numerous species including swine, lambs, chickens, Japanese quails, certain monkeys and even in the human.

Animal Nutrition

In mammals, depending on the species, failure to grow, abnormal formation of hair, nails and bone, anemia, anorexia, testicular atrophy, decreased size of the accessory sex glands, skin lesions and teratological defects are the most prominent consequences. A decrease in food intake is an early sign of the deficiency state, and force-feeding produces signs of ill health. Zinc supplementation promptly restores appetite. Zinc deficiency has been said to affect adversely the gustatory system and wound healing (Prasad, 1967) but much controversy surrounds these claims. In zinc deficient rats hypoplasia of the coagulating and prostate glands, seminal vesicles and hypospermia is observed, reversed by the administration of the metal. When testicular degeneration takes place, it is irreversible. Cadmium destroys testicular tissue when administered subcutaneously to rats, but when injected together with a large excess of zinc acetate this prevents testicular degeneration. This antagonism between cadmium and zinc is not limited to the testes, since zinc deficiency in chickens becomes more severe when cadmium is administered simultaneously, and higher concentrations of zinc are required to overcome the deficiency state.

Zinc-deficient rats characteristically develop hyperuricemia and the activities of pancreatic enzymes are

decreased, as are intestinal and kidney alkaline phosphatase activities. Addition of zinc to homogenates of the
intestine and kidney from zinc-deficient animals fully
restores alkaline phosphatase activity (Iqbal, 1970).
Synthesis of the apoenzyme apparently is unimpaired, but
enzyme activity is lacking, entirely due to inadequate
concentrations of zinc to form the active holoenzyme.
Pancreatic carboxypeptidase activity has been shown to
decrease quite specifically and rapidly in response to
zinc deficiency and return to normal on repletion (Mills
et al., 1967). Changes in the activities of a number of
dehydrogenases in the bone, kidney, intestine, esophagus
and testes of zinc-deficient rats and swine have been
noted (Prasad et al., 1967), apparently because these
tissues turn over rapidly and are most sensitive to zinc
depletion. Carbonic anhydrase activity usually remains
normal.

The RNA polymerase and thymidine kinase activities of
growing tissues are decreased as is the incorporation of
thymidine (Prasad and Oberleas, 1974). The vitamin A
concentrations of plasma are lowered, apparently as a
result of a defect in the mobilization of the vitamin
from liver into the circulation (Smith et al., 1973).
Hyperkeratinization of the epidermis and parakeratosis of
the esophagus are the most striking histological findings in zinc-deficient rats, similar to other animal
species (Follis, 1966). This may relate in part to the
prominent abnormal sulfur and mucopolysaccharide metabolism of zinc-deficient animals (Hsu and Anthony, 1970).

Offspring of rats made zinc-deficient for even short
periods of time during their pregnancy develop major
congenital malformations involving the skeleton, brain,
heart, eyes, gastrointestinal tract, and lungs, emphasiz-
ing the importance of zinc to normal growth and development (Hurley and Mutch, 1973) (see above). Interestingly,
the feeding of calcium-deficient diet can offset the

deleterious effects of zinc deficiency (Tao and Hurley, 1975). It appears that — through bone resorption — the concomitant calcium deficiency increases the availability of skeletal zinc stores which are not readily mobilized normally. Abnormalities and decrease in plasma proteins have been reported in zinc-deficient rats, chicks, swine and the Japanese quail. Table V summarizes and compares the morphologic chemical and enzymatic alterations of zinc-deficient animals with those in plants and micro-organisms.

In hogs, zinc deficiency may occur spontaneously or can be induced experimentally; the resultant disease is called Porcine Parakeratosis. It is characterized by dermatitis, diarrhoea, vomiting, anorexia, severe weight loss and eventual death (Figure 6). The disease apparently can occur spontaneously but has been attributed to the practice of adding bone meal to the diet to accelerate growth and ossification. Experimentally, the disease is aggravated by addition to the diet of large amounts of calcium; thus, parakeratosis in hogs is a *conditioned* zinc deficiency (Ershoff, 1948). The mechanism of the

Fig. 6. Drawing showing normal swine below and parakeratosis above.

TABLE V

Zinc Deficiency in Different Phyla

	Morphologic Changes	Chemical Compositions		Decrease in Enzyme Activity
		Decrease	Increase	
Microorganisms	Growth retardation Increase in cell size	Protein RNA (ribosomal) Pyridine nucleotides	DNA Amino Acids Polyphosphates Phospholipids ATP Organic Acids	Alkaline phosphatase Alcohol dehydrogenase Lactate dehydrogenase Tryptophan desmolase
Plants	Stunted Growth Small abnormal leaves Chlorotic mottling Decreased fruit production	Auxin Ethanolamine	Amino Acids	Tryptophan desmolase Carbonic anhydrase Aldolase Pyruvate carboxylase
Animals	Growth retardation Fetal wastage Birth anomalies Testicular atrophy Parakeratosis Coarse, sparse hair	Hemoglobin Serum protein	Uric Acid	Alkaline phosphatase Alcohol dehydrogenase Pancreatic proteases Leucine aminopeptidase Thymidine kinase Lactate and malate dehydrogenase NADH diaphorase

Ca-Zn antagonism is unknown. On supplementation of the diets of hogs with zinc carbonate both induced parakeratotic swine — whether afflicted spontaneously or experimentally — recover promptly. A similar disease has been reported in cattle.

Human Nutrition

A syndrome of severe iron deficiency, anaemia, hepatosplenomegaly, hypogonadism, hyperpigmentation and dwarfism in Iranian and Egyptian men was described in 1963 but attributed to primary zinc deficiency (Prasad, 1967). In the Iranian dwarfs geophagia was common; it was suggested that excessive consumption of a cereal diet containing large amounts of phytate might have caused zinc deficiency, since phytate inhibits iron and zinc absorption (see above). Egyptian dwarfs who exhibited a similar syndrome showed additional evidence of primary pituitary hypofunction accompanied by reduced activities of gonadotrophic and growth hormones. The plasma zinc content and that of the red blood cells and hair was decreased as was the 24-hour exchangeable zinc pool, while plasma turnover of zinc was increased. Sweat and urinary zinc excretion and serum alkaline phosphatase activities were decreased. Treatment with supplemental zinc salts resulted in striking responses in growth and development of secondary sex characteristics, exceeding those in response to a balanced diet alone. Serum zinc concentrations rose to normal. The administration of zinc did not reverse the anaemia, which was attributed to the concomitant iron deficiency of these patients, thought to be due to parasitic infestations. Such findings have recently been confirmed in Iranian women and school children. This zinc deficiency syndrome is now being recognized in the United States and elsewhere, particularly in young individuals with intestinal malabsorption of prolonged duration during adolescence (MacMahon *et al.*, 1968; Solomons *et al.*,

1976), or as a consequence of chronic, total parenteral
nutrition (Tucker *et al.*, 1976). Because of marginal or
deficient dietary intake, some children in the United
States have been reported to exhibit decreased body zinc
stores, as gauged by the content of zinc in hair, although
it must be pointed out that this remains a questionable
and unproven index of zinc deficiency. Such children
grow poorly and are amnaesic (Hambidge *et al.*, 1972).
Serum zinc is consistently decreased during pregnancy.
However, much of the decrease appears to be associated
with diminished binding of zinc to albumins, although a
relationship of this finding to marginal zinc nutritional
status has not been established (Giroux *et al.*, 1976).

Zinc in Human Disease States

Serum zinc concentrations are decreased below the
range of normal values in acute and chronic infections,
e.g., pneumonia, bronchitis, erysipelis and pyelonephri-
tis, but are restored upon recovery. A protein which is
released from leukocytes and stimulates the flux of zinc
from the plasma to liver has been thought to mediate the
zincemia associated with acute inflammatory stresses and
has been given the name *leukocytic endogenous mediator*
(Pekarek *et al.*, 1972). It appears also to have mediated
some of the generalized host responses to an inflammatory
stimulus, e.g., the synthesis of acute phase reactants
by the liver.

Serum zinc concentrations are decreased in myocardial
infarction and accompany the well-known changes in a
variety of enzyme activities and increase in the copper
concentration of serum in this disease. Various malig-
nancies are also accompanied by decreased serum zinc
values, although a uniform pattern is not discernible.
The leukocyte zinc content of patients with refractory
anaemia is increased while that of leukocytes in acute
and chronic lymphatic and myelogenous leukemia is

decreased to 10 per cent of the normal value (see above).
This phenomenon is apparently independent of the maturity
of the cells but may relate to the decreased leukocyte
alkaline phosphatase activities observed in some of these
conditions. Excessive zinc excretion has been noted in
leukemia and Hodgkins Disease, although the basis is not
immediately apparent. The zinc content and carbonic anhydrase activity of red cells correlate significantly in
individuals afflicted with anaemia, polycythemia vera,
secondary polycythemia, leukemia, and congestive heart
failure.

Dietary and conditioned zinc deficiency are potential
problems in patients with alcoholism and liver disease,
chronic renal disease, rheumatoid arthritis, inflammatory bowel disease and malabsorption syndrome. Patients
with chronic hemolytic anaemias such as sickle cell disease (Prasad *et al.*, 1975) exhibit decreased plasma and
erythrocyte zinc concentrations and zincuria. The growth
failure and hypogonadism in these patients may be related.
Post-surgical patients may also develop zinc deficiency,
accompanied by impaired wound healing (Sandstead *et al.*,
1974). Zinc chelating agents such as penicillamine and
histidine can induce zinc deficiency.

Zinc in Post-alcoholic Cirrhosis

The discovery of zinc in liver alcohol dehydrogenase
stimulated the study of zinc metabolism in human liver
disease. Marked abnormalities in the metabolism of this
metal were found to occur in post-alcoholic cirrhosis
(Vallee *et al.*, 1956). The concentration of zinc in the
serum of patients with a severe degree of this disease is
decreased markedly. Most significantly, in patients in
hepatic coma, concentrations of less than 30 μg per 100
ml suggest an ominous prognosis. Liver tissue of patients
who die of post-alcoholic cirrhosis contain only half the
normal content of zinc and iron, while calcium, magnesium,

aluminium, manganese and copper concentrations are normal. Significantly, patients with post-alcoholic cirrhosis excrete abnormally large quantities of zinc in their urine, 1000 ± 200 µg per 24 hours. Administration of zinc sulfate in physiologic quantities to such patients tends to restore the normal excretory pattern in urine, accompanied by a tendency toward restoration of normal liver function (Vallee *et al.*, 1957).

It has been suggested that the low serum zinc concentration in cirrhosis may reflect a change in the synthesis, degradation, metal content, or specific activity of the zinc-containing enzymes which are cardinally involved in intermediary and protein metabolism. It cannot be stated, however, whether these abnormalities of zinc metabolism constitute primary or secondary manifestations of the disease although they emphasize a relationship of the metal to this entity.

Most recently, an unusual variant of human liver dehydrogenase, π-alcohol dehydrogenase, has been isolated and identified as a zinc metalloenzyme. It is not found in all human livers examined. It is most effective when substrate concentrations are very high (Li *et al.*, 1977). Studies of this enzyme in patients with post-alcoholic cirrhosis should be of special interest in furthering understanding of the subcellular pathology and cytology of this disease.

In this regard, it is of interest that effects of chronic ethanol ingestion have been studied recently in rats. Liver, plasma and muscle zinc content all decrease progressively. The liver is depleted most severely, particularly its mitochondrial compartment, effects seen as early as two weeks after the onset of alcohol feeding. The findings clearly indicate that alcohol deranges hepatic zinc metabolism long before the onset of cirrhosis (Wang and Pierson, 1975).

Acrodermatitis Enteropathica

An inherited zinc-deficiency disorder has been described recently, characterized by alopecia, dermatitis, diarrhoea, photophobia, psychological changes and failure to thrive (Moynahan and Barnes, 1974) (Figure 7). The disease is transmitted as an autosomal recessive trait.

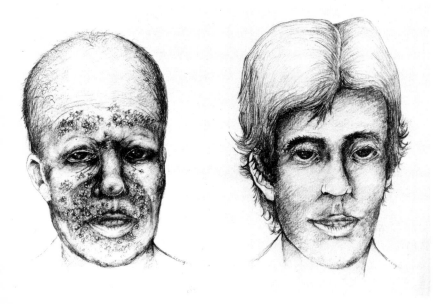

Fig. 7. Drawing of patient with *Acrodermatitis enteropathica* prior to treatment (left) and subsequent to successful therapy on the right.

Both the patients and their family members exhibit low plasma zinc concentrations (Hirsh *et al.*, 1976), although two instances of normal concentrations have now been observed. The primary defect is thus far unknown, although an abnormality of zinc absorption has been postulated but not proven. Daily supplements of zinc in amounts equal to the normal daily intake, approximately 15-25 mg per day, are curative. This would seem to be the first human disease in which zinc reverses all pathological lesions and accompanying clinical signs and symptoms. This discovery may justly be thought a milestone in the biology and pathology of this hitherto elusive metal.

Zinc Toxicity

Compared to copper, lead, mercury, and arsenic, zinc is relatively non-toxic. However, *inhalation* of zinc oxide fumes in high concentration, as may occur in industrial settings, produces an acute illness of relatively short duration characterized by chills, fever, cough salivation, headache, leukocytosis and pulmonary infiltrates (see above). Constant exposure produces tolerance, but intermittent exposure results in recurrence of the illness.

Poisoning due to *ingestion* of zinc may occur when foods have been stored in galvanized containers. Signs and symptoms of toxicity are nausea, vomiting, abdominal cramps, diarrhoea and fever. In experimental animals, feeding of massive doses of zinc produces anaemia, retarded growth and eventual death.

Acknowledgements

This work was supported by Grant-in-Aid GM-15003 from the National Institutes of Health of the Department of Health, Education and Welfare.

References

Anderson, R.A., Bosron, W.F., Kennedy, F.S. and Vallee, B.L. (1975). *Proc. Nat. Acad. Sci. USA* **72**, 2989.

Atsuya, I., Alter, G.M., Veillon, C. and Vallee, B.L. (1977b). *Anal. Biochem.* **79**, 202.

Atsuya, I., Kawaguchi, H. and Vallee, B.L. (1977a). *Anal. Biochem.* **77**, 208.

Atsuya, I., Kawaguchi, H., Veillon, C. and Vallee, B.L. (1977c). *Anal. Chem.* **49**, 1489.

Auld, D.S., Atsuya, I., Campino, C. and Valenzuela, P. (1976). *Biochem. Biophys. Res. Commun.* **69**, 548.

Auld, D.S., Kawaguchi, H., Livingston, D.M. and Vallee, B.L. (1974a). *Biochem. Biophys. Res. Commun.* **57**, 967, 1974a.

Auld, D.S., Kawaguchi, H., Livingston, D.M. and Vallee, B.L. (1974b). *Proc. Nat. Acad. Sci. USA* **71**, 2091.

Auld, D.S., Kawaguchi, H., Livingston, D.M. and Vallee, B.L. (1974c). *Federation Proc.* **33**, 1483, 1974c.

Auld, D.S., Kawaguchi, H., Livingston, D.M. and Vallee, B.L. (1975). *Biochem. Biophys. Res. Commun.* **62**, 296.

Bertrand, G. (1912). *8th Int. Congr. Appl. Chem.* **28**, 30.
Blair, A.H. and Vallee, B.L. (1966). *Biochemistry* **5**, 2026.
Blatti, S.P., Ingles, C.J., Lindell, T.J., Morris, P.W., Weaver, R.F.,
    Weinberg, F. and Rutter, W.J. (1970). *Cold Spring Harbor Symposium
    Quant. Biol.* **35**, 649.
Bosron, W.F., Kennedy, F.S. and Vallee, B.L. (1975). *Biochemistry* **14**,
    2275.
Branden, C.I., Jörnvall, H., Eklund, H. and Furugren, B. (1975). *In*
    'The Enzymes' (P.D. Boyer, ed.), Academic Press, New York, p. 104.
Busch, K.W. and Vickers, T.J. (1973). *Spectrochim. Acta* **288**, 85.
Christian, G.D. and Feldman, F.J. (1970). 'Atomic Absorption Spectro-
    scopy', Interscience.
Cold Spring Harbor Symposium on Quantitative Biology, Vol. 38, 1973.
Coleman, E.J. (1974). *Biochem. Biophys. Res. Commun.* **60**, 641.
Collier, E.S. and Drinker, K.P. (1926). *J. Ind. Hyg.* **8**, 257.
Colman, P.M., Jansonius, J.N. and Matthews, B.W. (1972). *J. Mol. Biol.*
    70, 701.
Cristol, P. (1922). Contribution à l'étude de la physio-pathologie
    du zinc et en particular de sa signification dans les tumeurs.
    Thèse Montpellier, 1922.
Drum, D.E., Harrison, J.H., Li, T.-K., Bethune, J.L. and Vallee, B.L.
    (1967). *Proc. Nat. Acad. Sci. USA* **57**, 1434.
Emerson, C.P. and Humphreys, T. (1970). *Develop. Biol.* **23**, 85.
Ershoff, B.H. (1948). *Phys. Rev.* **28**, 107.
Falchuk, K.H. (1977). *N. England J. Med.* **296**, 1129.
Falchuk, K.H., Fawcett, D.W. and Vallee, B.L. (1975a). *J. Cell. Sci.*
    17, 57.
Falchuk, K.H., Krishan, A. and Vallee, B.L. (1975b). *Biochemistry* **14**,
    3439.
Falchuk, K.H., Mazus, B.L., Ulpino, L. and Vallee, B.L. (1976). *Bio-
    chemistry* **15**, 4468.
Falchuk, K.H., Ulpino, L., Mazus, B. and Vallee, B.L. (1977). *Biochem.
    Biophys. Res. Commun.* **74**, 1206.
Falin, L.I. and Gromzewa, K.E. (1939). *Amer. J. Cancer* **36**, 233.
Follis, R.H., Jr. (1966). *In* 'Zinc Metabolism' (A.S. Prasad, ed.),
    Springfield, Charles C. Thomas, p. 129.
Fraser, R.S.S. (1975). *Eur. J. Biochem.* **50**, 529.
Fuwa, K., Pulido, P., McKay, R. and Vallee, B.L. (1964). *Anal. Chem.*
    36, 2407.
Fuwa, K. and Vallee, B.L. (1963). *Anal. Chem.* **35**, 942.
Giroux, E., Schecter, P.J. and Schoun, J. (1976). *Clin. Sci. Mol.
    Med.* **51**, 545.
Grant, P.T. and Coombs, T.L. (1970). *Essays in Biochem.* **6**, 69.
Guthrie, J. (1964). *Brit. J. Cancer* **18**, 130.
Hambidge, K.M., Hambidge, C., Jacobs, M. and Baum, J.D. (1972).
    Pediat. Res. **6**, 868.
Harless, E. (1847). *Müller's Arch. f. Anat. Physiol.* **148**.
Harris, M.I. and Coleman, J.E. (1968). *J. Biol. Chem.* **243**, 5063.
Herskovitz, T., Averill, B.A., Holm, R.H., Ibers, J.A., Phillips,
    W.D. and Weiher, J.F. (1972). *Proc. Nat. Acad. Sci. USA* **69**, 2437.
Hirsh, F.S., Michel, B. and Strain, W.H. (1976). *Arch. Dermatol.*
    112, 475.
Hoch, F.L. and Vallee, B.L. (1952). *J. Biol. Chem.* **195**, 531.
Holmquist, B., Kaden, T.A. and Vallee, B.L. (1975). *Biochemistry* **14**,
    1454.

Hsu, J.M. and Anthony, W.L. (1970). *J. Nutr.* **100**, 1189.
Hurley, L.S. and Mutch, P.B. (1973). *J. Nutr.* **103**, 649.
Hurley, L.S. and Shrader, R.E. (1972). *In* 'Neurobiology of the Trace Metals Zinc and Copper' (C.C. Pfeiffer, ed.), Academic Press, New York. p. 7.
Iqbal, M. (1970). *Enzymol. Biol. Clin.* **11**, 412.
Kaden, T.A., Holmquist, B. and Vallee, B.L. (1974). *Inorg. Chem.* **13**, 2585.
Kawaguchi, H., Atsuya, I. and Vallee, B.L. (1977). *Anal. Chem.* **49**, 266.
Kawaguchi, H. and Vallee, B.L. (1975). *Anal. Chem.* **47**, 1029.
Keilin, D. and Mann, T. (1940). *Biochem. J.* **34**, 1163.
Kennedy, F.S., Hill, H.A.O., Kaden, T.A. and Vallee, B.L. (1972). *Biochem. Biophys. Res. Commun.* **48**, 1533.
Kotsiopoulos, S. and Mohr, S.C. (1975). *Biochem. Biophys. Res. Commun.* **67**, 979.
Labbé, H. and Nepveux, P. (1927). *Le Progres Med.* **577**.
Latt, S.A. and Vallee, B.L. (1971). *Biochemistry* **10**, 4263.
Li, T.-K. (1977). *Adv. Enzymol.* Vol. 45, 427.
Li, T.-K., Bosron, W.F., Dafeldecker, W.P., Lange, L.G. and Vallee, B.L. (1977). *Proc. Nat. Acad. Sci. USA* **74**, 4378.
Lindskog, S. and Malmstrom, B. (1962). *J. Biol. Chem.* **237**, 1129.
Lindskog, S. and Nyman, P.O. (1964). *Biochim. Biophys. Acta* **85**, 141.
Lucus-Lenard, J. and Beres, L. (1974). *In* 'The Enzymes' (P.D. Boyer, ed.), Vol. X, p. 53, Academic Press, New York.
MacMahon, R.A., Parker, M.L.M. and McKinnon, M.C. (1968). *Med. J. Austr.* II, **210**.
McLennan, A.G. and Keir, H.M. (1975a). *Biochem. J.* **151**, 227.
McLennan, A.G. and Keir, H.M. (1975b). *Biochem. J.* **151**, 239.
Michailowsky, I. (1928). *Virchow's Archiv. Pathol. Anat.* **267**, 27.
Mills, C.F., Quarterman, J., Williams, R.B., Dalgarno, A.C. and Panic, B. (1967). *Biochem. J.* **102**, 712.
Morrison, G.H. (1965). 'Trace Analysis', Wiley & Sons, New York.
Moynahan, E.J. and Barnes, P.M. (1974). *Lancet* **2**, 399.
Nemer, M. and Infante, A.A. (1965). *Science* **150**, 217.
Neuberger, A. and Tait, G. (1964). *Biochem. J.* **90**, 607.
Ochoa, S. and Mazunder, R. (1974). *In* 'The Enzymes' (P.D. Boyer, ed.), Vol. X, p. 1, Academic Press, New York.
Parisi, A.F. and Vallee, B.L. (1970). *Biochemistry* **9**, 2421.
Pekarek, R.S., Wannemacher, R.W., Jr. and Beisel, W.K. (1972). *Proc. Soc. Exp. Med.* **140**, 685.
Pinta, M. (1970). 'Detection of Trace Elements', Humphrey Science Publishers.
Pogo, A.O., Littau, V.C., Allfrey, V.G. and Mirsky, A.E. (1967). *Proc. Nat. Acad. Sci. USA* **57**, 743.
Prasad, A.S. (ed.) (1966). 'Zinc Metabolism', Charles C. Thomas, Springfield.
Prasad, A.S. (1967). *Federation Proc.* **26**, 181.
Prasad, A.S. and Oberleas, D. (1974). *J. Lab. Clin. Med.* **83**, 634.
Prasad, A.S., Oberleas, D., Wolf, P. and Horwitz, J. (1967). *J. Clin. Invest.* **46**, 549.
Prasad, A.S., Schoomaker, E.B., Ortega, J., Brewer, G.J., Oberleas, D. and Oelshlegel, F.J., Jr. (1975). *Clin. Chem.* **21**, 592.
Prask, J.A. and Plocke, D.J. (1971). *Plant Physiol.* **48**, 150.
Preyer, W.T. (1866). 'De Haemoglobiono Observationes et Experimentia,

p. 27, M. Cohen and Sons, Bonn.

Quiocho, F.A. and Lipscomb, W.N. (1971). *Adv. Prot. Chem.* **25**, 1.

Raulin, J. (1869). *Am. Sci. Natl. Botan. et Biol. Vegetale* **11**, 93.

Riordan, J.F. and Vallee, B.L. (1974). *Adv. Exp. Med. Biol.* **48**, 33.

Roeder, R.G. and Rutter, W.J. (1970). *Biochemistry* **9**, 2543.

Roeder, R.G. and Rutter, W.J. (1969). *Nature* **224**, 234.

Rosenbusch, J.P. and Weber, K. (1971). *Proc. Nat. Acad. Sci. USA* **68**, 68.

Sandstead, H.H., Lanier, J.C., Shepard, G.H. and Gillespie, D.D. (1974). *Amer. J. Clin. Nutr.* **23**, 514.

Schenk, W.G. (1975). 'Analytical Atomic Spectroscopy', Plenum Press, New York.

Scrutton, M.C., Wu, C.W. and Goldthwait, D.A. (1971). *Proc. Nat. Acad. Sci. USA* **68**, 2497.

Simpson, R.T. and Vallee, B.L. (1968). *Biochemistry* **7**, 4343.

Slater, J.P., Mildvan, A.S. and Loeb, L.A. (1971). *Biochem. Biophys. Res. Commun.* **44**, 37.

Slavin, W. (1968). 'Atomic Absorption Spectroscopy', Interscience, New York.

Slavin, W. (1971). 'Emission Spectrochemical Analysis', Wiley & Sons, New York.

Smith, J.C., McDaniel, E.G., Fan, F.F. and Halsted, J.A. (1973). *Science* **181**, 954.

Solomons, N.W., Rosenberg, I.H. and Sandstead, H.H. (1976). *Am. J. Clin. Nutr.* **29**, 371.

Springgate, C.F., Mildvan, A.S., Abramson, R., Engle, J.L. and Loeb, L.A. (1973). *J. Biol. Chem.* **248**, 5987.

Stadtman, E.R., Shapiro, B.M., Ginsburg, A., Kingdon, D.S. and Denton, M.D. (1968). *Brookhaven Symp. Biol.* **21**, 378.

Tao, S.H. and Hurley, L.S. (1975). *J. Nutr.* **105**, 220.

Tate, W.P. and Caskey, C.T. (1974). *In* 'The Enzymes' (P.D. Boyer, ed.), Vol. X, p. 87, Academic Press, New York.

Tucker, S.B., Schroeter, A.L., Brown, P.W. and McCall, J.T. (1976). *JAMA* **235**, 2399.

Vallee, B.L. (1955). *Adv. Prot. Chem.* **10**, 317.

Vallee, B.L. (1976). 'Cancer Enzymology' (J. Schultz and F. Ahmad, eds.) Academic Press, New York, p. 159.

Vallee, B.L. (1969). *Clin. Chim. Acta* **25**, 307.

Vallee, B.L. (1973). *In* 'Metal Ions in Biological Systems' (S.K. Dhar, ed.), Vol. 40, Plenum Press, New York, p. 1.

Vallee, B.L. (1959). *Physiol. Rev.* **39**, 443.

Vallee, B.L. and Coleman, J.E. (1964). 'Comprehensive Biochemistry', Vol. 12, (M. Florkin and E.H. Stotz, eds.), Elsevier Publishing Co., Amsterdam, p. 165.

Vallee, B.L. and Latt, S.A. (1970). *In* 'Structure-Function Relationships of Proteolytic Enzymes' (P. Desnuelle, H. Neurath and M. Ottesen, eds.), Academic Press, New York, p. 144.

Vallee, B.L. and Neurath, H. (1955). *J.Biol. Chem.* **217**, 253.

Vallee, B.L. and Wacker, W.E.C. (1976). Metalloproteins and Metalloenzymes, *in* 'Handbook of Biochemistry and Molecular Biology' (G.D. Fasman, ed.), Third Edition, 'Proteins', Vol. II, CRC Press Inc., Cleveland, p. 276.

Vallee, B.L. and Wacker, W.E.C. (1970). *In* 'The Proteins' (H. Neurath, ed.), Vol. 5, Second Edition, Academic Press, New York.

Vallee, B.L. and Williams, R.J.P. (1968). *Proc. Nat. Acad. Sci. USA* **59**.

Vallee, B.L., Fluharty, R.G., Gibson, J.G. (1949). *Acta Union Intern. Contre Cancer* **6**, 869.

Vallee, B.L., Riordan, J.F. and Coleman, J.E. (1963). *Proc. Nat. Acad. Sci. USA* **49**, 109.

Vallee, B.L., Wacker, W.E.C., Bartholomay, A.F. and Hoch, F.L. (1957). *New Engl. J. Med.* **257**, 1055.

Vallee, B.L., Wacker, W.E.C., Bartholomay, A.F. and Robin, E.D. (1956). *New Engl. J. Med.* **255**, 403.

Veillon, C. (1976). Optical Atomic Spectroscopic Methods *in* 'Trace Analysis-Spectroscopic Methods for Elements' (J.D. Winefordner, ed.), Vol. 46, Wiley & Sons, New York, p. 123.

Veillon, C. and Vallee, B.L. (1978). 'Methods In Enzymology', in press.

Wacker, W.E.C. (1962). *Biochemistry* **1**, 859.

Wacker, W.E.C., Kornicker, W. and Pothier, L. (1965). *Abstr. Amer. Chem. Soc. 150th Meetings*, **88C**.

Wang, J. and Pierson, R.N., Jr. (1975). *J. Lab. Clin. Med.* **85**, 50.

Weser, A., Seeber, S. and Warnecke, P. (1969). *Biochim. Biophys. Acta* **179**, 422.

Winefornder, J.D. (ed.) (1976). 'Trace Analysis', Wiley & Sons, New York.

Yagura, T., Hanagisawa, M. and Iwabuchi, M. (1976). *Biochem. Biophys. Res. Commun.* **68**, 183.

3   SOME ASPECTS OF STRUCTURE AND FUNCTION IN COPPER CONTAINING OXIDASES

Bo G. Malmström

*Department of Biochemistry and Biophysics, Chalmers
Institute of Technology and University of Göteborg,
Fack, S-402 20 Göteborg, Sweden*

Copper Proteins as Unique Coordination Compounds

It has become common in recent years to use the term *bio-inorganic chemistry* to designate studies of the rôle of metals in biochemistry, particularly their function in metalloenzymes. Progress during the last two decades in this area of research has, however, revealed that metallo-proteins frequently display unusual coordination proper-ties which are difficult to describe on the basis of know-ledge of the simpler compounds generally investigated by inorganic chemists. The unique properties of the metal ions are imparted by the protein, which may create an en-vironment in the binding site, for example, in regard to stereochemistry or local di-electric constant, which it is quite hard to reproduce in a small molecule in aqueous solution. Thus, an understanding of the unusual coordina-tion chemistry of metalloenzymes must be based on a de-tailed knowledge of protein structure. Bioinorganic chemistry is consequently dependent on *bioorganic chem-istry*.

The unique coordination provided by the protein ligands in metalloenzymes imparts properties to the metal ion facilitating its rôle in the catalytic reaction. It is the purpose of this article to illustrate this statement by a review of present knowledge of a number of copper-

containing oxidases. The major known copper proteins are
listed in Table I. This review will largely be limited to
the group of copper enzymes designated as blue oxidases

TABLE I

Main Classes of Copper-Containing Proteins

Protein	Function
Hemocyanin	Oxygen carrier
Superoxide dismutase[a] (earlier names: erythrocuprein, hepatocuprein etc.)	O_2^- detoxification
Azurin	Electron carriers
Stellacyanin	
Umecyanin	
Plastocyanin	
Phenol *o*-monooxygenase (tyrosinase)	Oxygenation
Dopamine β-monooxygenase	
Monoamine oxidases (pig liver)	Oxidases (reducing
Diamine oxidases	O_2 to H_2O_2)
Galactose oxidase	
Laccase	Oxidases (reducing
Ascorbate oxidase	O_2 to $2H_2O$)
Ceruloplasmin[b]	
Cytochrome *c* oxidase[c]	Terminal oxidase

[a]Contains Zn as well as Cu.
[b]Has a weak oxidase activity of unknown physiological function
(suggested to be a ferroxidase or a Cu-transport protein).
[c]Contains haem as well as Cu.

(Malkin and Malmström, 1970; Malmström *et al.*, 1975).
These are the most complicated of copper proteins as they
contain three distinct types of copper. Since the copper
coordination in some of the simpler proteins may show
similarities to a given copper type in the blue oxidases,
some properties of other copper proteins will also be
considered.

The first clear demonstration that copper-containing

oxidases have a unique coordination of the metal came
from EPR investigations that I carried out together with
T. Vänngård in the late 1950s (Malmström et $al.$, 1959;
Malmström and Vänngård, 1960). We found that the EPR spectra of laccase and ceruloplasmin showed exceptionally low
hyperfine structure constants compared to a large range
of other copper complexes, including natural as well as
artificial copper proteins, such as erythrocuprein (superoxide dismutase) and Cu^{2+}-carbonic anhydrase. It had, of
course, been known earlier that these proteins also have
unusually strong blue colours (extinction coefficient of
the order of 10^3 $M^{-1}cm^{-1}$) but inorganic chemists had suggested that the colour must be due to charge-transfer
bands involving Cu^{1+} (Orgel, 1958; Williams, 1963). As
each oxidase molecule contains several ions of copper, it
was possible that the colour and EPR signals involved
different ions. Broman et $al.$ (1962) showed, however,
that the blue colour is indeed due to the same Cu^{2+} as
gives the unique EPR signal. In the same study we found
that the Cu^{2+} EPR signal only corresponds to about 50% of
the total copper content. Thus, the possibility remained
that the specific bonding responsible for the unique spectral properties involved an interaction between Cu^{2+} and
Cu^{1+}. This possibility was, however, ruled out by the
finding of Broman et $al.$ (1963) that blue electron-
transfer proteins, such as azurin, which contain a single
Cu^{2+}, display the same unique spectral properties as the
oxidases.

It was noted in our early EPR investigations (Broman
et $al.$, 1962) that both laccase and ceruloplasmin in
addition to the unique EPR signal contains superimposed
signals of a more normal type. A detailed quantitative
study of these signals with laccase later established
that they correspond to one Cu^{2+} ion each (Malmström et
$al.$, 1968). Consequently laccase was shown to contain
three distinct forms of copper: two EPR-detectable Cu^{2+},

distinguished by their EPR spectra, and two EPR-nondetectable copper ions. It has later been shown that all blue
oxidases contain these three types of copper but that the
stoichiometries are not the same as in laccase (Malmström
et al., 1975).

In the following the spectroscopic, magnetic and redox
properties of the three types of copper will first be
discussed in relation to the bonding of the metal in the
proteins. This knowledge will then be used in conjunction
with kinetic findings in an attempt to understand the
rôle of the various forms of the metal in the catalytic
reaction. The survey will stress the basic principles
which are common to the blue oxidases. Only a selection
of all available data, necessary for a presentation of
the principles, will be reviewed. Readers desiring a more
complete documentation of the properties of the individual proteins are referred to two recent, extensive reviews (Fee, 1975; Malmström *et al.*, 1975).

The Three Forms of Copper: Spectroscopic, Magnetic and Redox Properties in Relation to Bonding

Type 1 Copper

The optical and CD spectra of the oxidized forms of a
fungal laccase are shown in Fig. 1. The beautiful blue
colour of the enzyme stems from the strong absorption
band at 610 nm, having an extinction coefficient of 4900
$M^{-1}cm^{-1}$. All blue proteins have absorption maxima with
similar extinction coefficients around 600 nm. The four
bands resolved in the CD spectrum can all be assigned to
type 1 Cu^{2+}, but an absorption band associated with the
EPR-nondetectable copper ions contributes to the strong
absorption in the near-ultraviolet region of the optical
spectrum. This band is quite broad, so that the optical
transition at 450 nm revealed in the CD spectrum is not
resolved in the optical spectrum. With proteins containing a single type 1 Cu^{2+}, such as azurin and stellacyanin,

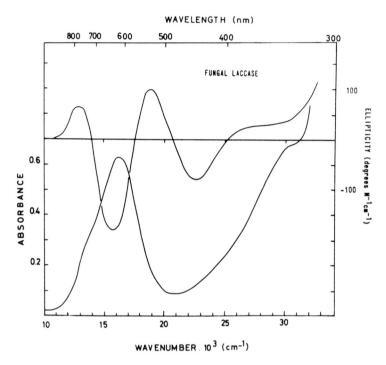

Fig. 1. Optical and CD spectra of fungal laccase (Falk and Reinhammar, 1972).

this band is clearly seen, however (Falk and Reinhammar, 1972). With many of the blue proteins more than four bands associated with type 1 Cu^{2+} can be observed. This, together with the high extinction coefficients, suggests that not only $d-d$ transitions but also charge transfer bands are involved (Falk and Reinhammar, 1972).

Experimental and simulated EPR spectra of a fungal laccase (Molitoris and Reinhammar, 1975) are given in Fig. 2. The simulation shows that the enzyme contains two Cu^{2+} ions with different EPR parameters. The spectrum assigned to type 1 Cu^{2+} is characterized by a very narrow hyperfine splitting, $|A_z|$ being 0.008 cm^{-1} compared to the range 0.015–0.020 generally found in Cu^{2+} complexes. The simulation also shows that there is a deviation from axial symmetry, as found in all blue proteins.

Despite the great similarities in the spectral properties of type 1 Cu^{2+} from all blue proteins, the oxidation-
reduction potentials show a relatively large variation
(Malkin and Malmström, 1970; Malmström et al., 1975). For
the oxidases the potentials range from 390 to 780 mV
(Reinhammar 1972; Deinum and Vänngård, 1973). Even the
lowest value falls in the upper potential range for Cu^{2+},
the value for fungal laccase (785 mV) probably being the
highest known for a stable Cu^{2+} complex.

Several attempts have been made to explain the unusual
properties of type 1 Cu^{2+} on the basis of structural
models. More than ten years ago Williams (1963) suggested
that the strong colour involves ligand-to-metal charge
transfer, probably with a cysteine sulphur as ligand. He
furthermore proposed that the symmetry around the Cu^{2+}
ion is nearly tetrahedral. A detailed structural model
incorporating these features has been proposed by Solomon
et al. (1976) on the basis of a large number of spectroscopic and chemical observations. They suggest that in
plastocyanin a Cu-S unit is found in a distorted tetrahedral environment with one peptide and two histidine
nitrogens as additional ligands. It should be noted that
all blue oxidases contain one cysteine residue, and that
cysteine is an invariant residue in the sequence of
several azurins (Malmström et al., 1975). An ESCA investigation by Solomon et al. (1975) also suggests sulphur coordination. The presence of one or more nitrogen ligands
is evidenced by ENDOR (Rist et al., 1970) and pulsed-EPR
measurements (Mims and Peisach, 1976). Evidence for two
histidine ligands in azurin (Hill et al., 1976), stellacyanin (K.-E. Falk, unpublished) and plastocyanin (Markley
et al., 1975) has been obtained by NMR.

We are now in the position that the various proposals
made for the type 1 Cu^{2+} site in the blue proteins can be
put to a direct test thanks to the beautiful crystallographic work of Colman et al. (1977) on plastocyanin from

the poplar tree. Unlike many other cases in the metallo-
protein field, the structure determination confirms most
of the essential features in the hypothetical models.
Thus, the symmetry around the metal ion is nearly tetrahedral, and cysteine sulphur as well as two histidine
nitrogens are ligands. The only unexpected feature is
that the fourth ligand is provided by a methionine sulphur instead of the peptide nitrogen suggested by Solomon
et al. (1976). Methionine can, however, not be a ligand
of the type 1 Cu^{2+} in all blue proteins, as stellacyanin
does not contain a methionine residue in its amino acid
sequence (Bergman et al., 1977). Available evidence would
suggest that the essential features of the unique bonding
is sulphur coordination and tetrahedral distortion, see
Fig. 2. Most likely all blue proteins also have at least
two nitrogen ligands. The fourth ligand, on the other
hand, may vary, and this variation can provide one basis
for the small changes in EPR parameters of type 1 Cu^{2+}
between different blue proteins.

Type 2 Copper

The EPR parameters associated with type 2 Cu^{2+} in all
blue oxidases are similar to those generally found for
simple Cu^{2+} complexes (Malkin and Malmström, 1970;
Malmström et al., 1975). A typical example is seen in the
simulated spectrum of Fig. 3. As the type 2 spectrum is
broader than that of type 1 Cu^{2+}, the low-field line is
entirely associated with type 2 Cu^{2+}. This line can,
therefore, be used for quantitative EPR studies of this
ion, for example, in attempts to establish its rôle in
the catalytic mechanism. So far it has not been possible
to associate any part of the visible or ultraviolet spectrum with type 2 Cu^{2+}. The distance between type 1 and
type 2 Cu^{2+} must be more than 8 Å, as the line width of
type 1 is the same in the absence of type 2 (Malkin et
al., 1969a), or vice versa that of type 2 is unchanged

Fig. 2a. The structure of the protein plastocyanin.

when type 1 Cu^{2+} is reduced (Brändén and Reinhammar, 1975).

In view of the normal EPR parameters there is no reason to assume that the type 2 Cu^{2+} site has an unusual stereochemical structure. The most characteristic unique property of type 2 Cu^{2+} is its ability to interact strongly with a number of anionic inhibitors, which include not only classic metal poisons such as N_3^- and CN^- but also the halides. In particular, F^- has been shown to be a strong inhibitor. The binding of F^- to type 2 Cu^{2+} can be observed directly on the basis of changes in the low-field EPR line (Malkin *et al.*, 1968). With fungal

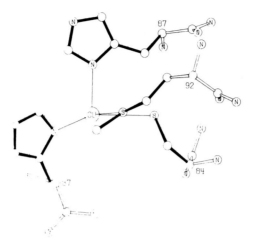

Fig. 2b. The binding site of the 'blue' copper (II) in plastocyanin.
We thank Prof. H.C. Freeman for these figures.

laccase the stability constant for the type 2 Cu^{2+}-F^-
complex is $>$ 10 μM^{-1}, compared to $<$ 10 M^{-1} for the Cu^{2+}
aquoion (Brändén *et al.*, 1973). The other blue oxidases
are also inhibited by F^- but the interaction is not as
strong as in fungal laccase (Brändén *et al.*, 1973).

There is only incomplete information on the protein
ligands to type 2 Cu^{2+}. Hyperfine structure in EPR spectra suggests that several nitrogens are ligands (Brändén
and Reinhammar, 1975; Malkin *et al.*, 1968). On the basis
of experiments with H_2O enriched in ^{17}O Deinum and
Vänngård (1975) have concluded that fungal laccase in its
resting state has H_2O or OH^- directly coordinated to type
2 Cu^{2+}. F^- expels this oxygen ligand (Deinum and Vänngård,
1975). The addition of F^- also leads to an increase in
pH suggesting that OH^- rather than H_2O is the ligand
(L.-E. Andréasson and B. Reinhammar, unpublished).

Type 3 Copper

All blue oxidases contain two or more copper ions
which are EPR-nondetectable, as was first shown with
ceruloplasmin and fungal laccase by Broman *et al.* (1962).

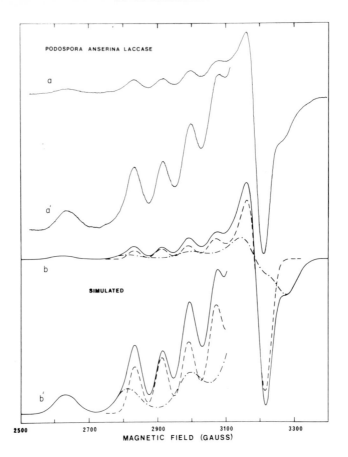

Fig. 3. Experimental and simulated EPR spectra at 9 GHz of laccase from the fungus *Podospora anserina* (Molitoris and Reinhammar, 1975).

These ions, nowadays generally referred to as type 3 copper, have been shown to be nonparamagnetic (Ehrenberg *et al.*, 1962; Moss and Vänngård, 1974), so that the absence of EPR absorption cannot be attributed to a dipolar broadening of signals from Cu^{2+} ions in close proximity. As the most common non-paramagnetic form of copper is Cu^{+}, the type 3 ions were long thought to be in this state. The situation changed, however, when it was found by Fee *et al.* (1969) that fungal laccase under anaerobic conditions can accept four electrons, i.e. the same number as the number of copper ions in the molecule. It was further-

more shown by Malkin *et al.* (1969b) that the two electrons
which are not used for reduction of EPR-detectable sites,
are accepted by a cooperative two-electron unit which has
a potential as high as that of type 1 Cu^{2+} and can be
monitored at 330 nm. For three reasons it was suggested
by Malkin and Malmström (1970) that this electron acceptor involves the type 3 copper ions: the stoichiometry,
the high oxidation-reduction potentials, and the 330 nm
absorption, not found in simple copper proteins. A reducible centre not involving type 1 or type 2 Cu^{2+} but having an absorption band at 330 nm has now also been demonstrated in the other blue oxidases (Malmström *et al.*,
1975).

Several models have been proposed for type 3 copper.
Byers *et al.* (1973) have suggested a pair of Cu^{+} ions
complexed to a disulfide bridge, but this model would
appear to be eliminated by the fact that there is no
increase in the number of sulphydryl groups on complete
reduction, as shown by Briving and Deinum (1975). Hamilton *et al.* (1973) favour Cu^{3+}. Even if Cu^{3+}-peptide complexes have been found (Margerum *et al.*, 1977), they are
all quite unstable, making this a less attractive possibility. Consequently the preferred model is an antiferromagnetically coupled Cu^{2+}-Cu^{2+} pair (Fee *et al.*,
1969; Malkin and Malmström, 1970).

The binding of F^- to type 2 Cu^{2+} changes the oxidation-
reduction potential of the type 3 coppers. This interaction suggests that the sites of these ions are close to
each other, even if a long-range interaction cannot be
excluded; some other evidence for a close proximity will
be considered in a later section. If the types 2 and 3
coppers are physically associated, this also means that
the type 3 ions are far from type 1 Cu^{2+}.

Catalytic Properties

The blue oxidases provide a facile pathway for four
electrons, derived from reducing substrates, and four
protons, derived from the solvent, to react with dioxygen,
forming two molecules of water. The problem of mechanism
entails a detailed description of the electron transfer
from substrate to dioxygen *via* the electron acceptors of
the enzyme. Anaerobic reduction experiments have shown
(Andréasson *et al*., 1973a; Malmström *et al*., 1969) that
type 1 Cu^{2+} is reduced in a second-order reaction involving a one-electron transfer from the substrate, which in
the case of organic substrates is oxidized to a free
radical (Broman *et al*., 1963). With fungal laccase the
rate constant is larger than 10^6 $M^{-1}s^{-1}$ with good substrates (Andréasson *et al*., 1973a; Malmström *et al*.,
1969), and it has been suggested that all electrons enter
the enzyme *via* the type 1 copper. It cannot, however, be
excluded that type 2 Cu^{2+} also can react directly with
the substrate, as further reduction of the enzyme is
dependent on substrate concentration (Andréasson and
Reinhammar, 1976). On the other hand, the type 3 copper
cannot be reduced directly, as shown by the fact that the
binding of F^- to type 2 Cu^{2+} prevents reduction of the
two-electron acceptor but not of type 1 Cu^{2+} (Andréasson
et al., 1973a).

The results from F^- inhibition suggest that type
2 Cu^{2+} is involved in the electron transfer from the primary acceptor, type 1 Cu^{2+}, to the two-electron acceptor,
the type 3 copper pair. EPR studies (Brändén and Reinhammar, 1975) with a rapid-freeze technique of the reduction of fungal laccase have provided direct evidence that
type 2 Cu^{2+} is reduced subsequent to the reduction of
type 1 Cu^{2+} but then reoxidized with a rate similar to
the rate of reduction of the two-electron acceptor. Consequently the type 2 copper plays an essential rôle in
the electron transfer between reducing substrate and

dioxygen. This does not exclude that it also can participate in the stabilization of intermediates, as suggested by its unusual affinity properties (Brändén et al.
1971).

Studies of the steady-state level of the 330 nm chromo-
phore (Carrico et al., 1971) have shown that type 3 copper
participates in electron-transfer reactions in the catalytic mechanism. Consequently the finding by Andréasson et
al. (1973a) that the anaerobic reduction of the two-
electron acceptor is considerably slower than the turnover
rate posed a dilemma. Detailed kinetic studies by
Andréasson and Reinhammar (1976) have now provided a solution to this enigma. They found that both tree and fungal
laccase exist in an inactive and an active enzyme form,
the equilibrium being towards the inactive form with the
fungal enzyme but becoming displaced towards the active
conformation during the catalytic reaction. It is thus no
longer necessary to postulate that dioxygen directly
affects the rate of intramolecular electron-transfer
steps, as suggested by Andréasson et al. (1973b).

It has been pointed out that one-electron reduction of
dioxygen to the hyperoxide radical would be expected to
be a slow process if the electron donor is the reduced
form of a high potential redox couple (George, 1965;
Malmström, 1973). In view of the high potential of all
copper sites in the blue oxidases a mechanism involving
four consecutive one-electron steps would not appear
likely. The presence of a cooperative two-electron unit
in these oxidases suggests that dioxygen is instead reduced directly to peroxide in a two-equivalent reaction,
thereby by-passing the energetically unfavourable formation of the hyperoxide radical. It has been proposed that
the presumed peroxide intermediate is then further reduced
in another two-electron step (Malmström, 1970). This
would, however, appear unlikely in view of the fact that
the complete reduction to two water molecules requires

not only additional electron and proton transfer but also
a breaking of the O–O bond. As there are no energetic
barriers to further reduction in one-electron steps, such
a mechanism would appear more plausible. Indeed it has now
been directly demonstrated that dioxygen reduction involves at least two one-electron steps, as a paramagnetic
oxygen intermediate has been found (Aasa *et al*., 1976a),
as shown in Fig. 4. Experiments with the oxygen isotope

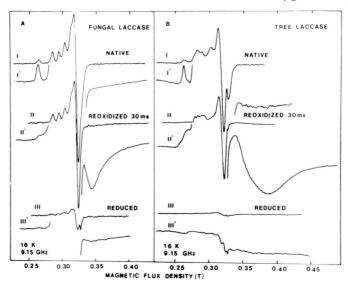

Fig. 4. EPR spectra recorded during reoxidation of fungal and tree
laccase, showing the formation of a paramagnetic intermediate (Aasa
et al., 1976a).

^{17}O have provided unambiguous evidence that this intermediate is derived from the dioxygen molecule participating in the reaction (Aasa *et al*., 1976b). The intermediate
was formed in these experiments during reoxidation of the
fully reduced laccase molecule. In this case the decay of
the intermediate occurs through electron transfer from
reduced type 2 copper. This is found to be a slow process,
which has the important consequence that it is easy to
trap the oxygen radical but unfortunately leaves doubt
about it being a true intermediate in the catalytic re-

action. In experiments with excess substrate it has, however, been shown that reduced type 1 copper rapidly transfers electrons to the intermediate (Aasa *et al.*, 1976a;
R. Brändén and J. Deinum, unpublished).

It is not possible to give a definite chemical identification of the radical intermediate. Stopped-flow experiments in which the reaction is followed at both 610 and
330 nm suggest that type 1 as well as type 3 copper have
been reoxidized when the intermediate is formed
(Andréasson *et al.*, 1976), but this is not a firm conclusion as the absorption spectrum of the oxidized form
of the two-electron acceptor in the intermediate state is
not known. A study of the decay of the intermediate shows,
however, a 1:1 correspondence between radical reduction
and type 2 reoxidation (R. Brändén and J. Deinum, unpublished), suggesting that the intermediate is at the O^-
reduction state. The unusual relaxation properties of the
intermediate are also consistent with it being the radical
O^- (Aasa *et al.*, 1976a).

With the isotope ^{17}O it has been shown by Brändén and
Deinum (1977) that when fully reduced laccase is oxidized
by dioxygen, one water molecule becomes coordinated to
type 2 Cu^{2+} (see Fig. 5) while the other one is released
into the solution (R. Brändén and J. Deinum, unpublished).
The water bound to type 2 Cu^{2+} does not exchange with bulk
water unless the enzyme is reduced. These results show
that the type 2 copper is part of the dioxygen reducing
site. It also suggests that this site is located in a
cavity which can be either closed or open depending on
the redox state of the different electron acceptors of
the protein, a property which may be important in stabilizing reactive oxygen intermediates. Recent experiments
by Brändén and Deinum (unpublished) have shown that protons from the bulk solution can interact with the intermediate, so that stabilization by the cavity being closed
does not prevent the entry of protons, required for the

formation of water from dioxygen.

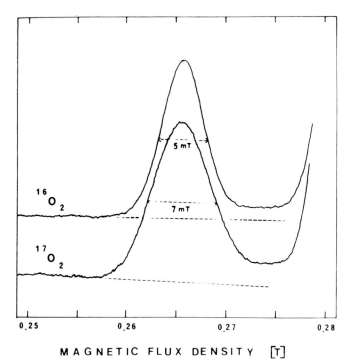

MAGNETIC FLUX DENSITY $[T]$

Fig. 5. Low-field part of the EPR spectrum of tree laccase reoxidized by $^{16}O_2$ and $^{17}O_2$, respectively (Brändén and Deinum, 1977).

Concluding Remarks

The importance of oxidases from a *biological* point of view is obvious as the reactions catalysed by these enzymes is the major source of energy in most living cells. The study of the mechanism of oxidase action has in addition provided knowledge that is of great interest from a purely *chemical* point of view. I introduced the present survey by stressing that metalloenzymes represent unique metal complexes in comparison with the simpler systems generally studied by coordination chemists. The review of structure and function in the blue, copper-containing oxidases has shown that these enzymes contain three distinct types of copper, all displaying unique features.

The type 1 Cu^{2+} has unusual spectroscopic and redox properties, while the type 2 Cu^{2+} shows a remarkably high affinity for certain anions, notably F^-. The type 3 copper exist as antiferromagnetically coupled pairs of Cu^{2+}, the coupling being uniquely strong so that the enzymes are non-paramagnetic even at room temperature. These pairs function as a strongly cooperative two-electron unit in the catalytic process. The unique properties are imparted to the metal ions by the nature and steric arrangement of the ligands provided by the protein. Only in the case of type 1 Cu^{2+}, however, do we have a detailed picture of the structural basis of the unique properties.

It is apparent that the unique metal ion coordination in the blue oxidases is intimately connected with the functional rôle of the ions in the catalytic mechanism. Thus, the tetrahedral distortion of the ligand arrangement around type 1 Cu^{2+} is undoubtedly responsible for this ion having an unusually high oxidation-reduction potential. Furthermore, its electronic structure, reflected in the spectroscopic properties, facilitates rapid electron transfer by an outer-sphere mechanism. The highly cooperative pair of type 3 copper ions allow a two-electron reduction of dioxygen directly to peroxide, thereby overcoming the kinetic inertness of the dioxygen molecule.

While the electron transfer steps involved in the mechanism of action of the blue oxidases are not understood in detail yet, our mechanistic understanding of these enzymes is a good deal more advanced than in the case of cytochrome *c* oxidase (Malmström, 1973), the major oxidase in most living cells. There are reasons to believe that there exist mechanistic analogies between cytochrome oxidase and the blue oxidases. Thus, the study of blue oxidases may also provide a guideline towards an understanding of one of the most complicated and important enzymes of bioenergetics.

References

Aasa, R., Brändén, R., Deinum, J., Malmström, B.G., Reinhammar, B. and Vänngård, T. (1976a). *FEBS Lett.* **61**, 115-119.

Aasa, R., Brändén, R., Deinum, J., Malmström, B.G., Reinhammar, B. and Vänngård, T. (1976b). *Biochem. Biophys. Res. Comm.* **70**, 1204-1209.

Andréasson, L.-E. and Reinhammar, B. (1976). *Biochim. Biophys. Acta* **445**, 579-597.

Andréasson, L.-E., Malmström, B.G., Strömberg, C. and Vänngård, T. (1973a). *European J. Biochem.* **34**, 434-439.

Andréasson, L.-E., Brändén, R., Malmström, B.G., Strömberg, C. and Vänngård, T. (1973b). *In* 'Oxidases and Related Redox Systems' (T.E. King, H.S. Mason and M. Morrison, eds.). Proceedings of the Second International Symposium, pp. 87-95. University Park Press, Baltimore.

Andréasson, L.-E., Brändén, R. and Reinhammar, B. (1976). *Biochim. Biophys. Acta* **438**, 370-379.

Bergman, C., Gandvik, E.-K., Nyman, P.O. and Strid, L. (1977). *Biochem. Biophys. Res. Comm.* **77**, 1052-1059.

Brändén, R. and Deinum, J. (1977). *FEBS Lett.* **73**, 144-146.

Brändén, R. and Reinhammar, B. (1975). *Biochim. Biophys. Acta* **405**, 236-242.

Brändén, R., Malmström, B.G. and Vänngård, T. (1971). *European J. Biochem.* **18**, 238-241.

Brändén, R., Malmström, B.G. and Vänngård, T. (1973). *European J. Biochem.* **36**, 195-200.

Briving, C. and Deinum, J. (1975). *FEBS Lett.* **51**, 43-46.

Broman, L., Malmström, B.G., Aasa, R. and Vänngård, T. (1962). *J. Mol. Biol.* **5**, 301-310.

Broman, L., Malmström, B.G., Aasa, R. and Vänngård, T. (1963). *Biochim. Biophys. Acta* **75**, 365-376.

Byers, W., Curzon, G., Garbett, K., Speyer, B.E., Young, S.N. and Williams, R.J.P. (1973). *Biochim. Biophys. Acta* **310**, 38-50.

Carrico, R.J., Malmström, B.G. and Vänngård, T. (1971). *European J. Biochem.* **22**, 127-133.

Colman, P.M., Freeman, H.C., Guss, J.M., Murata, M., Norris, V.A., Ramshaw, J.A.M., Venkatappa, M.P. (1978). *Nature* **272**, 319-324.

Deinum, J. and Vänngård, T. (1973). *Biochim. Biophys. Acta* **310**, 321-330.

Deinum, J.S. and Vänngård, T. (1975). *FEBS Lett.* **58**, 62-65.

Ehrenberg, A., Malmström, B.G., Broman, L. and Mosbach, T. (1962). *J. Mol. Biol.* **5**, 450-452.

Falk, K.-E. and Reinhammar, B. (1972). *Biochim. Biophys. Acta* **285**, 84-90.

Fee, J.A. (1975). *In* 'Structure and Bonding' (J.D. Dunitz, P. Hemmerich, R.H. Holm, J.A. Ibers, C.K. Jørgensen, J.B. Neilands, D. Reinen and R.J.P. Williams, eds.), Vol. 23, pp. 1-60. Springer-Verlag, Berlin, Heidelberg, New York.

Fee, J.A., Malkin, R., Malmström, B.G. and Vänngård, T. (1969). *J. Biol. Chem.* **244**, 4200-4207.

George, P. (1965). *In* 'Oxidases and Related Redox Systems' (T.E. King, H.S. Mason and M. Morrison, eds.). Proceedings of a Symposium held in Amherst, Mass. USA, Vol. I, pp. 3-36. John Wiley & Sons, Inc., New York.

Hamilton, G.A., Libby, R.D. and Hartzell, C.R. (1973). *Biochem. Biophys. Res. Comm.* **55**, 333-340.

Hill, H.A.O., Leer, J.C., Smith, B.E., Storm, C.B. and Ambler, R.P. (1976). *Biochem. Biophys. Res. Comm.* **70**, 331-338.

Malkin, R. and Malmström, B.G. (1970). *In* 'Advances in Enzymology' (F.F. Nord, ed.), Vol. 33, pp. 177-244. John Wiley & Sons, Inc.

Malkin, R., Malmström, B.G. and Vänngård, T. (1968). *FEBS Lett.* **1**, 50-54.

Malkin, R., Malmström, B.G. and Vänngård, T. (1969a). *European J. Biochem.* **7**, 253-259.

Malkin, R., Malmström, B.G. and Vänngård, T. (1969b). *European J. Biochem.* **10**, 324-329.

Malmström, B.G. (1970). *Biochem. J.* **117**, 15-16P.

Malmström, B.G. (1973). *Quarterly Rev. Biophys.* **6**, 389-431.

Malmström, B.G. and Vänngård, T. (1960). *J. Mol. Biol.* **2**, 118-124.

Malmström, B.G., Mosbach, R. and Vänngård, T. (1959). *Nature* **183**, 321-322.

Malmström, B.G., Reinhammar, B. and Vänngård, T. (1968). *Biochim. Biophys. Acta* **156**, 67-76.

Malmström, B.G., Finazzi Agrò, A. and Antonini, E. (1969). *European J. Biochem.* **9**, 383-391.

Malmström, B.G., Andréasson, L.-A. and Reinhammar, B. (1975). *In* 'The Enzymes' (P.D. Boyer, ed.) Vol. XII, Third Edition, pp. 507-579. Academic Press, New York and London.

Margerum, D.W., Wong, L.F., Bossu, F.P., Chellappa, K.L., Czarnecki, J.J., Kirsey, Jr., S.T. and Neubecker, T.A. (1977). *In* 'Bioinorganic Chemistry - II' (K.N. Raymond, ed.), Advances in Chemistry Series 162, pp. 281-303. American Chemical Society, Washington.

Markley, J.K., Ulrich, E.L., Berg, S.P. and Krogmann, D.W. (1975). *Biochemistry* **14**, 4428-4433.

Mims, W.B. and Peisach, J. (1976). *Biochemistry* **15**, 3863-3869.

Molitoris, H.P. and Reinhammar, B. (1975). *Biochim. Biophys. Acta* **386**, 493-502.

Moss, T.H. and Vänngård, T. (1974). *Biochim. Biophys. Acta* **371**, 39-43.

Orgel, L.E. (1958). *In* 'Metals and Enzyme Activity' (E.M. Crook, ed.) Biochem. Soc. Symposia 15, pp. 8-20. Cambridge University Press, London.

Reinhammar, B. (1972). *Biochim. Biophys. Acta* **275**, 245-259.

Rist, G.H., Hyde, J.S. and Vänngård, T. (1970). *Proc. Natl. Acad. Sci. USA* **67**, 79-86.

Solomon, E.I., Clendening, P.J., Gray, H.B. and Grunthaner, F.J. (1975). *J. Am. Chem. Soc.* **97**, 3878-3879.

Solomon, E.I., Hare, J.W. and Gray, H.B. (1976). *Proc. Natl. Acad. Sci. USA* **73**, 1389-1393.

Williams, R.J.P. (1963). *In* 'Molecular Basis of Enzyme Action and Inhibition' (P.A.E. Desnuelle, ed.). Proceedings of the Fifth International Congress of Biochemistry in Moscow 1961, Vol. IV, pp. 133-150. Polish Scientific Publishers, Warsaw.

4 MOLYBDENUM IN PROTEINS

José J.G. Moura and António V. Xavier

Centro de Química Estrutural das Universidades de Lisboa, I.S.T., Lisboa 1 (Portugal)

Introduction

Molybdenum is one of the less available elements of the earth's crust and even from sea water (Green, 1959; Mason, 1966; Calvin, 1969). However it is an element which is indispensable to life as we know it. Indeed, the molybdenum containing enzymes are crucial to the nitrogen cycle being involved both in nitrogen fixation and in the reduction of nitrate.

An inspection of the solution chemistry of molybdenum ions helps us to understand why it is that molybdenum has such an unique role among the transition metals. Some factors are

(i) The availability of relatively low-potential multi-redox states ($Mo(VI) \rightarrow Mo(III)$) which can be stabilized in an aqueous medium by ligands generally found in biological systems (nitrogen, sulphur, oxygen donors);

(ii) Molybdenum can be used to store several electrons;

(iii) Molybdenum has the capacity to act in atom-transfer which in conjunction with (i) makes it a good relay for converting reducing power to atom-transfer (Williams and Wentworth, 1974);

(iv) Molybdenum forms high coordination number complexes with ligands such as sulphur;

(v) Molybdenum forms oxo-complexes, e.g. $Mo=O$.

Thus molybdenum can be an element of choice for enzymes which catalyse processes such as oxidation-reduction
and the transfer or storage of electrons.

Besides the mentioned reactions of the nitrogen cycle
(via two enzymes: nitrogenase and nitrate reductase),
molybdenum is known to be an essential constituent of
other enzymes: molybdenum hydroxylases, sulphite oxidase
and formate dehydrogenase. Other molybdenum containing
proteins, for which no physiological role has yet been
established have been isolated from *Desulphovibrio* sp.
(sulphate reducers) (Moura *et al.*, 1976, 1977, 1978;
Hatchikian, 1978).

Molybdenum proteins are complex structures in which
molybdenum is associated with other redox centres such
as flavin, iron-sulphur centres and haem groups. In Table
I the different redox centres found in molybdenum enzymes
are listed. A complete list and characterization of the
known molybdenum proteins is presented in Table II. The

TABLE I

Redox Centres Associated with Molybdenum

Molybdoprotein	Fe/S	Flavin	Haem
Molybdenum hydroxylases	X	X	
Sulphite oxidase			X
D. gigas Mo-protein	X		
Nitrogenase	X		
Nitrate reductase	(X)	(X)	(X)
Formate dehydrogenase	(X)	(X)	(X)?

X - present
(X) - Not always present simultaneously.

aim of this article is to describe recent work on these
molybdenum enzymes stressing the value of the spectroscopic techniques that have been used to probe the role
of molybdenum. In this aspect electron paramagnetic resonance (EPR) has been the most useful tool.

Detailed information has been obtained by EPR for

three classes of proteins: molybdenum hydroxylases,
D. gigas Mo-protein and nitrate reductase. Mössbauer as
well as EPR spectroscopy has given much knowledge about
the iron-sulphur centres of nitrogenase. We shall compare
these studies as well as review recent work on formate
dehydrogenase. The participation of tungsten as a biologically active metal as well as the isomorphous substitution
of molybdenum (by tungsten and vanadium) are discussed.
As the enzymes usually contain several redox centres we
shall attempt an understanding of their interaction and
of the mechanism of electron transfer between them.

Some of the molybdenum containing proteins have been
extensively reviewed (for molybdenum hydroxylases and
sulphite oxidase see Bray (1975) and references therein;
for nitrogenase see Dalton and Mortenson (1972), Eady and
Postgate (1974), Burris and Orme-Johnson (1974), Zumft
and Mortenson (1975), Eady and Smith (1976) and Orme-
Johnson and Davis (1977); for formate dehydrogenase see
Ljungdahl and Andreesen (1975), Ljungdahl (1976) and
Thauer *et al.* (1977)). The role of molybdenum in biological systems has been subject to several recent reviews
(Spence, 1969; Williams and Wentworth, 1974; Bowden,
1975; Bray, 1976 and Wentworth, 1976).

Molybdenum in Biological Systems

As no X-ray data are available for any molybdoprotein
and since molybdenum does not contribute to the absorption spectrum of enzymes EPR remains the only powerful
tool for studying the metal ion inside the protein structure.* The line shapes and g-values of the EPR signals
can give important information about the species responsible for the signals in the enzyme. The study of the

---

*Recently the technique of X-ray absorption edge spectroscopy was
applied to the problems of the nature of molybdenum in the Mo—Fe
protein of nitrogenase from *C. pasteurianum* using a comparison
of experimental data with that from model compounds (Cramer *et al.*,
1976; Smith, 1976).

TABLE II

Molybdenum Containing Enzymes

Mo-protein		Reaction catalyzed	Redox centres	Substrate		Molecular weight	Occurrence	References
				donor	acceptor			
Molybdenum hydroxylases	Xanthine oxidase	RH→ROH	2Mo 8Fe	Purines aldehydes	O_2	275,000 (dimer)	Mammals Birds Insects Bacteria	Bray (1975) and references therein
	Xanthine dehydrogenase		(4×(2Fe,2S)) 2FAD	Purines aldehydes	Ferredoxin NAD dyes Ferry-cyanide			
	Aldehyde oxidase			Aldehydes	O_2		Liver (rabbit, pig)	
	Sulphite oxidase	$SO_3^= \rightarrow SO_4^=$	2Mo 2 haems b$_5$	$SO_3^=$	O_2	115,000 (dimer)	Liver (rat, pig, bovin, chicken)	
Molybdenum proteins involved in N_2 metabolism	Nitrogenase (MoFe protein)	$N_2 \rightarrow NH_3$	2Mo 24-32 Fe (Fe/S centres)	N_2 CH:CH MeCN	Ferredoxin dyes	200,000-230,000 (four subunits of two types 50,000 and 60,000 each)	*Clostridium pasteurianum* *Azotobacter vinelandii* *Klebsiella pneumoniae*	Orme-Johnson and Davis (1977) and references therein

Enzyme	Reaction	Donor	Composition	Acceptor	Molecular weight	Source	Reference
Nitrase reductase	$NO_3^- \rightarrow NO_2^-$	$FMNH_2$	0.4Mo 8Fe, 8S (Fe/S centres)	NO_3^-	160,000	*Micrococcus dinitrifi-cans*	Forget (1971)
			1.5Mo 20Fe, S(?) (Fe/S centres)		320,000	*E. coli* K$_{12}$	Forget (1974)
		NADH	2Mo FAD Fe(cyt b$_{552}$)	NO_3^-	230,000	*Neurospora crassa*	Garret and Nason (1969)
Mo-protein *D. gigas* (analogous proteins have been found in *Desulphovibrio* sp.)	unknown	(dithion-ite)	1Mo 12Fe, 12S (6× 2Fe, 2S)	(O_2)	120,000	*Desulpho-vibrio gigas*	Moura et al. (1976, 1977)
Formate dehydro-genases	$CO_2 \rightarrow HCOO^-$	NADP Ferre-doxin	Mo, Mo/W Se, (Fe/S centres)	CO_2	≃ 250,000	*Clostridium pasteuri-anum*	Thauer et al. (1977) and references therein
						Clostridium Thermo-aceticum	
						Clostridium formico-aceticum	
						E. coli	
						Methano-coccus vannielli	
						Vibrio succino-genes	

conditions under which a signal appears and disappears can give an indication of the reaction mechanism involved and has been used to determine the mid-point redox potential of the molybdenum centre (see 'Mid-Point Potentials of the Redox Centres', below). It is also possible to obtain information about the interaction between molybdenum and other centres using EPR (see 'Interactions between redox centres., below).

Oxidation states from VI to III can be expected for molybdenum in aqueous media. The sequence of reduction of Mo (VI) gives rise to two paramagnetic species, Mo(V) and Mo(III) which can be studied by EPR, and the diamagnetic states Mo(VI) and Mo(IV), which are EPR silent. Changes in oxidation state from VI to IV have been detected in xanthine oxidase (Bray, 1975), *D. gigas* Mo-protein (Moura *et al.*, 1978) and sulphite oxidase (Kessler *et al.*, 1974). The oxidation state III has been postulated for nitrate reductase (Forget and DerVartanian, 1972; DerVartanian and Forget, 1975, Orme-Johnson *et al.*, 1976), and suggested for nitrogenase (Bowden, 1975). There is a wealth of EPR data for Mo(III) and Mo(V) model compounds which allows useful comparison (Bowden, 1975; Goodman and Raynor, 1970; Marov *et al.*, 1972).

The fact that molybdenum has a tendency to form dimeric compounds (oxo-bridged species) has led to a possible model for the active centres of molybdoenzymes at least in the resting state (Spence, 1969; Williams and Wentworth, 1974; Wentworth, 1976). However Massey *et al.* (1970) have shown that in xanthine oxidase these models are not applicable, since two molecules of the competitive inhibitor, allopurinol, bind to 2 Mo(IV) with no difference in binding constant from one to the other. Also dissociation of nitrogenase into dimers (Smith, 1976 and references therein) shows no loss of activity and the dimers contain one Mo atom each, favouring the presence of mononuclear molybdenum, though this remains a somewhat controversial area.

Molybdenum Hydroxylases

An excellent and recent review has been published on molybdenum hydroxylases (Bray, 1975). This group of enzymes includes the molybdoproteins that catalyse the general reaction RH $\xrightarrow[+OH^-]{-2e \quad -H^+}$ ROH. This hydroxylation reaction has low specificity since the reducing substrate (RH) can be a purine, an aldehyde, a quaternary heterocyclic, etc., and the oxidizing substrate can be oxygen (when the enzyme is called an 'oxidase') or another electron acceptor such as a ferredoxin, NAD, or an artificial acceptor (when the enzyme is a 'dehydrogenase'). We shall consider here xanthine oxidase as a model for this group of enzymes, as it is the most studied (Bray and Swann, 1972; Massey, 1973; Bray, 1974; Bray, 1975, and references therein; Bray *et al.*, 1975; Lowe *et al.*, 1976; Cammack *et al.*, 1976).

Xanthine oxidase is a complex enzyme that contains two molybdenum atoms, 2 FAD and 8 iron-atoms per dimeric unit of 275,000 molecular weight. Two types of iron-sulphur (2Fe, 2S) centres, Fe/S I and Fe/S II, are present and each monomeric unit contains one of each type of group.

Molybdenum Oxygenized milk xanthine oxidase has generally no EPR signal and it is assumed to contain Mo(VI). However an EPR signal can be developed under reducing conditions (see Bray and Swann, 1972). Different types of signals have been obtained in xanthine oxidase EPR studies and were classified as very rapid, rapid, slow, inhibited, and resting II signals (Bray, 1974, 1975, 1976; Dalton *et al.*, 1976; Barber *et al.*, 1976 and Lowe *et al.*, 1976). The experimental conditions under which the signals develop are listed at the bottom of Table III. This Table also describes the parameters and the origins of the various EPR signals. The signals are relatively similar (g_{av}

J.J.G. MOURA and A.V. XAVIER

TABLE III

Parameters of Molybdenum EPR Signals in Molybdoproteins

Mo-protein	g-value		A_{av} values
	g_{av}	g_z	(^{95}Mo) (G)
Xanthine oxidase[a] MO(V)			
Very rapid[b]	1.977	2.025	34
Rapid[c]	1.973-1.974	1.989-1.994	
Inhibited[d]	1.973	1.953	
Slow[e]	1.965	1.956	
Resting II[f]	1.973	1.980	
Sulphite oxidase[g] Mo(V)			
High pH	1.965	1.984	none
Low pH	1.977	2.000	10
D. gigas Mo-protein[h] Mo(V)			
'Slow' type[i]	centered around 1.97		
M. denitrificans[j] nitrate reductase			
Mo(V)[k]	1.985($g\perp$)	2.045($g/\!/$)	
Postulated Mo(III)[l]	1.999($g\perp$)	2.023($g/\!/$)	
E. coli K_{12}[m] nitrate reductase			
Mo(V)	centered around 1.988		
Postulated Mo(III)	2.008($g\perp$)	2.032($g/\!/$)	
E. coli k_{12}[n] nitrate reductase			
Mo(V) first species[o]	1.983	1.999	
Mo(V) second species[p]	1.978	1.995	
Model Compounds[q]			
Quinoline-8-thiol Complex-Mo(V)	1.981	2.019	34

[a] Bray, 1975.
[b] Active enzyme (reduction at high pH with xanthine).
[c] Active enzyme (reduction with any substrate).
[d] Enzyme treated with HCHO or CH_3OH.
[e] Enzyme treated with cyanide (desulphoenzyme) and reduced for long periods.
[f] Mo(V) signal resistant to both oxidation and reduction obtained after conversion of the enzyme to the desulpho form and reaction with ethylene glycol or other reagents (Lowe *et al.*, 1976). Resting

Table III continued

I type signals have been observed in native xanthine dehydrogenase from *Veillonella alcalescens*, also resistant to oxidation with EPR parameters g_z = 2.010, g_{av} = 1.983, present in variable amounts from preparation to preparation.

[g]Cohen *et al.*, 1971; Bray, 1975.
[h]Moura *et al.*, 1978.
[i]Enzyme as prepared but reduced for long periods (e.g. 20–30 minutes).
[j]Forget and DerVartanian, 1972.
[k]Enzyme as prepared.
[l]Enzyme as prepared and chemically reduced with dithionite.
[m]DerVartanian and Forget, 1975.
[n]Bray *et al.*, 1976.
[o]Enzyme as prepared.
[p]Enzyme as prepared and reduced with small amounts of dithionite, or fully reduced with dithionite and slightly oxidized with nitrate.
[q]Marov *et al.*, 1972; Bray, 1975.

values in the range 1.971 ± 0.006). Only the very rapid and the rapid signals are observed during the turnover time of the enzyme, and are considered to be catalytically relevant (Bray, 1975). The conversion of very rapid to rapid signals seems to be related to the uptake of a proton. The interaction of this proton with the molybdenum gives rise to the hyperfine structure observed in the EPR signal of molybdenum. It is seen in all the Mo signals with the exception of the very rapid (Bray, 1975) and resting II signals (Lowe *et al.*, 1976). The interacting protons of the rapid and slow signals are exchangeable with solvent water molecules but in the inhibited species the proton is not exchangeable (Bray, 1975). The possibility that the interaction is due to the formation of a hydride has been ruled out since Bray and Vänngård (1969) estimated a distance of *c*. 3 Å between the molybdenum centre and the proton.

The molybdenum signals generally correspond to only one chemical species but the rapid signal represents the superposition of closely related spectra due to different complexes with substrates (Bray and Vänngård, 1969; Pick and Bray, 1969; Bray and Swann, 1972).

Iron-Sulphur Centres Two different iron-sulphur systems (Fe/S I and Fe/S II) have been distinguished in the
reduced state of the enzyme on the basis of different
g-values and relaxation rates of their EPR signals (Lowe
et al., 1972, Edmondson *et al.*, 1973). EPR quantification
and comparison with plant type ferredoxin suggests that
each centre is of the (2Fe,2S) type. This is supported by
CD and ORD measurements (Garbett *et al.*, 1967, Bayer *et
al.*, 1971).

Mössbauer and magnetic susceptibility measurements have
also been used in probing the iron-sulphur centres in
xanthine oxidase (Johnson *et al.*, 1969; Ehrenberg and
Bray, 1965). Spectral details and conditions under which
the iron-sulphur EPR signals are observed are summarized
in Table IV.

Desulphovibrio gigas *Mo-protein*

D. gigas Mo-protein contains one molybdenum atom, 12
iron-atoms per molecule and has a molecular weight of
120,000 daltons (Moura *et al.*, 1977, 1978). No subunits
or dimeric forms have been found and no FAD was detected.
Its physiological role has not yet been determined. However molybdoproteins associated with iron-sulphur centres
have also been isolated from *Desulphovibrio africanus* and
Desulphovibrio salexigens (Hatchikian, 1978). These two
proteins are not yet well characterized.

Molybdenum The quantification of molybdenum in the
protein was obtained by thermo-neutron activation analysis
which is a particularly sensitive technique. For the
molybdenum determination the 140.5 KeV photopeak, due to
$^{99}Mo-^{99}Tc$ was used. Details of the experimental conditions
are given by Moura *et al.* (1977). The gamma spectrum of
an irradiated sample is shown in Fig. 1.

Here again the most useful tool for the study and
characterization of the molybdenum in the protein is EPR.
The isolated protein at 77K shows a small signal around

TABLE IV

Parameters of Iron-Sulphur Centres EPR signals in Molybdoproteins

Fe/S[a] centre	g value			T[b]/K
	g_x	g_y	g_z	
Xanthine oxidase[c]				
Fe/S I	1.899	1.935	2.022	40
Fe/S II	1.97	2.07	2.12	25
D. gigas Mo-protein[d]				
Fe/S IA	1.93	1.94	2.02	77
Fe/S IB	1.93	1.94	2.02	77
Fe/S II	1.89	n.m.[e]	2.06	47
M. denitrificans[f] nitrate reductase				
Fe/S (native)	around 2.016			
Fe/S I	1.881	1.947	2.057	
Fe/S II	Same g-values as Fe/S I, with additional features at 2.031 and 1.926			
E. coli K_{12}[g] nitrate reductase				
Fe/S (native)	around 2.005			
Fe/S I	1.861	1.889	2.047	
Fe/S II	1.948		2.030	

[a] EPR signal of reduced enzymes except for native nitrate reductase.
[b] Temperature below which the signal can be observed.
[c] Bray, 1975.
[d] Moura *et al.*, 1978.
[e] Not measured due to overlapping with Mo signal.
[f] Forget and DerVartanian, 1972.
[g] DerVartanian and Forget, 1975.

g = 1.97 (Fig. 2a) which is attributed to a small amount of modified protein containing a stable form of Mo(V), similar to the Resting I signal observed in *Veillonella alcalescens* xanthine dehydrogenase (Dalton *et al.*, 1976). This accounts for 10% of the total molybdenum content. Upon reduction with dithionite a more intense signal at g = 1.97 appears (Fig. 2b). The reduction was slow when compared with the reduction of the iron-sulphur centres

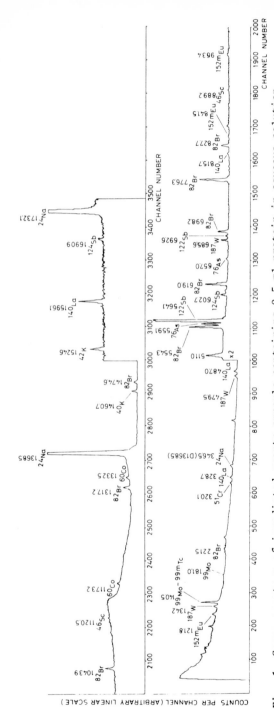

Fig. 1. Gamma spectrum of irradiated quartz ampoule containing 0.5 ml protein in aqueous solution $(9.3 \times 10^{-6} \pm 0.6 \times 10^{-6}$ mol Mo.l$^{-1})$ recorded 92 h after the end of 14 h irradiation. The counting time was 300 min and the intensity is shown as counts per channel on an arbitrary linear scale. The numbers of the spectrum are the energies (KeV) of the photo peaks. Most of the peaks were also detected in the irradiated empty ampoule, particularly ^{122}Sb, ^{124}Sb, and ^{76}As, but not the peak due to ^{99}Mo$-^{99m}$Tc (Moura et al., 1977).

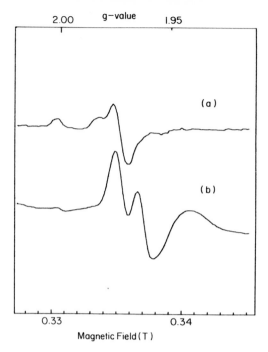

Fig. 2. Spectra around g = 1.97 due to molybdenum in *D. gigas* Mo-
protein: (a) 'Resting' signal from the native enzyme, recorded at a
gain setting of 5.000; (b) 'Slow' moybdenum signal from the reduced
protein, recorded at a gain setting of 1.500. Other experimental conditions were: temperature 77K, microwave power 0.25 mW, frequency
9.25 GH$_z$, modulation amplitude 1.0 mT (Moura *et al.*, 1978).

(Figure 3). Continuation of the reduction results in the
disappearance of the Mo signal. This behaviour is reversible. The pattern described is in agreement with the following sequence of reduction: Mo(VI)(diamagnetic) $\rightleftharpoons$ Mo(V)
(paramagnetic) $\rightleftharpoons$ Mo(IV)(diamagnetic).

    Iron-sulphur Centres The presence of six (2Fe, 2S)
type clusters was strongly suggested by the optical
(visible and CD) and EPR spectra (Moura *et al.*, 1976).
Since molybdenum does not contribute to the visible spectrum and no FAD is present the visible spectrum is directly comparable with proteins with known (2Fe, 2S)
centres (Hall *et al.*, 1974). Only some minor differences
can be seen. The CD spectrum is also typical of (2Fe, 2S)

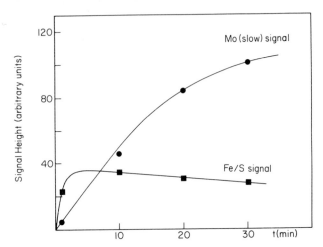

Fig. 3. Increase of EPR signal amplitudes with reduction time for Fe/S centre I and Mo in *D. gigas* Mo-protein. The protein was poised at -594 mV in the presence of dye mediators and samples were collected at different time intervals at 77K. Other conditions of measurement were: microwave power 1 mW, frequency 9.06 GHz, modulation amplitude 0.4 mT (Moura *et al.*, 1978).

centres being very similar to the characteristic spectra of plant ferredoxins and of xanthine oxidase (Garbett *et al.*, 1967).

The presence of (2Fe, 2S) centres can be confirmed by monitoring the EPR spectrum of the reduced protein in 80% dimethylsulphoxide (DMSO)(Moura *et al.*, 1976). Although in the native state of a protein any EPR spectrum is influenced by the polypeptide chain, in DMSO solution the protein is unfolded and the spectrum is more directly characteristic of the iron-sulphur centre (Cammack, 1975). For the (2Fe, 2S) centres a signal is detected at 77 K. The EPR spectrum of *D. gigas* Mo-protein in 80% DMSO (Fig. 4) is very similar to that of plant ferredoxin and xanthine oxidase (Cammack, 1975) confirming that the iron-sulphur clusters are of the (2Fe, 2S) type.

Further information on the iron-sulphur centres was obtained by EPR measurements of the reduced state of the protein at different temperatures. Fig. 5 shows the EPR

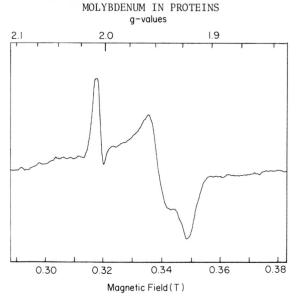

Fig. 4. EPR spectra of reduced *D. Gigas* Mo-protein in 80% DMSO solution reduced with dithionite and measured at 77 K, power 20 mW. The sample was prepared as described by Cammack and Evans (1975) (Moura *et al.*, 1976).

spectrum of the protein reduced with dithionite. Again as for xanthine oxidase different (2Fe, 2S) centres are detectable. The signal due to the Fe/S I is readily observed at 77 K with g_x = 1.93, g_y = 1.94, and g_z = 2.02. As the temperature is decreased further signals due to the Fe/S II centre develop at g_x = 1.89 and g_z = 2.06. The picture is here even more complex than for xanthine oxidase since the EPR redox titration (see Section below) shows a greater differentiation. The Fe/S I centres are composed of two components of identical temperature dependence but with different redox potential, Fe/S IA (-260 mV) and Fe/S IB (-440 mV) (Moura *et al.*, 1978).

Comparison with the xanthine oxidase spectral temperature dependence (Table IV) shows that the relaxation of *D. gigas* Mo-protein Fe/S centres is slower, indicating a closer proximity of the centres in xanthine oxidase.

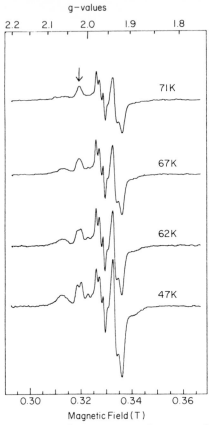

Fig. 5. EPR spectra of *D. gigas* Mo-protein at various temperatures as shown. The sample was reduced with dithionite before freezing. Other conditions of measurement were: microwave power 1 mW, frequency 9.06 GHz, modulation amplitude 0.4 mT (Moura *et al.*, 1978).

Nitrogenase

Nitrogenase catalyzes the reduction of nitrogen to ammonia and is composed of two proteins: an iron-protein (Fe protein) and a molybdenum-iron-protein (MoFe protein) (Smith, 1976; Orme-Johnson and Davis, 1977). Recently Shah and Brill (1977) reported the isolation of an iron-molybdenum containing cofactor (FeMo-co) from the MoFe protein. The EPR signals of the FeMo-co *Azotobacter vine-landii*, *Clostridium pasteurianum* and *Klebsiella pneumoniae* are identical. Recent EPR and Mössbauer studies (Smith and

TABLE V

Composition of nitrogenase MoFe protein and FeMo-cofactor[a]

	Number of metal atoms/molecule
MoFe protein	
molybdenum	2
iron	24
M_{EPR}	8-10 (42.5 ± 3)[b]
D	9-11 (14.0 ± 1)
Fe^{2+}	3-4 (38.5 ± 3)
S	1 (5.0)
FeMo-co	
molybdenum	1
iron	6-8[c]

[a]Table composed from Münck *et al.* (1975), Rawlings *et al.* (1977) and Münck, personal communication.

[b]Numbers between brackets refer to percentage of total absorption based on Mössbauer spectroscopy.

[c]8 atoms of iron based on chemical analysis of FeMo-co (Shah and Brill, 1977) and 6 atoms of iron based on quantification of the EPR signal (Rawlings *et al.*, 1977).

Lang, 1974; Münck *et al.*, 1975) indicate the presence of four spectroscopic classes of iron environments in the MoFe protein (Table V). Two of the clusters existing in the FeMo-co (Rawlings *et al.*, 1977) constitute the M_{EPR} clusters and undergo reversible redox reactions during catalytic turnover of the enzyme system. They represent the previously observed EPR active centres in intact nitrogenase with g-values 4.32, 3.65 and 2.01 (Orme-Johnson and Davis, 1977). They seem to contain six iron atoms and one Mo each.

Oxidation of the native state M_{EPR} ($S = 3/2$) with oxygen removes one electron giving a diamagnetic state ($S = 1$) (Münck, personal communication). Under nitrogen fixation conditions the M_{EPR} clusters go through a one electron reduction giving an integer spin state (probably $S \geqslant 1$ (Münck *et al.*, 1975). Since there is labile sulphide

and Fe in the FeMo-co unit as well as in nitrogenase itself, there are probably iron-sulphur centres present
which give rise to these changes. However there is evidence indicating that the spatial arrangements within the
clusters are different from the typical (2Fe, 2S) and
(4Fe, 4S) units (Rawlings *et al*., 1977). Thus *c*. 50% of
the Fe and all of the molybdenum are associated in an
unknown structure.

Two other types of centres labelled D and Fe^{2+} (containing *c*. 50% of the total iron) are sub-components of
a second type of metal cluster and have been characterized
by Mössbauer (Münck *et al*., 1975, Orme-Johnson and Davis,
1977, Debrunner *et al*., 1977). These studies suggest that
the D centres contain low-spin ferrous iron. This iron
seems to be in a diamagnetic complex, but participating in
spin-coupled clusters with the Fe^{2+} centre. The iron of
the Fe^{2+} centres is high-spin ferrous in character and
is coordinated by sulphur. It seems to be incorporated in
a diamagnetic complex, possibly involved with the D type
centres (in spin-coupled clusters). Since oxidation yields
a spin state $S \geqslant 5/2$ these clusters each contain one $1Fe^{2+}$
and 3D type iron atoms. The ratio of 3:1 observed be-
tween the D and Fe^{2+} groups suggests the presence of
4 × (4Fe, 4S) units which are somewhat distinct from typical ferredoxin (4Fe, 4S) clusters (Münck *et al*., 1975).
Component S, only observed by Mössbauer spectroscopy, is
identified as low spin ferrous and could be a contaminant.

Nitrate Reductase

EPR studies on molybdenum of nitrate reductase isolated
from *Pseudomonas aeroginosa* (Fewson and Nicholas, 1961),
Micrococcus denitrificans (Forget and DerVartanian, 1972),
and *Escherichia coli* K_{12} (DerVartanian and Forget, 1975;
Bray *et al*., 1976) have been reported.

Molybdenum Some controversy is present in the literature about the interpretation of molybdenum EPR signals of

E. coli K$_{12}$ nitrate reductase. DerVartanian and Forget
(1975) observed that at low temperature the molybdenum
signal is not detectable due to the intensity of the
iron-sulphur signals around g = 2. It is likely that the
Mo signal is very easily saturated at low temperatures.
However at higher temperatures (> 40 K) the iron-sulphur
centres are too broad to be detected and at 80 K Der-
Vartanian and Forget (1975) observed an assymetric EPR
signal at $g_\perp$ = 1.985 and $g_\parallel$ = 2.045 which was attributed to
Mo(V) in a monomeric state. Support for the fact that
Mo(V) is present in the native enzyme is the decrease observed in the signal when ferricyanide is added, (Mo(V) →
Mo(VI)). Quantification of the molybdenum signal accounts
for only 15% of the molybdenum content. Enzymatic reduction results in the complete disappearance of the Mo(V)
signal. Further reduction using dithionite gives a signal
with g = 2.008 and g = 2.031 which was postulated to
arise from Mo(III) in a different ligand environment
(Forget and DerVartanian, 1972). This signal was also observed for the *M. denitrificans* enzyme and as the intensities of both Mo(V) and Mo(III) (postulated) signals are
identical, these authors propose that one molybdenum
species is converted into the other. The reaction of nitrate reductase with nitrate (oxidising substrate), nitrite or azide (competitive inhibitor) results in an alteration of the Mo(V) signal with consequent shift of the
g-value (g in *M. denitrificans* nitrate reductase changes
to 2.090) suggesting that the substrate binds to molybdenum or that somehow it alters the molybdenum environment (DerVartanian and Forget, 1975).

Small shifts of the g_x value (from 1.999 to 2.003)
were also observed in *E. coli* K$_{12}$ nitrate reductase on
addition of nitrate (Bray *et al.*, 1976) but no conclusions
were made about the specificity of this effect. Bray *et*
al. (1976) observed a well defined EPR spectrum of Mo(V)
in the *E. coli* K$_{12}$ enzyme, which accounted for 10% of the

total molybdenum content.* It gave a rhombic type of spectrum corresponding to a single chemical species (g_x =
1.999, g_y = 1.985, g_z = 1.964 and g_{av} = 1.983). Comparison
of the EPR spectra in ^1H_2O and ^2H_2O showed that interaction with an exchangeable proton occurred, since in ^1H_2O
splitting is observed in the molybdenum signal (Bray *et*
al., 1976). At high pH the splitting due to proton interaction was not observable (Vincent and Bray, umpublished
data in Bray, 1976). This is a phenomenon similar to that
observed in milk xanthine oxidase, for the rapid and slow
signals (Bray, 1976) and for sulphite oxidase (Cohen *et*
al., 1971). Addition of small amounts of dithionite to the
enzyme, in the absence of nitrate, had little effect on
the intensity of the Mo(V) signal but caused the appearance of a new species which Bray *et al.* (1976) presume to
be due to Mo(V) also, with g_x = 1.995, g_y = 1.982, g_z =
1.956 and g_{av} = 1.978. With further addition of dithionite
both signals disappear, but both species appear again when
small amounts of nitrate were added. On further oxidation
with nitrate the second Mo(V) species disappeared and only
the original signal was observable with a slight change in
spectral parameters observed in the presence of nitrate.

Bray *et al.* (1976) rationalize their results in terms
of two different molybdenum species both undergoing reduction with dithionite and oxidation with nitrate. The
first signal would be due to the Mo(V) state in the native
enzyme and could be reduced to the Mo(IV) state. A second
species in a Mo(VI) state should appear as a Mo(V) transient during both reduction and reoxidation. The g-values
reported by Bray *et al.* (1976) are not as high as those
reported by DerVartanian and Forget (1975). These high
values seem to be outside the expected range for molyb-

---

*In milk xanthine oxidase the relative low intensity of the molyb-
denum signal was suggested to be due to equilibria between the different oxidation states of the metal (IV to III) (Olson *et al.*, 1974,
Bray *et al.*, 1976).

denum bound to biologically available ligands (Meriwether
et al., 1966). Bray *et al.* (1976) do not support the presence of a Mo(III) state and their results seem to indicate that nitrate reductase is more like other molybdenum
containing enzymes.

Iron-Sulphur Centres The work of Forget and DerVartanian (1972) and DerVartanian and Forget (1975) illustrates
also the functional role of iron-sulphur centres in the
active site of the enzymes. Since the EPR spectra of the
enzymes isolated from *E. coli* K_{12} and *M. denitrificans*
are identical we shall consider here the last one. A complex g = 2.016 EPR signal due to non-heme iron of unknown
symmetry was detected at 15 K in the native enzyme (Forget
and DerVartanian, 1972). This signal was readily observed
at temperatures lower than 40 K. The enzymatic reduction
of nitrate reductase by addition of *D. gigas* hydrogenase,
under hydrogen and in the presence of nitrate, resulted in
the complete disappearance of the initial signal at g =
2.016 and subsequent appearance of a reduced iron-
sulphur signal (Type I) with g-values at 2.057, 1.947 and
1.881. Further reduction of the enzyme accomplished chemically with dithionite, resulted in the appearance of a
complex signal with g-value around 1.9. This last signal
can be rationalized in terms of a single reduced iron-
sulphur centre (Type II) having approximately the same
g-values as Type I but with additional details at g =
2.031 and 1.926. These results are almost superimposable
with those observed for the *E. coli* enzyme but here the
single resonance at g = 1.881 is split into two peaks
at 1.889 and 1.861. Integration of the signals shows that
the g = 2.016 Type I and Type II reduced centres are equi-
molecular.

About 32% of the total iron content could be accounted
for. The fact that Type II centres are only developed when
the enzyme is reduced with dithionite and not enzymatically (then only Type I are visible) indicates that the

two types of centres have quite different redox potentials.

Similar EPR signals have been observed in homogeneous proteins, e.g. ferredoxin I from *A. vinelandii* (Sweeney *et al.*, 1975) and ferredoxin II from *D. gigas* (Cammack *et al.*, 1977) and were attributed to the (4Fe, 4S) centre clusters in an oxidation state which has three Fe(III) and one Fe(II) (the C⁻ state of the 'Three States Hypothesis' (Carter *et al.*, 1972)). Since optical spectra show that on reduction the absorption increases progressively from 600 to 280 nm, revealing a weak maximum at 410 nm, a plausible interpretation of the spectroscopic studies is that some of the iron-sulphur clusters can undergo a two electron reduction from the C^- to the C^{3-} states (both C^- and C^{3-} states are paramagnetic giving EPR spectra with g-values respectively g = 2.02 and g = 1.94), whereas the other iron-sulphur clusters use the C^{2-}/C^{3-} oxidation states. It is not possible however to assign Type I or Type II centres to the centre initially giving the g = 2.016 signal, or to say if all are independent clusters.

Isomorphous Replacement of Molybdenum

Tungsten and molybdenum are elements of the same group (Group VI of the Periodic Table) and have very similar solution chemistry. Their atomic and ionic radii and elec-tronegativies are almost identical. Binding constants with ligands usually found in biological systems are similar (Sanderson, 1967). The lower redox states (III → V) are less accessible for tungsten than for molybdenum (Mitchell, 1973, Sykes, 1976) and this fact has been used as an explanation of why molybdenum is generally preferred in biological systems (Callis and Wentworth, 1977). In order to probe the properties of the metal ion (Dennard and Williams, 1966) replacement of molybdenum in enzymes by tungsten has been carried out in nitrogenase, nitrate reductase and sulphite oxidase (Bennemann *et al.*, 1973, Guerrero and Vega, 1975, and Johnson *et al.*, 1974, respectively). Although in general an inactive enzyme is obtained weak activities were reported for nitrogenase and hepatic sulphite oxidase (Ljungdahl, 1976 and references therein). In this last case, while the native enzyme is reduced by sulphite *in vitro* within 1 minute the tungsten substituted enzyme shows only 30—40% reduction after 30 minutes. Full reduction was accomplished with dithionite (W(VI) $\rightarrow$ W(V)), giving an EPR signal with a line shape similar to that of Mo(V) and a g = 1.87. In the case of nitrate reductase tungsten (which inhibits the enzymatic activity) was shown to form an analogue of nitrate reductase which was inactive with respect to nitrate reduction but it retained the ability to act as a NADH dehydrogenase (Notton and Hewitt, 1973).

Several reports suggest an antagonistic interaction between tungsten and molybdenum (Wentworth, 1976). For example, oral administration of WO_4^{2-} to rats at levels of 1 to 100 ppm in drinking water, results in a decrease of xanthine and sulphite oxidase activities and consequent decrease of the EPR signal (Johnson *et al.*, 1974). However, administration of 1 ppm of MoO_4^{2-} in the presence of 100 ppm of WO_4^{2-} causes no observable inhibition. Restoration of both sulphite and xanthine oxidase activities could also be accomplished by intraperitonial injection of MoO_4^{2-}. The conditions were such that the protein synthesis was inhibited, indicating that the restoration of the activities is due to a displacement of W by Mo even in a hundred fold less concentrated solution. However, reconstitution of the molybdenum enzyme *in vitro* was not possible and tungsten replacement of molybdenum in xanthine oxidase was never detected (Johnson *et al.*, 1974).

Isomorphous replacements have also been tried in nitrogenase synthesis (Smith, 1976). Experiments on the competitive effect of tungstate versus molybdate uptake indicate that molybdenum may induce the synthesis of nitrogenase in both *K. pneumoniae* and *C. pasteurianum* and of

the MoFe protein in *A. vinelandii*. This induction can be
accomplished by tungsten (not vanadium) in the case of
A. vinelandii, but resulting only in the formation of
apoprotein (Smith, 1976, and references therein). However, when *A. vinelandii* was grown under nitrogen-fixing
conditions, on a medium containing tungsten instead of
molybdenum, the metal was incorporated into a very unstable nitrogenase which was partially purified (Bennemann, 1973). Since the level of activities of the extracts was the same as that of cells from unsupplemented
medium it was probably due to residual molybdenum (Bennemann, 1973; Nagatani and Brill, 1974; Smith, 1976).

Recent work on model complexes of methyliminodiacetic
and L-cysteine with Mo(VI) and W(VI) (Callis and Wentworth, 1977 and references therein) imply, without consideration of stereochemical constraints (Sykes, 1976),
that these elements would have similar affinity to apo-
enzymes and carriers whether or not a sulphur donor atom
is involved. The model studies show that transport across
a membrane could not explain differential uptake between
the two metals when in the same oxidation state (Callis
and Wentworth, 1977). The fact that Mo(VI) is more easily
reduced than W(VI) could provide a way to distinguish
between them. However there is no evidence for the presence of a selective mechanism of uptake for molybdenum
based on redox state changes.

Formate Dehydrogenase

Formate dehydrogenases catalyze the reduction of
$CO_2 \rightarrow HCOO^-$ and have been isolated from different sources
(for a complete review see Thauer *et al.*, 1977). They form
a very heterogeneous group of enzymes since they differ
significantly in the reducing substrate (NAD, NADP, ferredoxin, cytochrome), in metal (molybdenum, tungsten, selenium, iron), flavin content, and in molecular weight
(250,000 dalton for prokaryotes and 100,000 for eukaryotes).

Here we shall be interested in the formate dehydrogenases from anaerobic bacteria, which contain molybdenum and/or tungsten.

This particular class of enzymes shows that tungsten is also a biologically important metal ion. Although molybdate or tungstate (and selenite) are required in the growth medium of *Clostridium thermoaceticum* and *Clostridium formicoaceticum*, the activity of the enzyme is increased when tungstate is used (Andreesen and Ljundgahl, 1973, Ljundgahl and Andreesen, 1975). Experiments using ^{185}W-tungstate in the presence of increasing amounts of molybdate shows that incorporation of tungsten into the formate dehydrogenases was not adversely affected by molybdate (even at concentrations 100 times greater than that of tungstate) indicating that in this case tungsten is the preferred metal (Ljungdahl and Andreesen, 1976). Formate dehydrogenases are the only enzymes which indicate that tungsten is a biologically active metal.

No EPR measurements on formate dehydrogenases have been reported. As indicated in Table VI even the presence of iron-sulphur centres is not clearly established although some suggestion of their presence comes from the similarity between the absorption spectra of this group of enzymes and those of nitrate reductase which is known to contain iron-sulphur centres (Forget, 1974). It is interesting to note that formate dehydrogenases containing Mo or W are irreversibly inactivated by cyanide, but the enzyme of the iron-sulphur flavin type (e.g. from *Pseudomonas oxalaticus* (with neither molybdenum nor tungsten) is reversibly affected. Although inactivation by cyanide is common to metallo-enzymes,* the irreversibility

---

*Cyanide reacts in different ways with the molybdenum containing enzymes (Bray, 1976 and references therein). With nitrate reductase the inhibition is competitive, and cyanide binds directly to the metal. Cyanide reacts with the molybdenum hydroxylases group of enzymes with a sulphur near the molybdenum centre and with release of thiocyanate. For nitrogenase, cyanide acts as a substrate, and little is known about the inhibition of sulphite oxidase.

TABLE VI

Characteristics of some formate dehydrogenases –
molybdo-tungsten proteins*

Source	Redox centres	Reducing agents	M.W.
Clostridium pasteurianum	Mo	Fd	$>$ 250,000
Clostridium thermoaceticum	Mo/W, Se, Fe/S**?	NADP	$\sim$ 290,000
Clostridium formicoaceticum	Mo/W, Se, Fe/S**?	not known NADP	—
Methanococcus vannielii	W, Se (?)	?	—
Eschirichia coli	Mo, Se, Fe/S**? Flavin?	Cyt b type	$>$ 230,000
Vibrio succinogenes	Mo	Cyt b type	—

*Table composed from Thauer *et al.*, 1977.
**Molybdenum or tungsten may be associated with iron-sulphur
centres since the absorption spectrum of *C. thermoaceticum* formate
dehydrogenase have been reported (Andreesen and Ljungdahl, 1974) to
be similar to that of *E. coli* nitrate reductase which is known to
contain iron-sulphur centres (Forget, 1974). Also the synthesis of
the *E. coli* formate dehydrogenase during the organism growth is iron
dependent (Fukuyama and Ordal, 1965).

observed for Mo/W enzymes suggests, by analogy with xanthine oxidase and aldehyde oxidase, that a persulphide
group is removed from the molybdenum environment with consequent release of thiocyanide (Massey and Edmondson,
1970).

Since the data available are limited it is difficult to
discuss the role of tungsten in the formate dehydrogenase
enzymes. However the fact that CO_2 can be inserted at room
temperature into the hexakis-(dimethylaminato)-tungsten-
CO_2-N complex has been suggested as a good model for subsequent reduction of CO_2 to formate (Thauer, 1972, Ljungdahl and Andreesen, 1975).

Vanadium

Molybdenum has a great affinity for sulphur ligands. This has been one of the bases of the suggestion that thiol groups are good candidates for molybdenum binding in proteins (Williams and Wentworth, 1974). The g-values of Mo(V) EPR signal also support the presence of sulphur ligands in the molybdenum environment (Dennard and Williams, 1966; Bray, 1974; Meriwether *et al.*, 1966, Marov *et al.*, 1972). These properties as well as the existence of multiredox states are only shared by one other early transition metal — vanadium (Mitchell, 1973). In fact, vanadium can replace molybdenum in nitrogenase with apparent maintenance of enzymatic activity (Smith, 1976) but not in nitrate reductase (Notton and Hewitt, 1973).

For some organisms such as *A. vinelandii*, *A. chroococcum*, and *Mycobacterium flavum*, vanadium can apparently substitute molybdenum in the same growth conditions, but slower growth rates are obtained (Benemann *et al.*, 1972). The activity of unsupplemented cells was very unstable and oxygen-sensitive. Vanadium grown cells gave extracts of better stability than extracts from molybdenum grown cells which are both stable and oxygen-tolerant (Smith, 1976 and references therein). However it is not established that vanadium is an active isomorphous replacement since a plausible explanation is that vanadium has a stabilizing structural role enabling the residual molybdenum present in the medium to function since the activity found can be accounted for by the residual molybdenum content (Nagatani and Brill, 1974; Benemann *et al.*, 1976).

Mid-Point Potentials of Redox Centres

To understand the electron transfer mechanisms between the redox centres present in complex enzymes it is important to make a complete study of the redox behaviour of these active centres. Direct measurements of mid-point potential have been reported for xanthine oxidase by

Cammack *et al*. (1976) and for *D. gigas* Mo-protein by
Moura *et al*. (1978).

The mid-point potential of the various groups (molybdenum, iron-sulphur centres, and FAD) have been measured
by potentiometric titration in the presence of dye-
mediators, removing samples at different redox potentials
for quantification of the reduced species (Mo(V), Fe/S_{red}
and FADH) by EPR spectroscopy. Details of the experimental
procedures have been previously described (Cammack *et al*.,
1976, Cammack *et al*., 1977, Moura *et al*., 1978).

The absolute values of the potentials determined by
this method enable a suggestion to be made of the reaction
mechanism of the enzyme with different substrate and artificial donors/acceptors of electrons. Comparison with
similar redox centres of other systems is then possible.

Since it is now believed (Olson *et al*., 1974; Bray,
1975) that reducing equivalents coming from substrate or
cofactors are rapidly distributed among the different
reducible centres, the relative rates and extent of reduction depends on the relative values of the mid-point potentials. The method also allows the study of possible devia-
tions from ideal behaviour, which might, for example,
result from interaction among the centres (Cammack *et al*.,
1976).

Xanthine oxidase   Oxidation-reduction potentials of
molybdenum, flavin and iron-sulphur centres in milk
xanthine oxidase were studied in detail by Cammack *et al*.
(1976), using potentiometric titration followed by EPR
measurements. The native enzyme as well as modified forms
of the enzyme (deflavo (alkylated)* and desulphoenzyme)
were studied and the values obtained are listed in Table
VII. The behaviour of the four redox-active centres in
functional xanthine oxidase (in Tris buffer) is almost
ideal. Cammack *et al*. (1976) studied in detail the effect

---

*Enzyme treated with iodoacetate, which modifies the FAD group
(Cammack *et al*., 1976 and references therein).

TABLE VII

Comparison of Mid-Point Reduction Potentials for D. gigas Mo-Protein (Moura et al., 1978) and Xanthine Oxidase (Cammack et al., 1976)

	Fe/S I	Fe/SII	E (mV) Mo(VI)/Mo(V)	Mo(V)/Mo(IV)	FAD/FADH	FADH/FADH$_2$
D. gigas Mo-protein*	-260 (centre A) -440 (centre B)	-285	-450 (slow)	-530 (slow)	—	—
Xanthine oxidase native enzyme*	-336	-255	-397 (rapid)	-450 (rapid)	-387	-223
Xanthine oxidase native enzyme**	-343	-303	-355 (rapid)	-355 (rapid)	-351	-236
Xanthine oxidase desulphoenzyme**	-340	-300	-400 (slow)	-480 (slow)	-344	-237
Xanthine oxidase deflavo(alkylated)enzyme**	-327	-288	-368 (rapid)	-385 (rapid)	—	—

*Tris buffer   **Pyrophosphate buffer

of different buffer systems. The results are not easily
interpretable.

D. gigas *Mo-Protein* The same method as that used for
xanthine oxidase was applied to the determination of the
mid-point potentials of the redox centres of D. *gigas* Mo-
protein. Even in the presence of redox mediators the reduction of the molybdenum was much slower than the iron-
sulphur centres and the molybdenum signal showed extremely
non-ideal behaviour, as was also observed for the 'slow'
molybdenum signal of desulpho-xanthine oxidase (Cammack
et al., 1976).

The redox titration curves are shown in Figure 6 and
mid-point potentials are listed in Table VII. An important conclusion of these studies comes from the fact that
the signal at g = 1.94 shows two stages of reduction, the
first at -260 mV and the second at approximately -440 mV.
Neither of these processes could be due to Fe/S II centres
since they were observed both at 77 K and at 24 K. The
two stages were labelled as Fe/S IA (-260 mV) and Fe/S IB
(-440 mV). Thus in this protein, three types of iron-
sulphur centres could be distinguished by means of temperature dependence, EPR measurements and redox titration
methods. Double integration of Fe/S I centres accounts for
3—4 of these centres per molecule. A comparison of mid-
point potentials of the redox centres of D. *gigas* Mo-
protein with xanthine oxidase (Table VII) shows that the
iron-sulphur centres in the D. *gigas* protein have a larger
range of available redox potentials while in xanthine
oxidase they are quite similar. The potentials of the
iron-sulphur centres in the proteins are within the range
of redox potentials observed for ferredoxins (Hall *et al.*,
1974). The reductions of molybdenum in xanthine and D.
gigas Mo-protein is that for the conversion Mo^{VI}(dia) $\rightleftharpoons$
Mo^{V}(para) $\rightleftharpoons$ Mo^{IV}(dia), as detailed above.

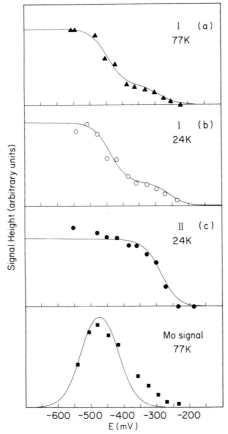

Fig. 6. Redox titration curves in *D. gigas* Mo-protein. EPR signal intensities plotted *vs*. redox potential: (a) Fe/S centre I measured by the height of the g = 1.94 signal; (b) The same feature, measured at 24 K; (c) Fe-S centre II measured by the height of the g = 2.06 signal at 24 K; (d) Mo(V) signal measured at 77 K (Moura *et al*., 1978).

Interactions between Redox Centres

The mechanism of electron transfer in proteins is a subject of great importance (Dickerson and Timkovich, 1975; Moore and Williams, 1976). Due to the large number of redox centres present it is expected that there will be some degree of interaction between them. The study of these interactions might prove useful in understanding the redox centres inside the proteins. Interaction between the centres have been detected by EPR methods for xanthine oxidase (Lowe *et al.*, 1972; Bray, 1978), *D. gigas* Mo-protein (Moura *et al.*, 1978) and nitrate reductase (Forget and DerVartanian, 1972).

Xanthine Oxidase

The EPR signal of FADH shows anomalous saturation be-haviour indicating that another paramagnetic centre is in its proximity (Beinert and Hemmerich, 1965). Flavin satu-ration studies lead to the postulation of a distance of 10–20 Å (Rajagopalan *et al.*, 1968, in this case for alde-hyde oxidase) to a non-identified paramagnetic centre. Although the nature of this paramagnetic centre is not completely established the most likely candidate is an iron-sulphur centre. In fact, EPR experiments with the deflavo (alkylated) enzyme show that no direct interaction exists between the Mo and FAD since the EPR line shape, the redox potential, and the g-values of the Mo-signal of the deflavo enzyme are not affected (Bray, 1974; Cammack *et al.*, 1976).

Lowe *et al.* (1972) reported the existence of spin–spin interaction between Mo(V) and reduced Fe/S I. This was achieved by the observation of the appearance of a split-ting in the Mo(V) EPR signal which could be correlated with the sharpening of the Fe/S I centre at temperatures lower than 40 K. Further characterization of the spin–spin interaction was achieved by the observation that no splitting of the Mo(V) signal is observed when the ex-perimental conditions are such that only molybdenum is reduced. As Bray (1975) pointed out this observation allows the separation of molecules containing only Mo(V) from those containing simultaneously Mo(V) and Fe/S I reduced centres.

Magnetic studies of the interaction between the differ-ent redox centres of xanthine oxidase allowed the deter-mination of limits for the distance from the Fe/S I centre

to the molybdenum atom (c. 20–25 Å) (Lowe and Bray, 1978).
As the molecule has a diameter of c. 40 Å there are
clearly no problems in accommodating these centres.

D. gigas *Mo-Protein*

Interaction between the Fe/S centres was shown to occur
in the *D. gigas* Mo-protein. The temperature dependence of
the EPR spectra (Fig. 5) shows that as the temperature is
lowered and the signal of the Fe/S II develops a splitting
of the g_z peak of Fe/S I (marked with an arrow) appears.
This is consistent with a spin–spin interaction between
the two iron-sulphur centres. Again the slow-reducing
Mo(V) signal of the protein at 77 K was much less easily
saturated with microwave power than the resting signal of
the unreduced protein (Moura *et al.*, 1978). As all the
Fe/S centres are reduced (paramagnetic state) before the
molybdenum it is possible that the relaxation of Mo(V) is
enhanced by the neighbourhood of the paramagnetic centres
in the protein. In fact, there is no spectral contribution from the Fe/S centres when the resting signal is observed. No estimation of the distance between the centres
has yet been made.

Nitrate Reductase

In the *E. coli* K_{12} nitrate reductase the Mo(V) signal
at g = 1.998 presents again unusual spin-relaxation properties since it is not saturated by a microwave power of
30 mW and at 100 mW power only 18% saturation is observed
(DerVartanian and Forget, 1975). The signal could be
readily detected at 25°K, suggesting an intense coupling
and/or interaction with a paramagnetic centre, for which
the most plausible candidates are the iron-sulphur centres.

Electron Transfer Mechanisms

The qualitative data obtained from interaction studies
and the estimation of the redox potentials of the differ-

ent centres suggest mechanisms for the electron transfer
within the proteins. Details on electron transfer from
one centre to the other can be useful for the understanding of complex electron transfer chains and this class of
proteins seems a good model since the electrons are transferred within a single complex where multiredox centres
are present (Bray *et al.*, 1975). Information on the way
substrates interact with enzymes may give additional
information about these mechanisms.

Xanthine oxidase

Bray (1974) suggested that the electron transfer between the redox centres and the interaction between the
groups are manifestations of the same phenomenon and
postulated a suitably oriented pathway inside the protein
(see also Lowe and Bray, 1978). These considerations and
the analysis of the redox potentials allow a picture of
electron transfer inside the protein to emerge (Bray,
1974, 1976) (see scheme in Fig. 8). Reducing substrates,
e.g. xanthine, interact via the molybdenum and transfer a
proton and an electron to the enzyme, while oxygen acts
as an oxidizing substrate via the flavin. The possibility
of other reducing equivalents coming to molybdenum, or to
the iron sulphur centres, is not excluded. Within the
enzyme all the potential internal electron transfer reactions are considered by Bray (1974) but experimental
data only support the interaction between molybdenum and
Fe/S I and the presence of a paramagnetic centre near the
FAD group (see previous section).

D. gigas *Mo-Protein*

In order to visualise the electron transfer mechanism
in the molybdoprotein from *D. gigas*, a kinetic study of
the reduction of this protein was carried out (Moura *et
al.*, 1978). All the iron-sulphur centres were reduced by
dithionite but with different rate constants (Fig. 7).

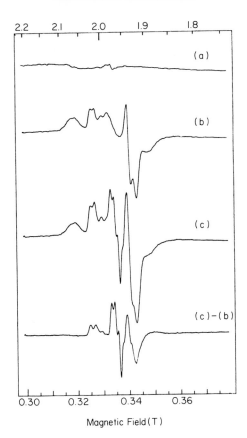

Fig. 7. Reduction of *D. gigas* Mo-protein by dithionite: (a) unreduced; (b) reduced for 1 minute; (c) reduced for 10 minutes; (d) difference between (c) and (b) obtained with a Nicolet 1020 digital oscilloscope. Conditions of measurements were: temperature 24 K, microwave power 20 mW, frequency 9.25 GHz, modulation amplitude 1 mT (Moura *et al.*, 1978).

After one minute of reduction (Figure 7b) the Fe/S II centres were completely reduced as seen by the features at g = 1.94 and 2.02. Complete reduction of the Fe/S I centres was achieved in less than 10 minutes. Spectral subtraction (Figure 7d) enables a visualization of all the spectroscopic states which are reduced between 1 and 10 minutes, Fe/S I and Mo (partially reduced).

The results are consistent with a mechanism by which the reduction of one of the centres (not molybdenum) with dithionite is slow, and the reducing electrons are rapidly distributed among the centres (Moura *et al.*, 1978). The Fe/S IA and Fe/S II (less negative redox potentials) would be reduced first followed by Fe/S IB and finally molybdenum.

It is a noteworthy fact that the Fe/S centres II and IA have very high redox potentials, similar to that of the FAD group in xanthine oxidase (Table VII). This may indicate that native *D. gigas* enzyme has replaced an FAD by Fe/S centres.

Other Molybdoproteins

Hepatic sulphite oxidase (Cohen *et al.*, 1971) and *Neurospora crassa* nitrate reductase (Garret and Mason, 1969) contain both haem and molybdenum. It is generally established that the electrons are transferred between these components during electron transfer from the reducing substrate to the oxidized one. Cohen *et al.* (1972) have shown that cytochrome *c* is a natural electron acceptor for the enzyme since it is reduced more rapidly than the haem cofactor. It provides an alternative pathway to oxygen, via cytochrome oxidase (see scheme in Fig. 8).

For *Neurospora crassa* assimilatory NADPH-nitrate reductase model compounds (Spence, 1969; Spence and Lawrence, 1976) support the electron transfer scheme proposed originally by Nicholas and Mason (1954) (see also Mason *et al.*, 1976) and shown in Fig. 8. A speculative scheme is shown both for *E. coli* K_{12} and *M. denitrificans* nitrate reductase.

Acknowledgements

This article was partly written during a leave of absence of the authors in the Freshwater Biological Institute, University of Minnesota which was supported by an NSF group-grant N$^{\circ}$.PCM76-82501, (A.V. Xavier) and the Calouste Gulbenkian Foundation (J.J.G. Moura). We should

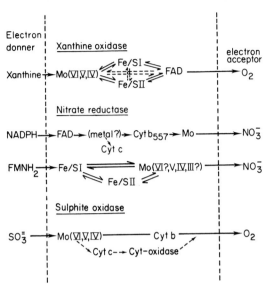

Fig. 8. Schematic representation of electron transfer in molybdo-proteins.

like to thank Professor J.J.R. Fraústo da Silva for help-ful discussion. We thank Drs. Eckard Münck and Robert C. Bray for letting us have work, respectively on nitrogen-ase and xanthine oxidase, prior to publication.

References

Andreesen, J.R. and Ljungdahl, L.G. (1973). *J. Bacteriology* **116**, 867.

Barber, M.J., Bray, R.C., Lowe, D.J. and Coughlan, M.P. (1976). *Biochem. J.* **153**, 297.

Bayer, E., Bacher, A., Krauss, P., Voelter, W., Barth, G., Bunnenberg, E. and Djerassi, C. (1971). *Eur. J. Biochem.* **22**, 580.

Benemann, J.R., McKenna, C.E., Lie, R.F., Traylor, T.G. and Kamen, M.D. (1972). *Biochim. Biophys. Acta* **264**, 25.

Benemann, J.R., Smith, G.M., Kostel, P.S. and McKenna, C.E. (1973). *FEBS Lett.* **29**, 219.

Bowden, F.L. (1975). *In* 'Techniques and Topics in Bioinorganic Chemistry', ed. C.A. McAuliffe, p. 205.

Brill, W.J., Steiner, A.L. and Shah, V.K. (1974). *J. Bacteriol.* **118**, 986.

Bray, R.C. (1974). *J. Less Common Metals* **36**, 413.

Bray, R.C. (1975). *In* 'The Enzymes', ed. Paul D. Boyer, Academic Press, New York, 3rd edn., Vol. XII, p. 299.

Bray, R.C. (1976). *In* 'Proceedings of the Climax Second International Conference on the Chemistry and Uses of Molybdenum', ed. P.C.H. Mitchell, Climax Molybdenum Co., Ltd., London, p. 271.

Bray, R.C. and Swann, J.C. (1972). *Structure and Bonding* **11**, 107.

Bray, R.C. and Vänngård, T. (1969). *Biochem. J.* **114**, 725.

Bray, R.C., Barber, M.J., Lowe, D.J., Fox, R. and Cammack, R. (1975). *In* 'Proceedings of the Tenth FEBS Symposium', Paris, Vol. **40**, 'Enzymes and Electron Transport Systems', ed. P. Desnuelle and A.M. Michelson, North Holland Publ. Co., Amsterdam, p. 159.

Bray, R.C., Vincent, S.P., Lowe, D.J., Clegg, R.A. and Garland, P.B. (1976). *Biochem. J.* **155**, 201.

Burris, R.H. and Orme-Johnson, W.H. (1974). *In* 'Bacterial Iron Metabolism', ed. Nielands, J.B., Academic Press, N.Y., p. 187.

Calvin, M. (1969). *In* 'Chemical Evolution', ed. Clarendon Press, Oxford.

Cammack, R. (1975). *Biochem. Soc. Trans.* **3**, 482.

Cammack, R. and Evans, M.C.W. (1975). *Biochem. Biophys. Res. Commun.* **67**, 544.

Cammack, R., Barber, M.J. and Bray, R.C. (1976). *Biochem. J.* **157**, 469.

Cammack, R., Rao, K.K., Hall, D.O., Moura, J.J.G., Xavier, A.V., Bruschi, M., Le Gall, J., Deville, A. and Gayda, J.P. (1977). *Biochim. Biophys. Acta* **490**, 311.

Cardenas, J. and Mortenson, L.E. (1975). *J. Bacteriol.* **123**, 978.

Cohen, H.J., Fridovich, I. and Rajagopalan, K.V. (1971). *J. Biol. Chem.* **246**, 374.

Cohen, H.J., Betcher-Lange, S., Kessler, D.L. and Rajagopalan, K.V. (1972). *J. Biol. Chem.* **247**, 359.

Cramer, S.P., Eccles, T.K., Kutzler, F.W., Hodgson, K.O. and Mortenson, L.E. (1976). *J. Amer. Chem. Soc.* **98**, 1287.

Dalton, H. and Mortenson, L.E. (1972). *Bacteriol. Rev.* **36**, 231.

Dalton, H., Lowe, D.J., Pawlik, R.T. and Bray, R.C. (1976). *Biochem. J.* **153**, 287.

Dennard, A.E. and Williams, R.J.P. (1966). *In* 'Transition Metal Chemistry', ed. R.L. Carlin, Marcel Dekker, New York, **2**, 115.

DerVartanian, D.V. and Forget, P. (1975). *Biochim. Biophys. Acta.* **379**, 74.

Dickerson, R.E. and Timkovich, R. (1975). *In* 'The Enzymes', ed. P. Boyer, Academic Press, p. 397.

Eady, R.R. and Postgate, J.R. (1974). *Nature* **249**, 805.

Eady, R.R. and Smith, B.E. (1976). *In* 'Dinitrogen Fixation', ed. R.W.F. Hardy, Wiley Interscience.

Edmondson, D., Ballou, D., VanHenvelen, A., Palmer, G. and Massey, V. (1973). *J. Biol. Chem.* **248**, 6135.

Forget, P. (1971). *European J. Biochem.* **18**, 442.

Forget, P. (1974). *European J. Biochem.* **42**, 325.

Forget, P. and DerVartanian, D.V. (1972). *Biochim. Biophys. Acta* **256**, 600.

Fukuyama, T. and Ordal, E.J. (1965). *J. Bacteriol.* **90**, 673.

Garbett, K., Gillard, R.D., Knowles, P.F. and Stangroom, J.E. (1967). *Nature* **215**, 824.

Garet, R.H. and Nason, A. (1969). *J. Biol. Chem.* **244**, 2870.

Goodman, B.A. and Raynor, J.B. (1970). *Advan. Inorg. Chem. Radiochem.* **13**, 135.

Green, J. (1959). *Bull. Geol. Soc. Amer.* **70**, 1127.

Guerrero, M.G. and Vega, J.M. (1975). *Arch. Microbiol.* **102**, 91.

Hall, D.O., Cammack, R. and Rao, K.K. (1974). *In* 'Iron in Biochemistry and Medicine', ed. A. Jacobs, and M. Worwood, Academic Press, London, p. 279.

Hatchikian, E.C. (1978). Unpublished results.

Hewitt, E.J., Notton, B.A. and Rucklidge, G.Y. (1976). *In* 'Proceed-
ings of the Climax Second International Conference on the Chem-
istry and Uses of Molybdenum', ed. P.C.H. Mitchell, Climax Molyb-
denum Co. Ltd, London, p. 276.
Johnson, E.C., Bray, R.C., Cammack, R. and Hall, D.O. (1969). *Proc.
Nat. Acad. Sci.* **63**, 1234.
Johnson, J.L., Rajagopalan, K.V. and Cohen, H.J. (1974a). *J. Biol.
Chem.* **249**, 859.
Jones, J.B. and Stadtman, T.C. (1976). *Symp. Microbiol. Production
and Utilization of Gases*, ed. H.G. Shlege, G. Gottschalk, and
N. Pfenning, Institut für Mikrobiologie, Göttingen, G.F.R.
Ljungdahl, L.G. (1976). *Trends in Biochem. Series* **1**, 63.
Ljungdahl, L.G. and Andreesen, J.R. (1975). *FEBS Lett.* **54**, 279.
Ljungdahl, L.G. and Andreesen, J.R. (1976). *Symp. Microbiol. Produc-
tion and Utilization of Gases*, ed. H.G. Schmegel, G. Gottschalk,
and N. Pfenning, Institut für Mikrobiologie, Göttingen, G.F.R.
Lowe, D.J. and Bray, R.C. (1978). *Biochem. J.* **169**, in press.
Lowe, D.J., Lynden-Bell, R.M. and Bray, R.C. (1972). *Biochem. J.* **130**,
239.
Lowe, D.J., Barber, M.J., Pawlik, R.T. and Bray, R.C. (1976). *Bio-
chem. J.* **155**, 81.
Marov, I.N., Belyaeva, V.K., Dubrov, Y.N. and Ermakov, A.N. (1972).
Russ. J. Inorg. Chem. **17**, 515.
Mason, B. (1966). *In* 'Principles of Geochemistry', 3rd edn., Wiley,
New York, pp. 45 and 195.
Massey, V. (1973). *In* 'Iron-Sulphur Proteins', vol. I, ed. W. Lowen-
berg, Academic Press, N.Y., p. 301.
Massey, V. and Edmondson, D. (1970). *J. Biol. Chem.* **245**, 6595.
Massey, V., Komai, H., Palmer, G. and Elion, G.B. (1970). *J. Biol.
Chem.* **245**, 2837.
Meriwether, L.S., Marzluff, W.F. and Hodgson, W.G. (1966). *Nature
(London)* **212**, 465.
Mitchell, P.C.H. (1973). *In* 'Proceedings of the Climax First Inter-
national Conference on the Cehmistry and Uses of Molybdenum', ed.
P.C.H. Mitchell, Climax Molybdenum Co. Ltd, London, p. 1.
Moore, G.R. and Williams, R.J.P. (1976). *Coord. Chem. Rev.* **18**, 125.
Moura, J.J.G., Xavier, A.V., Bruschi, M., Le Gall, J., Hall, D.O.
and Cammack, R. (1976). *Biochem. Biophys. Res. Common* **72**, 782.
Moura, J.J.G., Xavier, A.V., Bruschi, M., Le Gall, J. and Cabral,
J.M.P. (1977). *J. Less Common Metals* **54**, 555. Also in 'Proceedings
of the Climax Second International Conference on the Chemistry
and Uses of Molybdenum', ed. P.C.H. Mitchell, Climax Molybdenum
Co. Ltd, London, p. 271.
Moura, J.J.G., Xavier, A.V., Cammack, R., Hall, D.O., Bruschi, M.
and Le Gall, J. (1978). *Biochem. J.*, in press.
Münck, E., Rhodes, H., Orme-Johnson, W.H., Davis, L.C., Brill, W.J.
and Shah, V.K. (1975). *Biochim. Biophys. Acta* **400**, 32.
Nagatani, H.H. and Brill, W.J. (1974). *Biochim. Biophys. Acta* **362**,
160.
Nagatani, H.H., Shah, V.K. and Brill, W.J. (1974). *J. Bacteriol.* **120**,
697.
Notton, B.A. and Hewitt, E.J. (1973). *In* 'Proceedings of the Climax
First International Conference on the Chemistry and Uses of Molybdenum', ed. P.C.H. Mitchell, Climax Molybdenum Co. Ltd, London, p. 228.

Olson, J.S., Ballou, D.P., Palmer, G. and Massey, V. (1974). *J. Biol. Chem.* **249**, 4363.

Orme-Johnson, W.H. and Davis, L.C. (1977). *In* 'Iron Sulphur Proteins', Vol. III, ed. W. Lovenberg, Academic Press, N.Y., p. 15.

Orme-Johnson, W.H., Jacob, G., Henzl, M. and Averill, B.B. (1976). Abstracts, Meeting of the American Chemical Society, New York, April, p. 137.

Pick, F.M. and Bray, R.C. (1969). *Biochem. J.* **114**, 735.

Rawlings, J., Shah, V.K., Chisnell, J.R., Briel, W.J., Zimmermann, R., Münck, K.E. and Orme-Johnson, W.H. (1977). Accepted for publication in *J. Biol. Chem.*

Sanderson, R.T. (1967). *In* 'Inorganic Chemistry', Rheinhold, New York, pp. 52, 74 and 78.

Shah, V.K. and Brill, W.J. (1977). *Proc. Nat. Acad. Sci. U.S.A.* **74**, 3249.

Shrauzer, G.N., Robinson, P.R., Moorehead, E.L. and Vickrey, T.M. (1976). *In* 'Proceedings of the Second International Conference on the Chemistry and Uses of Molybdenum', ed. P.C.H. Mitchell, Climax Molybdenum Co. Ltd, p. 246.

Smith, B.E. (1976). *In* 'Proceedings of the Second International Conference on the Chemistry and Uses of Molybdenum', ed. P.C.H. Mitchell, Climax Molybdenum Co. Ltd, p. 237.

Smith, B.E. and Lang, G. (1974). *Biochem. J.* **137**, 169.

Spence, J.T. and Lawrence, G.D. (1976). *In* 'Proceedings of the Second International Conference on the Chemistry and Uses of Molybdenum', ed. P.C.H. Mitchell, Climax Molybdenum Co., Ltd, p. 259.

Stiefel, E.I., Newton, W.E. and Pariyadath, N. (1976). *In* 'Proceedings of the Second International Conference on the Chemistry and Uses of Molybdenum, ed. P.C.H. Mitchell, Climax Molybdenum Co., Ltd, London, p. 265.

Sweeney, W.V., Rabinowitz, J.C. and Yoch, D.C. (1975). *J. Biol. Chem.* **250**, 7842.

Sykes, A.G. (1976). *In* Proceedings of the Climax Second International Conference on the Chemistry and Uses of Molybdenum', ed. P.C.H. Mitchell, Climax Molybdenum Co. Ltd, London, p. 201.

Thauer, R.K. (1972). *FEBS Lett.* **27**, 111.

Thauer, R.K., Fuchs, G. and Jungerman, K. (1977). *In* 'Iron-Sulphur Proteins', Vol. III, ed. W. Lovenberg, Academic Press, p. 121.

Watt, G.D., McDonald, J.W. and Newton, W.E. (1976). *In* 'Proceedings of the Second International Conference on the Chemistry and Uses of Molybdenum, ed. P.C.H. Mitchell, Climax Molybdenum Co. Ltd, London, p. 216.

Wentworth, R.A.D. (1976). *Coordin. Chem. Rev.* **18**, 1.

Williams, R.J.P. (1970). *Quarterly Reviews* **24**, 331.

Williams, R.J.P. and Wentworth, R.A.D. (1974). *J. Less Common Metals* **36**, 405.

Zumft, W.G. and Mortenson, L.E. (1975). *Biochim. Biophys. Acta* **416**, 1.

Note added to proof

Recently a new possibility as to the structure of the active site of nitrogenase has arisen through the use of EXAFS, Extended X-ray Absorption Fine Structure. The study of this enzyme which was always known to have an unusual molybdenum binding when compared with the more general run of molybdenum proteins indicates that the molybdenum may be bound in a cluster with three iron atoms and four sulphur groups, $MoFe_4S_4$. The study of models of this cluster has been carried out in at least three groups and the physical parameters of the models accord closely with those of nitrogenase. Of immediate interest will be the redox properties and capacity of the cluster, and the possibility of an expansion of the coordination of the molybdenum atom so as to bind N_2 or other molecules which are reduced by nitrogenase. This work has not yet appeared in print but a full description has been given by Dr. K.O. Hodgson, S.P. Cramer, C.I. Stiefel and W.E. Newton. A second possibility would involve the unit $Fe(S)_2Mo(S)_2Fe$.

Examination of the cofactor which can be removed from nitrogenase shows that it is a unit of molecular weight around 2,000, that it contains approximately eight iron, one molybdenum and at least five sulphur atoms. The associated organic material is unknown at the moment (May 1978). This information has been obtained by several groups. The relationship of this cofactor to the apparently common cofactors from other molybdenum enzymes is also unknown but they do not contain iron.

5 HIGH REDOX POTENTIAL CHEMICALS IN BIOLOGICAL SYSTEMS

R.J.P. Williams and J.J.R.F. Da Silva

The Inorganic Chemistry Laboratory, Oxford, England and Centro de Química Estrutural, Universidades de Lisboa, I.S.T., Lisboa 1, Portugal

Introduction

The majority of present day living organisms require rather strict environmental conditions: temperature from 0 to about 40°C, pressure of the order of 1 atm, salinity up to about 4%, pH in the range 4 to 9, redox potentials in the region of stability of water [-0.4 to +0.8 volts at pH = 7 vs. the standard H^+/H_2 electrode].

In a few cases these limits can be exceeded to a marked extent - Table I - and it is known that, during the history of the Earth, conditions existed which were quite distant from the present ones without hampering the development of life, although they imposed constraints upon the *forms* of life that could exist. Indeed, one essential feature of evolution must be the adaptation of living organisms to changing environmental conditions, the fight against the damaging effect of some new factors. The adaptation can either take the form of the development of defences of various types against those changes or, whenever possible, the turning of the new (adverse) factor into a favourable one. The result is the survival of the fittest. We may perhaps conclude that the present day operating protective and utilising devices against (un-desirable) external factors are still one step behind the *most effective* devices, since these are developed as a

TABLE I

Minimum and maximum values of environmental factors of present day life

Factor	Lower limit	Upper limit
Temperature	-18°C (fungi, bacteria)	+104°C (sulphate reducing bacteria under 1000 atm hydrostatic pressure)
Redox potential	-450 mV at pH 9.5 (sulphate reducing bacteria)	+850 mV at pH 3 (iron bacteria)
pH	0 (Acontium velatum, fungus D, Thiobacilus thiooxidans)	13 (Plectonema nostocorum)
Hydrostatic pressure	Essentially zero	1400 atm (deep-sea bacteria)
Salinity	Double distilled water (heterothropic bacteria)	Saturated brines (Dunaliella, halophyllic bacteria, etc.)
Water activity	0.65-0.70 (Aspergillus glaucus)	Essentially 1

Source: J.R. Valentyne in G. Mamikunian and M.H. Brigs (eds.), 'Current aspects of exobiology', Pergamon Press, London 1965, pp. 1-12.

delayed reaction to an external disturbance and slowly adapted to it.

In general this adaption will include:

a) The utilisation of new conditions for growth - by the evolution of new enzymes or other compounds to take advantage of the change.

b) The protection against possible damage by the new conditions, again by the evolution of enzymes or other chemicals which eliminate the threat imposed by the change in environment.

However, no system could be expected to be perfect and some inevitable consequences arise from processes (a) and (b) above: *new* competition from other newly adapted organisms - e.g. for animals competition from newly evolved viruses, bacteria, protozoa, etc. - and *new* threats from *new* chemicals produced even by the host itself as side

products of steps (a) and (b). In this article we shall
tackle these general problems by taking the particular
case of the introduction of oxygen into the atmosphere
showing how it was related to new biochemistry of both
metals and non-metals in their high oxidation-potential
states.

Oxygen

The emergence of oxygen in biology was a particularly
important change in the environmental conditions, as it
introduced a whole range of effects under both headings
(a) and (b). New chemicals which could only be synthesised
in oxidising media with or without biological catalysts,
e.g. nitrates, made their appearance, and a far more
powerful source of energy became available to those organisms unable to use light directly through the evolution of new enzyme systems catalysing O_2 reactions with
sugars.

Thus the survival of the fittest depended first upon
the evolution of ways of burning bound hydrogen with oxygen and many organisms have evolved such a system. The
reaction clearly has many possible side-products:

$$O_2 \rightarrow O_2{}^{\cdot -} \rightarrow H_2O_2 \rightarrow OH^{\cdot} \rightarrow H_2O$$

and while the reactions of O_2 in the absence of catalysts
are quite restricted, those of $O_2{}^{\cdot -}$, H_2O_2 and $OH^{\cdot}$ are
much more indiscriminately destructive of organic materi-
als. Thus it was immediately necessary to devise further
tricks to get rid of these unwanted intermediates, see
Fig. 1, or to guide oxygen utilisation.

Unwanted though the intermediates may be to the hosts
directly they can be put to use in protective devices for
they are 'killer' chemicals (Fig. 1). Thus oxidative reactions could be made to function in the favourable ways

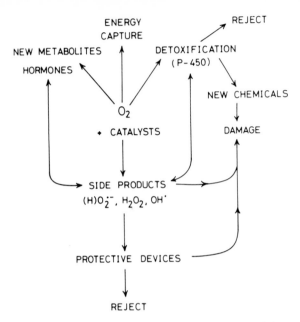

Fig. 1. A scheme showing the reactions of oxygen and its products.

$$O_2 + \text{bound-}H_2 \rightarrow \text{energy} + H_2O \qquad -(1)$$

$$O_2 \rightarrow \text{intermediates} \rightarrow \text{protective devices} \quad -(2)$$

Clearly it was valuable to evolve these two systems in separate spatial units (to avoid suicide) and advanced cells today have mitochondria for energy capture ($O_2 \rightarrow H_2O$) and various protective organelles such as peroxide-generating vesicles ($O_2 \rightarrow H_2O_2$) for attack on foreign bodies. As we shall see, quite primitive as well as advanced animals use the peroxide generating system in their protective response.

Oxygen can also be used in a further way in the preparation of new valuable chemicals and in the synthesis of intermediates (Fig. 1). Typical chemicals which are produced in the oxidative reactions of oxygen are

(i) Small transmitters and hormones, adrenaline and 5-hydroxy-tryptamin, derived by oxidation from simple amino-acids.

(ii) Protective colouring devices, e.g. melanocytic particles, and protective plastic coats of polyphenols and polyindoles, e.g. of insects.

(iii) Protective browning after physical damage — e.g. the browning of severed vegetables.

As stated above, the oxidative pathways, once available, could be used in detoxification (Fig. 1) supplementing detoxification by reduction. The most well known of these systems is the P-450 series of cytochromes, which are capable of hydroxylating a vast range of organic chemicals starting from molecular oxygen and using its partial reduction:

$$O_2 + \text{bound-}H_2 + RH \rightarrow ROH + H_2O$$
$$\downarrow$$
$$\text{destroyed}$$

The excess of many drugs and hormones are destroyed in this way. There are also more specific enzymes which use oxygen to remove excess of transmitters and hormones, e.g. the amine oxidases.

All the above effects, except for the introduction of the intermediates $O_2^{\cdot-}$, H_2O_2 and $OH^\cdot$ in some reactions, are of immediate advantage to the organism. However the evolution of the new enzymic activities introduced a new threat. The chemicals which are destroyed by *oxidative* metabolism pass through reactive intermediates such as epoxides and free radicals. These chemicals and others to which we refer below are mutagens. Mutagens in themselves are not necessarily harmful to an adapting organism for they increase the ability of the organism to evolve. This is mutagenic attack on a germ cell. By way of contrast mutagenic attack on a *somatic* cell can lead to grievous diseases such as cancer, when the individual of the species has reduced life expectation. [*Note*: a somatic cell is a body cell of a multi-cellular organism and its mutation

during the life of the individual organism is clearly
deleterious especially if this mutagenic change produces
a cell (stem-cell) capable of unrestricted growth. It
remains unclear that cancer in general is deleterious to
the species as a whole, as it removes very selectively
the older members, and the chemicals which cause it could
also be a source of increased evolutionary rate. Maybe
this is why the immune protection against cancer is
relatively feeble: perhaps cancer carries little evolutionary disadvantage.]

The introduction of oxygen brought with it the possible
production of a vast variety of new environmental chemicals formed from other non-metals such as NO, SO_2, halogens, etc. which are themselves very reactive and harmful to life. Once again while these chemicals can arise
by environmental chemical changes, evolution has found a
way to employ this same chemistry

$$O_2 + halide \rightarrow halogen + H_2O$$

in protective response devices. And yet again these very
reactions introduce new hazards of the halogenated compounds, as mutagenic agents, and they can also be used
in the production of new valuable compounds such as hormones, etc. e.g. iodo-tyrosine derivatives, thyroxines.
We note that man's oxidative industrial processes are only
another way of introducing and using these damaging
chemicals.

It can be seen that each change in the chemical (or
physical) system is a threat to life, but that this threat
can often be turned to an advantage. The usual price for
gaining the advantage however is the introduction of a
new threat ... and so on, and so on. Thus the very act of
evolving protection systems evolves an aggressive system
which can attack also the host, and from such developments
we suffer the auto-immune diseases. The parallel with the

atomic energy industry is very close for this represents
an evolved use of the environment which carries with it a
very grave risk. It is by the appropriate balance of advantage and risk that life evolves.

In the present paper we attempt to organise and unify
in a chemically coherent pattern the rather large mass
of data which is interwoven in the complex process of exploiting the potentialities of oxygen and thereby high
redox potential chemicals, especially inorganic chemicals,
for metabolic use, energy capture, and detoxification, as
well as in the development of protective devices against
the many secondary consequences of its use, while advanc-
ing some suggestions for the reasons underlying the
'logics' of the systems.

The Redox Properties of Oxygen

The inter-relationship of the couples derived from
oxygen and hydrogen are shown in the oxidation state
diagram, Fig. 2. This is a diagram of potential for re-

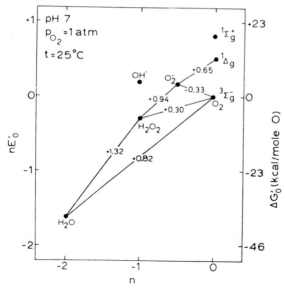

Fig. 2. The Oxidation State Diagram at pH = 7 for Oxygen. (After
Koppenol *et al.*, 1976).

action; reaction itself is then limited by activation free
energies and the presence of catalysts. While O_2, $O_2^{\cdot-}$ and
even H_2O_2 require catalysts, $OH^{\cdot}$ does not. O_2 in the singlet state is also very reactive. The immediate interest
of this oxidation state diagram is its relationship to
diagrams for other non-metals, present in large quantities
and open to attack, and other metals and non-metals available in small amounts which can act as catalysts. We note
that oxygen and its intermediates can generate redox potentials of the order of +1.0 volts, and that partial reduction of O_2 produces the highest potentials.

Other High-Potential Non-Metals

The introduction of oxygen into the environment in the
presence of water at pH = 7 changed the maximum equilibrium redox potential $E^{o'}$ due to the atmosphere from a value
associated with sulphur, $H_2S \rightarrow S$ ($E^{o'}$ = 0.0), to one
linked to oxygen, $H_2O \rightarrow O_2$ ($E^{o'}$ = +0.8). In this environment many elements change their redox states quite strikingly. For example some ammonia will be oxidised to NO
and NO_2 so that NO_2^- and NO_3^- become available in water.
In this section we look at the changes in oxidation state
of the non-metals which the presence of oxygen can induce
with or without assistance from biological processes and
in the next section we examine the same changes for the
metal elements.

Highly electronegative and readily available elements

The distinctive features of these electronegative
elements are:
  (a) They readily form simple anions, e.g., OH^-, O^{2-},
RS^-, F^-, Cl^-, Br^-, I^-
  (b) In combination with other atoms (and in biology we
are especially interested in compounds formed with carbon)
they form polar bonds such as $\frac{\backslash}{/} C^+ - X^-$
  The polarity of the bonds means that these compounds

are good reagents for the transfer of particular fragments, e.g.

$$R.CH_2.halogen + R_2N-H \rightarrow R_2N-CH_2.R + H.halogen$$

$$\begin{matrix} CH_2 \\ | \quad \backslash \\ | \quad \quad O + R_2NH \rightarrow R_2N-CH_2-CH_2OH \\ | \quad / \\ CH_2 \end{matrix}$$

(c) All their higher oxidation states are highly oxidising with respect to the element and to the negative anion. Fig. 3 summarizes the redox data on the electronegative elements, relatively to the O_2/H_2O standard couple, which allows a clear visualisation of the observations.

Now, such high redox potentials mean that as well as giving rise to highly reactive chemicals for substitution reactions (as in (b)), these elements generate oxidation systems in which the thermodynamics of the reaction are very favourable to oxidation, only restricted, in some cases, by kinetic limitations. It is this last limitation that makes the evolution of new enzymes so important in biology.

There are some other implications as far as biology is concerned. We note that oxidised nitrogen and halogens, e.g. NO_2, ClO_4^-, ClO_3^-, ClO_2^-, ClO^-, Cl_2, Br_2, I_2, etc. are all killers of biological material; indeed, they are used in disinfectants, antiseptics and weed killers. We will return to their hazardous nature again later. Note however that they are also a potential source of energy as is oxygen, and a potential source of protection against invaders.

Sulphur and selenium high oxidation states are not such powerful oxidants but even they can oxidise almost all organic materials, going themselves to the elemental

state. More generally, however, they are used in biology
in mild redox reactions, and carbon excepted, are indeed
the non-metal elements that can be used for this purpose.
The elements phosphorus and silicon are always in their
highest oxidation state in biological systems and have no
value in redox reactions, Fig. 3.

(d) A further feature of this part of the Periodic
Table is that in the higher oxidation states, the respective elements are extremely acidic and generate strong
acids, e.g. $HClO_4$, HNO_3, H_2SO_4, HCl etc. Hence, apart from
their oxidising action, nitrogen and sulphur fumes produce
dangerously acid conditions. Note

$$SO_2 \xrightarrow{O_2} SO_3 \xrightarrow{H_2O} H_2SO_4$$

$$NO \xrightarrow{O_2} NO_2 \xrightarrow{H_2O} HNO_3$$

Thus some of the hazards of SO_2 and NO pollution derive
from the presence of oxygen and water.

From a comparison of the values presented in Fig. 3 it
may be concluded that many of the oxygenated compounds of
the halogens and nitrogen could only be synthesised in
reactions involving free radicals. This is a hazard in
itself. Once formed they are thermodynamically unstable
relatively to reduction to the lower oxidation states and
cannot survive in biological media except if kinetically
inert. Many, such as NO_2^-, ClO^- are not unreactive. On the
other hand these compounds can be important synthetic
reagents forming new types of bonds and functional groups,
although in many instances catalysts are required.

Carbon Redox Potentials

High redox potential species are not known in the
simple two-electron redox reactions between hydrocarbons,

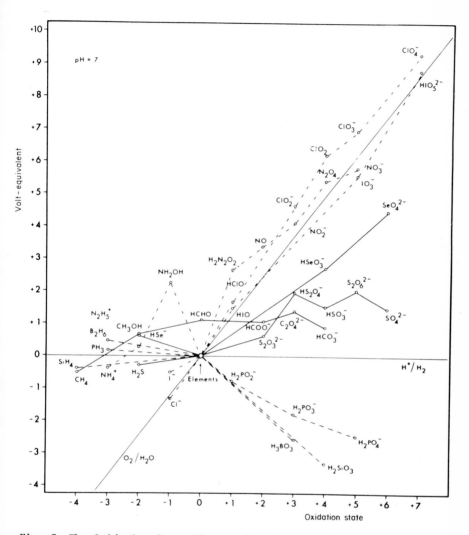

Fig. 3. The Oxidation State Diagram for the major Non-metals, pH = 7.

alcohols, ketones and acids for example in the Kreb's
cycle. On the other hand one-electron reactions of carbon
compounds to radicals give potentials greater than +1.0
volt. Typical examples are the radicals of phenols and
indoles, Table II. Radicals of electron-poor compounds
such as benzene and alkyl chains are of even higher potential and are not observed in biology. The radicals of

TABLE II

Redox Potentials of Some Carbon Compounds

Compound	Product	Redox Potential (volts)
Simple Phenol	Radical	$\sim +1.0$
Simple Indole	Radical	$\sim +1.0$
p-Diphenol	Quinone	$\sim +0.4$
Methane	Carbon Dioxide	$+0.1$
Methane	Glucose	$+0.3$
Methane	Methanol	$+0.5$
Ethanol	Acetaldehyde	$+0.2$
Water	Oxygen	$+0.8$

Potentials Relative to the $H_2 \rightarrow 2H^+$ electrode at pH = 7 (which has a potential of -0.5.

nitrogen, sulphur and selenium chemistry are not of a much higher redox potential than the two-electron couples of the oxidation states of these elements.

Metal Systems of High Redox Potential

Whereas the major non-metals must have been major bulk elements of biology the rarer non-metals and especially the transition metal ions were the catalysts from the beginning. The advent of oxygen and its simple metabolites due to reduction 'required' the evolution of new catalytic centres capable of utilising these oxidation states of oxygen and the higher oxidation states of the other non-metals. As a striking example, before we examine the redox potentials of the metals in order to see which oxidation states might be of increased value at potentials up to 0.8 v, we note that all copper enzymes in biology are high potential catalysts. Could it be that copper was very largely introduced into biology with and in response to the introduction of oxygen?

One characteristic feature of transition metals is that changes in oxidation states of 1 unit are relatively easy (compare the case of non-metals when changes of 2 units are the rule), but to get to oxidation states greater than III for the later transition metals involves

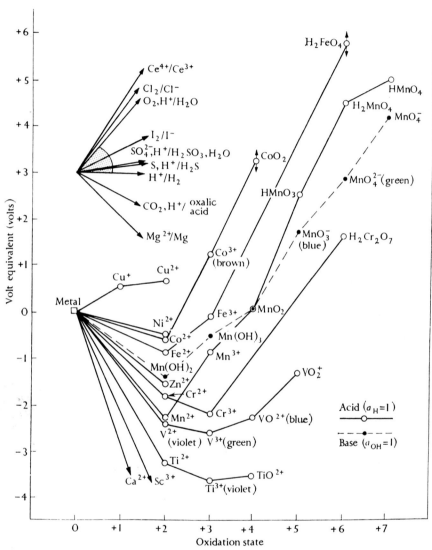

Fig. 4. The Oxidation State Diagram for the major Metals, at low and high pH.

a rather large energy change — Fig. 4, Williams (1973). In Fig. 4 we present the oxidation state diagram for a number of common biological metals relative to the O_2/H_2O couple at pH = 7 to stress the oxidising power of the several ion couples compared with that of oxygen.

There are very few examples of systems which have higher redox potential than O_2/H_2O. For example $VO^{2+}/$ $V(III)$, $Mn(III)/Mn(II)$, $MnO_2/Mn(II)$, $Co(III)/Co(II)$ and $Cu(III)/Cu(II)$ have this property. Hence the oxidised forms of some of these couples cannot exist in common biological media unless complexed by ligands which stabilize them selectively or, as is the case of VO^{2+}, in special compartments at rather low pH (e.g. in the 'vanadocytes' of ascidia where the medium is 2.5M H_2SO_4). $Mn(III)$ is obtained on oxidation of $Mn(II)$ complexes, e.g. those of polyaminocarboxylic acids, by dissolved oxygen and occurs in several enzymes such as the super-oxide dismutases and pyruvate carboxylase; complexed $Co(III)$ is, of course, the active centre of vitamin B_{12}; $Cu(III)$ has been postulated as a possible state in the mechanism of copper enzymes but direct evidence for this state is still lacking except in peptide complexes, Bossu *et al.* (1977).

The oxidation diagram also shows that the occurrence of $Mn(IV)$, $Fe(IV)$, and $Co(IV)$ (bonded to an oxygen atom) is possible but more likely if stabilised by suitable ligands. We have excluded from consideration the complexes of the oxygen molecule with metals, which can have very high redox potentials too, as these will be treated as low oxidation state metal complexes of O_2.

Summary of Redox States

We see now that the introduction into life chemistry of oxygen led not only to new chemistry involving oxygen itself but also to an extension of the chemistry of all elements from the redox range H_2/H^+ (-0.5v) to H_2S/S_n (~0.0v) to the new range H_2/H^+ (-0.5v) to H_2O/O_2 (+0.8v). This extension was almost concomitant with the evolution of new catalysts which, naturally enough for redox chemistry, are based upon transition metals. However the dif-

ferent transition metals are used differently by biology
as is shown in broad outline by Table III. We would like
to know if this specialisation of metal function is a
required feature for optimal efficiency or is accidental.
If the former is true then man has a lot to learn from
biological systems. An analysis of this problem can be
broken down into two parts. (a) The description in some
detail of the mechanisms of biological redox reactions
using the letter M at first to identify any non-specified
metal involvement. (b) An examination of the properties
of individual metals to see if in the environment of biology (i.e. bound by possible biological ligands) only
certain metals can generate the required intermediates
uncovered in (a). Alternatively the observed use of metals
in biology is haphazard when no underlying pattern should
emerge.

Putting together High-Potential Metal and Non-Metal Couples in Biology

Introduction

In this section we examine the ways in which biological
systems handle available high redox potential substrates
such as O_2 and NO_2^-, using metal ion catalysts, and how
they make other high redox potential systems from these
reactants by using oxidation reactions often in combinations. In these oxidations there are three distinct mechanisms of metal/non-metal interaction which can occur -
an inner sphere mechanism of electron transfer, an outer
sphere attack, or using combined oxygen as the attacking
group. The inner sphere attack requires a metal ligand
complex ML, the outer sphere attack is redox transfer at
a distance (by tunnelling M.....L e.g. one-electron
transfer), and the attack using oxygen can be any one of
the following: MO_2 takes e from L (one-electron outer-
sphere), MO.L or MO_2.L gives O or O_2 to L (two or four
electron, virtual inner-sphere reaction of the oxygen in

TABLE III

Requirements of Catalysts for High Potential Reactions

Reaction	Required Intermediates	Metals used in Biology
O_2 - binding	MO_2 complex (M_2O_2)	Fe (myoglobin) hemerythrin Cu haemocyanin
O_2 - reduction to H_2O	MO_2 complex, MO?	Fe(a_3)
O_2 - evolution from H O	MO_2 unstable, MO?	Mn
O_2 - reduction to H_2O plus oxidation of RH	(a) one-electron reaction (b) MO_2 complex	Cu, Fe (laccase)
O_2 - half reduction to H_2O 0-atom insertion	MO_2 complex MO relatively stable	Fe(P-450)
O_2 insertion	MO_2 and MR(?) Complex	Fe
$O_2 \cdot^-$ Disproportionation	$MO_2 \cdot^-$ complex	Fe, Cu, Mn superoxide-dismutases
H_2O_2 Disproportionation or use in oxidation	MO_2H^- complex, MO relatively stable	Fe catalase and peroxidase
Amine oxidation	Amine complexes (?)	Cu
NO_2^- Reduction	$M.NO_2^-$ complex (?)	Fe
Halide Oxidation	see H_2O_2 reactions	Fe
NO_3^- Reduction	$M.NO_3^-$ complex, MO	Mo
N_2 Reduction	$M.N_2$ complex (?)	Mo
RSH Oxidation	(?)	Se
Electron-transfer	one-electron reactions	usually Fe, (Cu)

RH = organic substrate.

which O or O_2 is inserted). Oxygen can be replaced by
another non-metal reagent such as hydrogen peroxide, sulphur or a halogen compound in two-electron reactions
involving inner sphere atom transfer. For example

$$M^+I^- + S \rightarrow M^-I^+ + S \rightarrow M^- + SI^+.$$

An alternative is oxidation which produces a one-electron
radical attack which can be outer sphere

$$M^+I^- \rightarrow M + I^\cdot; \quad I^\cdot + S \rightarrow SI^\cdot$$

Biological systems have come to use all these mechanisms
and different catalysts are used in the different reactions. We shall concentrate mainly upon oxidations
using oxygen.

Oxygen and Metal Ions in Biology

As has been mentioned before, the functions of oxygen
in biology can be grouped under three main headings;
production of new metabolic products, energy capture,
and the production of new protective devices. In the
latter case there are different aspects to be considered,
e.g. (i) the formation of protective polymers from aromatic oxidations, (ii) the elimination (detoxification)
of unwanted chemicals by hydroxylation (followed by conjugation with certain compounds to give easily excretable
species) and (iii) the destruction of foreign invaders,
such as viruses, protozoa, bacteria or just foreign
particles, although here H_2O_2 rather than O_2 is generally
used. Related to these aspects one must also consider the
problems of storage, transport, retention and release of
oxygen, which are basic to the mechanisms of utilization
of this chemical.

In what follows we will consider each of the above
aspects separately to stress the different types of re-

action which we believe are particularly relevant and
later we shall enquire about the selection by biology of
specific (metal) catalysts for the various reactions involved. The sequence of topics chosen is dictated by the
convenient development of the arguments rather than by
the logical ordering of functions as outlined above.

Oxygen binding, uptake and storage

We note that almost every use of oxygen involves the
binding of O_2 to a metal atom in an initial step and the
distribution of the bound form within a large organism.
We must then ask about the conditions which must be obeyed
for this to be possible.

TABLE IV

O_2-Binding Model Complexes

Cobalt	Wide range of pentacoordinate cobalt(II) species usually nitrogen coordination
Iron	Many (Mostly) porphyrin compounds
Manganese	Very few examples of porphyrin type
Copper	Very few examples many questionable

See G. McLendon and A.E. Martell, Coordination Chemistry Reviews
(1976), **19**, 1-39.

There are several chemical model complexes which can
bind O_2 in a reversible manner — Table IV — and we find
that it is those metal ions which have Lewis base donor
(*d*-electron) properties which can combine with oxygen and
they do it in low oxidation states (high oxidation states
oppose the donation of electrons). At the same time if the
release of oxygen is to be controlled then the overall
scheme of reaction required is:

$$M + O_2 \underset{\text{slow (controlled)}}{\overset{\text{fast}}{\rightleftarrows}} MO_2$$

This can be met by a change in the nature of M on binding O_2 such as that found on going from a high-spin to a low-spin state

$$M + O_2 \quad \rightarrow \quad MO_2$$
$$\text{(high spin)} \quad \text{(low-spin)}$$

It is iron that is the obvious choice for uptake and transport in this type of reaction. Another possible way of controlling the on/off reactions uses *dimeric* centres in which the metal donates an electron virtually completely, Lontie and Witters (1972) and Vannerte and Suberbuhler (1974)

$$O_2 + 2\ Cu(I) \underset{\text{intermediate}}{\overset{\text{fast}}{\rightleftharpoons}} (Cu(I))_2O_2 \underset{\text{slow}}{\overset{\text{slow}}{\rightleftharpoons}} (Cu(II))_2O_2{}^{2-}$$

The control mechanism might be based on the change of oxidation state and conformation at the metal site but the charge re-distribution can be effectively relayed to the surrounding protein preventing O_2 release until a reaction cycle is complete by reduction of the metal ion. This reaction mechanism is known for copper proteins. We know that iron can pick up O_2 reversibly in the similar reaction, Okamura and Klotz (1973),

$$O_2 + 2Fe(II) \rightleftharpoons (Fe(III))_2\ O_2^{2-}$$

Oxygen can be retained also at a single metal centre as peroxide with a two-fold redox step,

$$M(II) + O_2 \rightleftharpoons M(IV)O_2{}^{2-} \underset{}{\overset{H^+}{(\rightleftharpoons}} M(IV).O_2H^-)$$

Such reactions have been reported for manganese as has the final possibility, Wechsler *et al.* (1975).

$$2Mn(II) + O_2 \underset{fast}{\overset{fast}{\rightleftarrows}} 2Mn(II).O_2 \overset{slow}{\rightleftarrows} 2MnO(IV)$$

Now we note the special requirements of the different reactions — spin state change, oxidation state change upon oxygen addition either in a one or two electron reaction, and formation of oxo-cations, MO.

This example illustrates the *requirements* for reversible oxygen uptake and the metals which are used. In the next sections we shall attempt to specify the requirements for catalysis of oxygen reactions. We note first the actual biological systems which pick up oxygen.

In biology there are the hemoglobins and myoglobins (single Fe centre), the haemocyanins (two copper centre) and the hemerythrins (two iron centre) for oxygen uptake. (Rather little metal ion selectivity is shown in forming oxygen complexes.) The choice of iron and copper rather than cobalt or manganese for these functions is not an obvious one except perhaps on the basis of their availabilities. Curiously the *release* of oxygen from plants when it is formed by oxidation of water is carried out by a manganese centre only, see later.

Catalysis of formation of protective polymers: one-electron oxidation

The introduction of oxygen in biology allowed the possibility of such reactions as

$$O_2 + RH \rightarrow R^{\cdot} + H(O_2)$$

where the redox potential of the RH/R$^{\cdot}$ couple can be as high as +1.0 volt, Table II. Typical radical reactions of this type, which could not occur earlier in evolution in reducing media, are the oxidation of phenols and indoles. These reactions are the basis of formation of melanine, the pigments which are responsible for the colours of skin and hair in mammals and for the plastic coats of insects, as well as of other protective polymers, e.g. in plants. Here the reaction requires a catalyst which must be able to undergo a rapid cycling process while going to a high redox potential, probably by an outer sphere path,

$$RH + M_{ox} \rightarrow M_{red} + R^\cdot + H^+$$

$$M_{red} + O_2 + 4H^+ \rightarrow M_{ox} + 2H_2O$$

If oxygen is the direct source of oxidising equivalents then the catalyst has also to be such that $R^\cdot$ does not react with O_2^{-} or H_2O_2 which might be formed on reduction of O_2 (if these intermediates are set free). It is required not only to form $R^\cdot$ but to initiate the polymerization through

$$R^\cdot + RH \rightarrow R_2H^\cdot \xrightarrow{RH} (R)_n$$

The reduction products of oxygen must then be kept bound and separated from $R^\cdot$ or be present in very low concentration. The very best procedure is to reduce O_2 to H_2O and the second best is make and remove H_2O_2, which can be turned to H_2O and O_2 by catalase, or to make and then disporportionate O_2^{-} by turning it to $O_2 + H_2O_2$. Direct use of H_2O_2 will be discussed later.

The requirements for the whole cycle are that there should be (i) a combination of the catalyst M with O_2; (ii) a one-electron oxidation of RH; (iii) a multi-electron reduction of O_2 preferably to H_2O and (iv) O_2^{-}, H_2O_2, and $OH^\cdot$ must be avoided if possible, or removed subsequently.

Now the redox potentials of RH, where RH is a phenol or indole (Table II) are very high whence the requirement under (ii) is for a *high* potential *one*-electron metal-redox centre. (We shall assume that only metal ions can do the reactions we describe.) However the requirements

in (i) and (iii) are for an initially lower potential
system which is converted to a multi-electron reaction
centre on reaction with O_2. It could be that all three
functions could not be served by a single metal centre
when we would have

$$
\begin{array}{c}
RH \\
R^{\cdot}
\end{array}
\;\rightleftharpoons\;
\begin{array}{c}
M_1^{+} \\
M_1
\end{array}
\xrightarrow[\text{4}e\ \text{steps}]{\text{connecting}}
\begin{array}{l}
M_2O_2 \quad \leftarrow M_2 + O_2 \\[1mm]
\qquad\qquad\ \uparrow \\[-1mm]
M_2(H_2O)_2 \rightarrow M_2 + H_2O
\end{array}
$$

A typical example of the potentials in a real enzyme
which is known to be able to carry out the reaction scheme
are given in Table V. The reactions are most commonly
carried out by copper centres.

TABLE V

Copper Redox Potentials in Proteins

Type of Site	Redox Potential (mVolts)
Blue Copper Type 1	+200 to +800
Copper Type 2	~+500
Copper Pair	~+500

See the reference in the article of B.G. Malmström
in this volume.

[Note that one-electron reactions are required elsewhere at *low* as opposed to *high* potential and there are a
vast range of iron centres which are used for this purpose
e.g. cytochromes, rubredoxins, ferredoxins and no other
metal than iron is used to our knowledge in *low* potential
reactions of this kind. The actual attack in all these
one-electron steps as well as those of high potential
would seem to be by outer-sphere reactions.]

The problems of the removal of H_2O_2 or $O_2^{\cdot -}$, should
they be formed, will be dealt with later.

In biological systems there is a range of compounds
which are oxidised by the above radical path using both

single metal-atoms and several metal atoms in enzymes.
Control can be based upon matching of the redox potentials
of the reactants and the catalyst. Earlier in this article
we have seen that aromatic compounds including phenols and
indoles have redox potentials of about +1.0 volt. The
above Table indicates that Cu(II)/Cu(I) couples are not of
this high potential. It is the substrates of medium redox
potential which are uniquely controlled by single-copper
atom catalysts. Higher potentials are only reached using
oxo-cations, see Fig. 4. Copper reactions in biology are
then associated with for example the oxidation of free
Fe(II) to Fe(III), and of enols (ascorbate).

The one-electron oxidation of simple phenols and indoles is carried out in plants (for certain) and probably
in animals by the peroxidases, single metal-site iron containing enzymes which go through oxo-cation intermediates,
E=1.0 volt. Referring now to Fig. 4, note that formation
of complexes, especially with the *anionic* centres of biological ligands, will make all the oxidation state plots
pivot around zero oxidation state (largely an entropy
effect) such that the iron redox couples from III to VI
roughly run parallel with the oxygen couple, Fig. 2.
Earlier elements, e.g. Cr, remain very reducing in low
oxidation states, and later elements are very oxidising
in oxidation states above III. The peculiarity of copper
is that *stable* combination largely with *neutral* ligands
permits its oxidation state diagram between I and II to
go to *increased* potentials, +0.5 volts. Biological evolution then selected between iron and copper for different
substrate oxidations using iron in single metal atom
centres at approximately 1.0 volt and copper in *single*-
metal atom enzymes, at approximately 0.5 volt, see later.
In multi-metal atom enzymes which bind oxygen potentials
rise to those of the oxygen intermediates, 1.0 volt, Fig.
2, and *multi*-copper centres then act at high potentials.

Detoxification of unwanted chemicals: Two-electron oxidation

A general way of producing required chemicals and of removing unwanted chemicals such as excess of certain hormones, e.g. sterols or drugs, is by oxidation, but it is again essential that this should not produce oxygen radicals. Instead, groups should be introduced which by themselves or through conjugation with other chemicals communicate water solubility to the product, for the chemical can then be excreted, while oxygen goes to water.

A possible scheme which is frequently found in biological systems is

$$RH + O_2 \rightarrow ROH + (O)$$

followed, if necessary, by

$$ROH + (sugar) \rightarrow RO\text{-sugar conjugate (excretion)}$$

As an example, one may quote the case of benzene, which is insoluble in water and difficult to eliminate. Through oxidation to phenol and conjugation with glucuronic acid, it gives a phenylglucuronide which forms soluble salts and is excreted by the kidneys

$$pK_a = 10 \qquad\qquad pK_a = 3.4$$

This reaction requires the hydroxylation of the compound RH for which an oxygen atom transfer (two electron oxidation) catalyst is necessary but the metal must be able to retain oxygen first

$$M + O_2 \rightarrow M(O_2) \rightarrow M(O) + (O)$$

and, subsequently, the following reaction must be possible:

$$RH + M(O) \rightarrow M + ROH$$

Hence, MO must have a high redox potential.

Taking into account the redox potentials of metals and non-metals we find that some oxocations of the transition metals are ideal intermediates for the purpose, as many of them have high oxygen-atom transfer (MO/M) redox potentials (Fig. 4) and in lower oxidation states form oxygen complexes readily. (Note flavins can undergo similar reactions.)

Now we can rationalise the steps of this system to requirements for

(i) O_2 uptake by the metal, M.

(ii) splitting of oxygen to MO + H_2O

(iii) an O-atom transfer to RH

(iv) a supply of two electrons from a reducing source to turn MO_2 to MO + H_2O

Once again it is found that this scheme is not carried out by one centre:

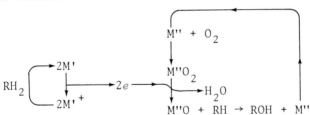

Two metal centres are used where M' undergoes two one-electron *low* potential steps. The metal M' need only be an electron transfer agent from flavin and NADH or NADPH, and its redox potential could be very low. It is invariably iron — see above. This reaction is not open to M" for there could then be the danger of radical production from O_2, e.g. OH·. The oxo-cations must be of sufficient redox potential to transfer oxygen to such molecules as benzene.

The most common metalloenzyme used in M" reactions is in fact an iron-enzyme, namely cytochrome P-450 oxidase, which hydroxylates several substrates undergoing a cycle of reactions which seems to be fairly well established, Gunzalus *et al.* (1975), and has recently received confirmation, Hrycay (1976).

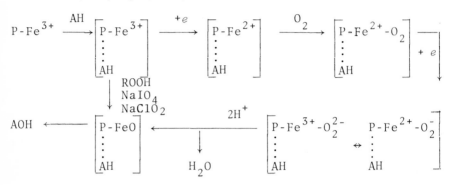

(Note the short-circuit using high-redox potential one-electron inorganic oxidants or organic hydroperoxides in place of oxygen and that there is a spin-state switch for the control of oxygen uptake and activation. We will come to this subject later again when describing some of the individual systems which operate in these processes). The reactions are inner sphere insertion of an oxygen atom into an R-H bond, perhaps by initial H-transfer to MO.

Production of Desirable Metabolites

Materials such as benzene and aliphatic hydrocarbons are useful as energy sources and carbon sources *only* if they can be converted to oxygen containing intermediates. The attack on such materials would not seem to have been possible before the advent of oxygen. Thus the anaerobic metabolic path

$$2(CHOH) \rightarrow CO_2 + CH_4$$

yielded the present sources of gas and oil. In an aerobic environment this loss of carbon from the biological
cycle does not occur. Thus apart from the use of oxygen
atom transfer reactions in detoxification they are useful in generating metabolites. There are non-haem and
P-450 like enzymes in bacteria for this purpose.

Quite different reactions are the milder hydroxylations
used in producing dihydroxy substituted benzene from
phenols e.g. catecholamines

$$(2e) \; + \; \text{(benzene ring with OH)} \; + \; O_2 \; \rightarrow \; \text{(benzene ring with OH, OH)} \; + \; H_2O$$

We observe that copper containing enzymes are very common
in these catalysed reactions which partly belong within
the one-electron systems of earlier sections. Even the
mechanism, inner sphere or outer sphere, is in doubt in
the case of these copper hydroxylating enzymes, such as
dopamine β-hydroxylase and the phenolases. A typical
possible mechanism for the enzymes is:

$$E(Cu^{2+})_2 \; + \; o\text{-diphenol} \rightarrow E(Cu^+)_2 \; + \; o\text{-quinone} + 2H^+$$
$$E(Cu^+)_2 \; + \; O_2 \rightarrow E(Cu^+)_2 O_2$$
$$E(Cu^+)_2 O_2 \; + \; \text{monophenol} \rightarrow E(Cu^{2+})_2 \; + \; o\text{-diphenol} + H_2O$$
$$E(Cu^+)_2 O_2 \; + \; o\text{-diphenol} \rightarrow E(Cu^{2+})_2 \; + \; o\text{-quinone} + 2 \; OH^-$$

Amine Oxidases

Detoxification from biologically produced amines is
brought about by the reactions

$$\backslash \!\! /CH - NH_2 \rightarrow \; >\!C = 0$$

which use molecular oxygen and some electron-transfer
components. The exact mechanism of the reaction is hard to
discern from the literature but we find that generally
copper centres form the catalysts.

Dioxygenases: Four electron oxygen-addition

We have described systems which insert *one* atom of oxygen into the molecule of the substrate; in other cases reactions such as the following can be desirable

$$RH_2 + O_2 \rightarrow R(OH)_2$$

These correspond to the insertion of the molecule (two atoms) of oxygen and the enzymes which catalyse such reactions are called dioxygenases.

Apparently the catalyst must be able to activate oxygen so as to do the reaction

$$(O_2) + RH_2 \xrightarrow{\text{M}} R(OH)_2 + M$$

An example is pyrocatechase, a non-heme iron enzyme whose mechanism is not well understood but it depends upon a single iron-atom enzyme centre.

Now usually the group RH_2 into which oxygen is inserted is a highly activated aromatic such as a resorcinol. In fact this is a very major biological pathway for it is involved in the production of lignins. Recent evidence indicates that the oxygen is in fact not activated by the metal but the oxygenation is brought about as follows

This type of O_2 attack requires a metal ion which binds phenols strongly at pH = 7, and later binds $O_2^{.-}$ but the metal ion need not cycle through oxidation states!

The requirements are very different from those in

previous sections for we now need

   (i) a metal to bind a substrate (not needed before)
which here is an anion, see Table VI.

   (ii) a metal to bind oxygen O_2 and possibly to reduce
it to an attacking group, $O_2^{\cdot -}$.

   (iii) a transfer or insertion step of $O_2^{\cdot -}$ into substrate.

Clearly the reaction must be at *one* centre, M_3, so that a
three centre reaction occurs

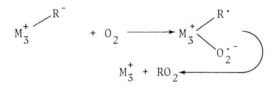

R could transfer an electron directly to O_2 without the
participation of M. The reactions are all inner-sphere.

Energy capture from O_2 reactions

   Biological systems are largely reducing systems for the
capture of oxidised C/H/O/N compounds in water. It follows
that energy can be gained from *controlled* oxidation reactions of these compounds with O_2 and the greatest energy
is gained if oxygen is reduced to two moles of water,
since a 2-electron step leads to H_2O_2 which is still
'energy rich'.

   The required reaction is now

$$O_2 + 2H_2 \text{ (bound hydrogen)} \rightarrow 2H_2O$$

and, in order to avoid chemical complications, it must
avoid free radical paths. No reduced O_2 intermediates
should be released. To a certain extent this is similar
to the problem of producing (R) in polymers, when we must
keep oxidant and reductant apart:

$$2RH_2 \rightarrow 2R + 4H^+$$

$$O_2 \rightarrow 2O^{2-}$$

$$2O^{2-} + 4H^+ \rightarrow 2H_2O$$

and we must avoid

$$RH \rightarrow R^{\cdot} + H^+$$

$$O_2 \rightarrow O_2^{-}, \ H_2O_2, \ OH^{\cdot}$$

$$RH + O_2 \rightarrow ROH, \ R(OH)_2$$

To achieve this we need separated systems of at least two sites:

$$\left.\begin{array}{c} 2\,RH_2 \\ \\ 2R + 4H^+ \end{array}\right] \ M(1) \ \xrightarrow{\ 4e\ } \ M(2) \ \left[\begin{array}{c} O_2 \\ \\ 2\,H_2O \end{array}\right.$$

The requirements of M(2) which reacts with O_2 are much as before except that it must give a 4-electron change. The M(1) couple is a *low* potential single electron reaction which requires iron, see above. The transfer of the $4e$ requires other metal centres of potential between -0.5 and +0.4.

In the above scheme, the potential drop from M(1) to M(2) in mitochondria is used for energy capture and it is known that this is done by coupling to an ADP/ATP system. Clearly, a high redox potential couple is required at M(2), which must also bind O_2 and retain $O_2^{\cdot-}$, H_2O_2 and $OH^{\cdot}$ while H_2O is formed. The following hypothetical metal carries out the reaction at a single centre in the scheme

$$M^{2+} + O_2 \rightarrow M^{2+}O_2 \xrightarrow{\ +e\ } M^{2+}O_2^{-} \xrightarrow{\ +e\ } M^{3+}O^{-} + H_2O$$

$$M^{2+} + H_2O \xleftarrow[2H^+]{\ +e\ } M^{3+}O^{2-} \quad \Big\downarrow{\scriptstyle +e}$$

Several problems now arise. Unlike the example of oxygen insertion (P-450) all steps in the reduction of O_2 to H_2O even the step Fe(III) to Fe(II), should be at as high a potential as possible to conserve energy. In fact the Fe(III)/Fe(II) potential of cytochrome oxidase is above +300 Mvolts in contrast to the system for oxygen-atom insertion (P-450) where the Fe(III)/Fe(II) potential is at -300 Mvolts. The requirement in the P-450 series was for a relatively long-lived oxo-cation, FeO, which may well force a low Fe(III)/Fe(II) potential.

The initial electron-acceptor from RH_2 in the energy conservation scheme must be at as low a potential as possible, less than -400 Mvolts, now close to that in the P-450 system. In both energy-generating and detoxification systems Fe/S proteins are used. Between the two ends of the energy-conserving electron transfer chains there must be an ordered series of intermediate redox potentials, Fe/S and haem centres, which drop electrons down the H_2/O_2 potential gradient so as to make ATP from ADP. Following the work of Williams and Mitchell the pathway involves the deliberate oxidation of organic hydrogen carriers, reduced flavin and quinone, at potentials of -400 and zero Mvolts respectively. The oxidation of the organic compounds and the reduction of O_2 are three steps where the redox reactions generate hydrogen ion concentration changes. By special arrangement of all the reactants in a membrane these changes appear as an energised proton gradient. It is this acidity gradient which is dissipated in the production of ATP. In photo-energy conversion the catalysts are very similar except for the steps involving the O_2/H_2O reaction and its reversal, see below. Copper is only involved in these systems as a medium potential one-electron carrier, i.e. much as before, in the reaction of O_2 itself which involves a four electron donor enzyme, cytochrome oxidase.

It is known that 2M can also take O_2 to H_2O in multi-centre enzymes but it is curious to note that in these reactions (e.g. of laccase) there is *no* energy capture. These reactions are inner sphere O_2 reactions and outer sphere successive one-electron transfer. Cytochrome oxidase achieves this cycle.

Medium Potential Multi-electron Transfers

A short aside is not out of order to the *medium* potential electron transfers which require many electrons. The reduction of oxygen requires a series of oxidation states of *high* potential e.g.

$$FeO_2 \rightarrow FeO_2^- \rightarrow FeO + H_2O \rightarrow FeOH \rightarrow Fe \rightarrow H_2O$$

$$VI \quad\quad V \quad\quad IV \quad\quad\quad\quad III \quad\quad II$$

$$\sim+0.8v \quad \sim+1.0v \quad\quad \sim+1.0v \quad \sim+0.0v$$

However in order to utilise fully the reaction potentials of other non-metal oxidation states there is a requirement to reduce at lower potential e.g. in the steps

$$SO_4^{2-} \rightleftarrows SO_3^{2-} \ (+0.2v) \quad NO_3^- \rightleftarrows NO_2^- \ (+0.4v)$$

and there is also the requirement of the overall reaction

$$N_2 \rightleftarrows 2NH_3 \ (-0.1v) \quad cf \ O_2 \rightleftarrows 2H_2O \ (+0.8v)$$

In biological systems all these steps use molybdenum. Though not within the topic of this article the association of molybdenum with the two major paths of nitrogen assimilation is a striking example of the chemical drive implicitly 'required' in evolution. Copper can not carry out these reactions as it forms no oxo-cation and iron is selected for high-potential reactions.

Detoxification of products of O_2 reduction

As well as handling O_2 for the above advantageous purposes the accidental or deliberate production of $O_2^{\cdot-}$,
OH$^{\cdot}$ and H_2O_2 from O_2 would supply dangerously aggressive
chemicals unless these can be eliminated. Again metal
catalysts are clearly required. A separate chapter of this
book by Dr. Hill describes the elimination and reactions
of superoxide dismutase. The only point of note here is
that the biological handling of $O_2^{\cdot-}$ is carried out by
three different metal ions Cu, Mn, and Fe all of which
can give the necessary high one-electron redox step for
disproportionation. Here we look at the reactions of H_2O_2
which is both deliberately produced, used (peroxidases),
and deliberately destroyed (catalases).

The Reactions of Hydrogen Peroxide

There are two possible ways of handling hydrogen peroxide. The first is a simple decomposition

$$2H_2O_2 \rightarrow 2H_2O + O_2$$

and such a reaction requires a metal which binds to H_2O_2,
probably as HO_2^-, and then undergoes a redox cycle in
which O_2 is produced and released.

$$M + H_2O_2 \rightarrow MO + H_2O$$
$$MO + H_2O_2 \rightarrow MO_2 + H_2O$$
$$MO_2 \rightarrow M + O_2$$

is a possible cycle. The requirement is now that the first
state of M of the metal does *not* bind oxygen, but it must
bind HO_2^- in an inner sphere reaction. Typically iron is
used.

The second path, a general oxidation, is very similar
to the use of oxygen in such reactions as described above.

$$M + H_2O_2 \rightarrow M(H_2O_2)$$
$$M(H_2O_2) + RH \rightarrow R^\cdot + (MH_2O_2)^- + H^+$$
$$M(H_2O_2)^- + RH \rightarrow R^\cdot + (MH_2O_2)^{2-} + H^+$$
$$2H^+ + M(H_2O_2)^{2-} \rightarrow 2H_2O + M$$

The requirements of the cycle are obvious, inner sphere uptake of HO_2^- and outer sphere reaction of RH. There are general features of these reactions to which we turn next.

Binding of Anions, $O_2^{\cdot-}$ and HO_2^-

Unlike the activation of oxygen the handling of superoxide and hydrogenperoxide involves an anion. The relative strength of binding of such simple anions by metal atoms can be estimated from the reaction of metal ions with OH^-, Table VI. The order is much the same for the binding of

TABLE VI

Stability constants (log K) for the formation of hydroxide complexes

Metal ion	log K
Cr^{3+}	10.0
Mn^{2+}	3.4
Mn^{3+}	13.9
Fe^{2+}	4.2
Fe^{3+}	11.5
Co^{2+}	4.2
Co^{3+}	14.3
Ni^{2+}	4.2
Cu^{2+}	6.6
Zn^{2+}	4.2

Data from 'Stability Constants' - The Chemical Society Special Publication No. 14 and 24 - London 1964 and 1971.

other anions such as nitrite, sulphite and phenolate and
similar metal oxidation states are involved in these reactions as are involved in the reactions of HO_2^- and $O_2^{\cdot-}$.
It is clear that trivalent cations, M^{3+}, are suitable but
that only Cu^{2+} of the divalent ions can bind such anions
strongly. Lack of metal selection is obvious in $O_2^{\cdot-}$ reactions. However in biology copper is not used with H_2O_2
for the reaction of copper with H_2O_2 is a one electron
step which releases radicals. The iron H_2O_2 reaction re-
tains the oxygen radicals, for the observed intermediates
are always of the kind MO (compounds I and II). Thus
perhaps it is not sufficient for the metal to bind some
of the anions RO^- (OH^-, HO_2^-) but it must also retain the
active radicals, in the oxo-cations. The difference from
the $O_2^{\cdot-}$ reaction is clear.

The reaction $2H_2O \rightarrow O_2$

This reverse reaction which is a part of photosynthesis
requires the reaction of a metal centre

$$(M)_n \xrightarrow{-4e} (M)_n^{4+}$$

Knowing that the product is oxygen the reaction is likely
to be

$$[M(H_2O)]_2 \longrightarrow M_2O_2 + 4H^+ + 4e$$

Thus there is a requirement for an unstable oxygen complex and clearly the metal must have a high redox potential. In biological systems it is known that manganese
fills this role. Thus overall in biology there are two
seemingly irreversible processes

$$4H^+ + 4e + M + O_2 \rightarrow M + 2H_2O \text{ (iron only)}$$

and $\quad M + 2H_2O \quad \rightarrow M + 4H^+ + 4e + O_2$ (manganese only)

Other High Redox Potential Couples

Nitrogen Couples Starting from the highest oxidation
state, NO_3^- is reduced by oxygen atom transfer

$$NO_3^- + M \rightarrow MO + NO_2^-$$

Although this couple has a redox potential of about +0.5v
it is not as high as that of NO_2^-. We have already mentioned that the activation of NO_3^- to NO_2^- is carried out
by oxygen atom abstraction at a molybdenum atom centre.
     Nitrite is reduced directly to ammonia in biology and
it is both a source of nitrogen for synthesis and a source
of energy, compare O_2. The reduction requires a metal site
to bind NO_2^- and a large pool of electrons ($NO_2^- \rightarrow NH_3$ is
a 6-electron reduction). Obviously the reaction resembles
the reduction of O_2 to water and the iron atom is an
obvious centre for the multi-electron process, see above.
In fact NO_2^- is reduced by a haem-iron enzyme; NO_2^- has
a very similar redox potential to that of O_2.

The Use of Selenium in Biology

     Apart from the bulk non-metals, O, N, S, C of biology
only the trace non-metal, selenium, seems to be used as
a redox catalytic centre. Such a curious feature poses
the question 'What can selenium do which can not be done
by metal ions as effectively?' The selenium enzymes are
listed in Table VII and its redox potentials are given in
Fig. 2. There is one possible answer to the question —
selenium undergoes *two-electron* reactions at a potential
close to 0.0 volts. The first row transition metals react
at much higher potentials, molybdenum reacts at rather
lower potentials in two-electron steps. The way in which
the three couples

TABLE VII

Selenoenzymes and the reactions they catalyze

Enzyme	Reaction Catalyzed
Formate dehydrogenase	$HCOOH + A \rightarrow A.H_2 + CO_2$
Glutathione peroxidase	$2GSH + H_2O_2 \rightarrow GSSG + 2H_2O$
Glycine reductase	$\underset{NH_2}{CH_2COOH} + R(SH)_2 + Pi + ADP \rightarrow CH_3COOH + NH_3 + R\big\langle\begin{smallmatrix}S\\|\\S\end{smallmatrix} + ATP$
Small seleno-protein of muscle	Unknown

GSH and GSSG are reduced and oxidized glutathione.
After T.C. Stadtman (1977), *Nutrition Reviews* **35**, 161-166.

$\begin{array}{ccc} H\text{-}O\text{-}O\text{-}H, & R\text{-}S\text{-}S\text{-}R, & R\text{-}Se\text{-}Se\text{-}R \\ \text{substrate} & \text{co-enzyme} & \text{enzyme} \\ & \text{glutathione} & \text{glutathione peroxidase} \end{array}$

can be used to act as a redox buffer despite the fact that H_2O_2 is a very powerful redox agent is then clearer. The direct reactions of sulphydryls with H_2O_2 is perhaps too slow. The exact intermediates in the selenium chemistry are not known.

The Halogens in Biology

At first sight the halogens are surprisingly little used in biological systems. There are iodine compounds in animals, thyroxines, and there are some chloro- and bromo-organic compounds in sea-plants. Some natural anti-microbial agents contain chlorine — chloro-mycins. However the major uses of the halogens are destructive. The reaction O_2 or $H_2O_2 + X^- \rightarrow X^+$ provides a very aggressive chemical Cl^+, Br^+ or I^+ which can destroy almost any organism. To oxidise Cl^- requires a higher potential than to oxidise I^- and there are special chloro-peroxidases, which are to be grouped with P.450 enzymes and for different reasons with lacto- and myelo-peroxidases, and more conventional plant peroxidases which oxidise I^-. These peroxidases will be discussed at length in the next section.

Summary

The above schemes include virtually all the require-ments for metals in catalysts in redox reactions associa-ted with O_2 and its products. We list in Table III the demands on the catalyst. A major feature of the metal complex must be that the metal is available and can be retained by biological systems. Secondly, the complex must control the redox potentials. Note however that in Table III one complex cofactor haem can act in the fol-lowing ways (a) O_2 — transport; (b) O-atom insertion; (c) high-potential electron transfer; (d) H_2O_2 activation; (e) H_2O_2 disporportionation, but different coordination chemistry is involved in each case. Thus selectivity rests not only in choice of metal ion and its retention and not only in the poise of redox potential, but also in the choice of liganding groups from the protein.

There is the strong suggestion that individual metal ions have been selected for special redox potential path-ways and that specific chemistry is involved in this selection. We turn in the final section to some possible rationalisations of this chemistry in biology.

The Choice of Metal Catalysts in Biology

Table III indicated the metal ions which have been selected for particular tasks in biological systems. We now seek the origin of this choice. The natural avail-ability of metals which are potential redox catalysts in the Periodic Table is $Fe \gg Mn > Cu = Co, Cr, V > Mo$, where we exclude metal ions which could not take part in redox reactions. The list contains several metals, Co, Cr, V which are not present in Table III and Mn only

appears once. Only the dominance of iron in Table III is
explicable by this availability order and then only in
part. Why is it that metals such as Mn and Co are so
little used and why is copper sometimes used and not iron?
The discussion will start from the ranges of redox states
of free metal ions and will then describe their ranges of
redox potential. The way in which protein ligands can
adjust the potentials will be examined as will the control exerted over rates of redox processes by such ligands. We shall also look at combinations of metals and
non-metals which can produce rather special redox reactions. We cannot give more than a gross simplification
of the reasons for biological selection.

Review of the Choice of Metal Ions in Biological Catalysis

A major consideration when judging the value of a
redox couple is the number of equivalents it can absorb.
Table VIII lists some of the common metals and their

TABLE VIII

The Redox Equivalence Span of Element Oxidation States

Element	M^n/M^{n+1}	M^n/M^{n+2}	M/M^{n+3}	M/M^{n+4}
Fe	2/3, 3/4, 4/5, 5/6	2/4, 3/5, 4/6	2/5, 3/6?	2/6?
Cu	1/2, 2/3?	1/3?	–	–
Mn	2/3, 3/4	2/4	–	2/6?
Co	1/2, 2/3	1/3?		
Se	(0/1)	0/2		0/4
Mo	2/3?, 3/4, 4/5, 5/6	2/4?, 3/5, 4/6	2/5?, 3/6	2/6?

Notes on biological systems:
Fe The redox states are seen in many oxygenases and peroxidases.
    Oxidation state 6 is the $Fe(II)O_2$ complex.
Cu The oxidation state 3 is not fully authenticated except in
    $Cu_2^+O_2$.
Mn The oxidation states 5 to 7 are very unlikely as they are known
    in anionic systems only, e.g. permanganate.
Co The oxidation states are those found in B_{12} complexes but Co(I)
    is not fully authenticated in biological systems.
Se In theory selenate (6) could occur but it is unlikely that states
    higher than the oxo-compound SeO occur in biology.
Mo The only doubtful state here is Mo(2).

redox equivalent ranges in their most simple complexes. Iron has the possible variation of redox states from II to V or VI if $Fe(II)O_2$ is included. Molybdenum can go from Mo(III) to Mo(VI) and Mo(II) is just possible. Some of the other metals have a maximum span over at most three oxidation states, Mn, Cu, Co, but there are questions as to the availability of Co(I) and Cu(III). Metals not included are Cr and Ni which are not known to be redox active in biology. Selenium is unusual in that it usually undergoes two-electron (atom) transfer rather than one-electron radical steps. See Fig. 5 for a general comparison of metal and non-metal redox states.

Elaboration of High- and Low-Potential Systems in Biology

The consideration of the control of redox equivalent changes must be coupled with that of control of the redox potential at which the redox reactions occur. In Table IX we place simple metal ion couples in compartments according to their potentials. The observations from the study of biological couples are listed in parentheses.

The quite separate functional possibilities of Fe, Mn, Cu and Mo are clear from Table VIII plus Table IX. Using only the simple metal complexes as a guide we see that

(a) A wide range of redox states all of which are of low potential is only possible for molybdenum.

(b) A wide range of redox states of high potential is only possible for iron or manganese.

(c) Oxygen complexes are relatively common, Cu, Mn, Fe, Co.

(d) Copper complexes do not give rise to low potential couples.

(e) Oxo-cations occur with Mn, Fe and Mo but not with Co or Cu.

But how and why has biology produced those limitations in Table III which are not so obvious from this inorganic chemistry? The answer obviously lies in the selection

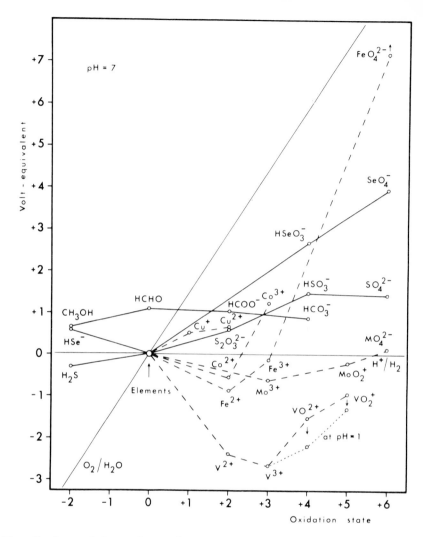

Fig. 5. General Comparison of Oxidation States of Some Metals and Non-Metals.

of ligand environment for the metal ions as much as in their intrinsic chemistry as free ions. Biological systems have two different approaches to metal binding. In the first the metal binds to a set of protein side-chains usually combinations of strong donors imidazole, thiolate and phenolate. The second type of centre is a specially built frame in a co-factor. Examples are Fe(haem),

TABLE IX

Redox Potentials of Simple Inorganic Complex Couples

Element	1/2	2/3	3/4	4/5	5/6	0/1
Fe	-	medium (low)	v. high	v. high	v. high	not observed
Cu	medium	v. high	-	-	-	not observed
Mn	-	high (medium)	high	-	-	not observed
Co	low	high (medium)	-	-	-	not observed
Mo	-	low	low	low	low	not observed
Se	medium	-	medium	-	medium	(medium)

Potentials in the range -0.5 to ±0.0 are called low and values above +0.4 are called high.

Co(corrin), and perhaps Mo (special small ligand?). The fact that given metal ions have been selected in biology for combination with special ligands strongly suggests that evolution has discovered some very favourable ligand/metal combinations rather than that it has just chosen metals *per se* for special tasks.

This selected chemistry can be simply illustrated by haem-iron coordination in oxidative enzymes. Ligands are required which stabilise the high oxidation state, FeO(IV), relative to Fe(III). It is interesting that the three well known iron catalysts which use FeO(IV) are

$$E^{o\prime} \ M^{2+}/M^{3+} \text{Mvolts}$$

(1) haem-Fe(III). RCO_2^- (?) (catalase)   -400
(2) haem-Fe(III). imidazole (peroxidase)   -250
(3) haem-Fe(III)/. RS^- (P-450)   -300

where $E^{o\prime}$ values refer to reduction to the Fe(II) state. These potentials are in marked contrast with those of most of the O_2 carrying haem complexes.

(4) haem-Fe(II). imidazole (Haemoglobin)   +100 to +200 Mv.
                         (myoblogin)

and for the cytochrome complex which converts O_2 directly to H_2O

(5) haem(a) - Fe(II). imidazole (cyt. a_3) +350 Mvolts

It would seem that if FeO(IV) is to be a stable intermediate then the Fe(II) state will be strongly destabilised with respect to Fe(III) and Fe(III) with respect to FeO, see Fig. 4. In the light of changes to Fig. 4 which we might reasonably expect from ligand binding to the metals we shall examine Table III again.

A Rationale of the Observed Potential in Biology

If metals are to act as redox catalysts in proteins then they must be permanently bound to their sites. We note that if Fe(II), Mn(II), Co(II) or Cu(II) are to be firmly held in a thermodynamic and a kinetic sense in biological systems they require powerful nitrogen or sulphur donor centres. Many of these ligands are anions and must preferentially stabilise higher oxidation states. Again the simple ionic higher oxidation states of iron or manganese cannot be reached by oxygen oxidation e.g. Fe(IV), Fe(V) or Fe(VI), but in anionic complexes these states are sufficiently stabilised for them to be reached by oxygen oxidation. Thus the ligands which stabilise very high oxidation states automatically give very low Fe(II)/Fe(III), Co(II)/Co(III) and medium Mn(II)/Mn(III) potentials. None of the catalyst centres which must use cobalt(II) or iron(II) together with high valence states can have high M(II)/M(III) potentials. It is then possible that haem iron in haemoglobin has a chosen medium Fe(II)/ Fe(III) potential (~ +200 mV) so that O_2 is not activated: the higher redox states are inadequately stabilised. In this environment low-spin $Fe(II)O_2$ is relatively stable.

In the environment of P-450 ($\sim$ -300 mV) Fe(II)O_2 is very much less stable relative to Fe(III)$O_2^{\cdot-}$ and FeO(IV). The reason that the reverse reaction

$$Fe(II) + 2H_2O \rightarrow Fe(II).O_2 + 2H_2$$

is not observed in biology even in cytochrome oxidase ($\sim$ +400 mV) is then that it may well be that although FeO can be obtained as a transient intermediate it is so unstable that evolution of O_2 is not possible.

The comparison with manganese chemistry is interesting. Mn(II) does not provide an oxygen carrier in biology neither does it give an (Mn(II))$_2O_2$ carrier. The chosen ligands for Mn(II) would appear to be such that when the reaction

$$2Mn(II) + 2H_2O \rightarrow (Mn(II))_2O_2 + 2H_2$$

oxygen is evolved immediately. When we compare iron and manganese the following comments are apparent

(a) Mn(II) forms weaker complexes with O_2 than does Fe(II)

(b) Mn(II) does not easily go low-spin

(c) Mn(III) is unstable in contrast with Fe(III), see Fig. 4.

(d) Mn(IV) is relatively stable in contrast with Fe(IV).

We are left with conclusions. It is easier to make two (adjacent) MnO(IV) than to make two FeO(IV); any iron complex which can give FeO(IV) will react in the Fe(II) form with O_2 to give stable O_2 complexes or to give Fe(III)$O_2^{\cdot-}$; it is possible to arrange redox state stabilities such that O_2 is less oxidising than MnO(IV)/Mn(II) in such a system Mn(III) may not be involved; manganese as Mn(II) may well stay high spin even with an O_2 ligand. Thus for O_2 uptake and reaction iron is to be preferred, but for O_2 generation from H_2O manganese is preferred. In such a way the cycle of reactions of water

$$h\nu + H_2O(+Mn) \rightarrow O_2$$

$$O_2(+Fe) \rightarrow H_2O + heat$$

can only go irreversibly in the direction from light to heat.

In the light of the above possibilities we can go forward to an examination of the chemistry of cobalt and copper in biology. A major change from manganese and iron chemistry is that while the states MO(IV) become of too high a potential to be reached M(I) is a potentially valuable oxidation state. It has not been proved that M(I) is used in cobalt, vitamin B_{12} biochemistry, but M(I) is a very usual biochemical state of copper. Referring to Fig. 4 we see now that the inability of this element to form an oxo-cation generates a grave disadvantage in catalysis. In the case of cobalt it may explain why this element is not used in conventional redox catalysis in biology although it is used in some very special M-carbon chemistry to generate carbon free radicals.

Copper chemistry is constrained very largely to Cu(I)/Cu(II) one-electron redox reactions in biology. The curious feature of the copper oxidation states is that Cu(I) forms complex ions of almost equal strength with those of Cu(II) with either nitrogen or sulphur ligands. Thus the redox potentials of copper complexes are often as high as or higher than the aquo ions. Copper then provides a ready source of one-electron couples of potential in the range +300 to +800 Mvolts. While it is possible to generate one-electron couples from other elements which would have about these potentials and using the ligands of proteins it is not at all sure that such complexes would be stable in their low oxidation states.

We have stressed earlier that the binding of the *single* metal ions by ligands, anions earlier in the transition metal series and neutral ligands for copper, causes a change in the redox diagram such that the O_2/H_2O couple closely matches the Fe(III) to Fe(VI) couples. Earlier first transition series metals are too reducing relative to the H_2/H_2O couple to be effective in multi-electron redox steps and later metals are too oxidising. The H_2/H_2O redox is however matched closely by molybdenum oxidation states for multi-electron reductions. The use of more than one metal-atom centre, cluster reactions, provides extra possibilities.

Biology has multi-atom clusters in the Fe/S systems for low-potential one electron transfer and possibly this is an early device for overcoming the Franck-Condon barrier while using high-spin states. They may also be used in hydrogenases and nitrogenase. The use of two metal-atom 'cluster' redox catalysts is observed for oxygen transport (Fe, Cu), and probably for oxygen release (Mn). In all cases the reaction would appear to give M_2O_2. The oxygen is not made into a good attacking group as it is bound at both ends even though it is effectively reduced to O_2^{2-}.

These oxygen-carriers can now be compared with the superoxide dismutases which have single metal ion centres, do not bind oxygen but release it, and can also use Fe, Cu, or Mn. We suspect that the single centres are very similar to those in the oxygen-carriers which have *two* metals. In each case the metal would appear to have a medium redox potential and ligands which keep the metal in the high-spin state, e.g. imidazole and some oxygen anion. All three metals are involved in one-electron reaction steps in both oxygen-carriers and superoxide dismutases; however the metals differ in the chemistry and stability of the different oxidation states.

The failure of these single centre to bind oxygen arises in part from the weakness of the ligands for both

iron and manganese which are such that (i) the potentials
of higher oxidation states remain higher than that of
oxygen, i.e. FeO(IV) and MnO(IV) are not sufficiently
stabilised (copper does not give oxo complexes); (ii) the
electron donor power of high-spin Fe(II) and Mn(II) is
inadequate to retain O_2; and (iii) the superoxide iron is
too reducing to stabilise Mn(III), Fe(III) or Cu(II) in
these complexes. The situation is strikingly different
from that in porphyrin complexes when the fifth ligand is
a good donor, imidazole, but is very similar to that where
the fifth ligand is a weaker donor e.g. in catalase where
it is thought to be a carboxylate group. This last enzyme
does not give low-spin complexes, will not bind oxygen and
in fact is used to decompose hydrogen peroxide and thus to
release oxygen. We now see that oxygen-binding and release
are under tight control in the one-metal centres. If two
of the above weak-ligand centres are put together in a
protein, overcoming mutual repulsion, then the reaction:
$2M^{2+} + O_2 \rightarrow 2M^{3+}.O_2^{2-}$ occurs and can yield a relatively
stable product as is well known in inorganic chemistry of
cobalt and iron. The manganese complex is inherently less
stable as Mn(III) is more strongly oxidising. Thus all
that we know of these biological systems suggests that
they are carefully adapted to make optimal use of the
inorganic chemistry.

There is a further cluster of metal atoms in some oxygen using copper enzymes where oxygen is reduced to water
directly. It now appears that three copper atoms are involved in a tight cluster of two plus an extra, type II,
copper. The redox potentials are those of O_2, Fig. 2.

We can contrast this three-copper reduction site for
oxygen, where there is also an extra distant copper atom
to react with substrate (making a four electron site in
all) with the four electron reduction of cytochrome oxidase ($O_2 \rightarrow H_2O$) at a single metal atom. In each case
oxygen atoms from di-oxygen are only released as water,

but the iron centre acts very differently from the copper
cluster:

$$Fe(II) \;+\; O_2 \rightarrow FeO_2 \overset{c}{\rightarrow} Fe^{III}O_2{}^{2-} \xrightarrow{2H^+} FeO \overset{H}{\rightarrow} Fe^{III}OH \overset{H}{\rightarrow} Fe(II)$$

| high-spin | low-spin | low-spin | low-spin | low-spin | high-spin |

fast uptake   ↕

                                        high-spin   release

All reactions are controlled on one site by keeping low-
spin complexes so that oxidising equivalents are retained.
The scheme for copper may well be as follows:

$$2Cu(I).\;Cu(II) + O_2 \rightarrow (Cu(I))_2O_2.\;Cu(II) \xrightarrow{2e} 2Cu(I).\;Cu(II).\;O_2H \rceil$$

$$2H_2O + 2Cu(I).\;Cu(II) \;\;\;\; \leftarrow \rfloor$$

    Reactions do not depend on the ground state geometry
alone but on the ease of reaching some excited electronic/
geometric state. The simplest reaction to consider is
electron transfer for which a catalyst should have a com-
promised ground-state geometry between the geometry de-
manded by each of the two oxidation states of the catalyst, an entatic state. This condition has been obtained
in several of the biological electron transfer catalysts
which have very low activation energies for electron
transfer, see the geometries of Table X. Evolution may
not demand zero free energy of activation since a transfer
catalyst of this kind would not be open to control.
    The reactions of the uptake and use of oxygen demand
open-sided catalysts — as found in all the metallo-enzymes
which function in the O_2 reactions. Control is exerted by
the redox potentials of various oxidation states, the approach of the substrate (close or remote) and by the
ordered addition of oxygen, external reducing equivalents,

TABLE X

Some Metal Centres of Redox Catalysts

Metal	Enzyme	Ligands	Geometry	Function
Cu	Blue Proteins	2 imidazole 1 thiolate 1 thio-ether	Flattened Tetrahedron	electron transfer (high E^O)
	Type II Oxidase	4 imidazoles	Tetragonal (Open)	Superoxide
Fe	Cytochromes	heme 1 imidazole 1 thio-ether	Distorted Octahedron	electron transfer (medium E^O)
	Cytochromes b	heme 2 imidazoles	Distorted Octahedron	electron transfer (low E^O)
	Ferredoxin	2 thiolate 2 sulphur	Distorted Tetrahedron	electron transfer (low E^O)
	P-450	heme thiolate	Open-sided to Closed, Switch	O_2- activation
	Peroxidase	heme imidazole	Open-sided	H_2O_2 activation
Co	Ribonucleotide Reductase	Corrin benzimidazole carbanion	Open sided to Closed, Switch	Reduction (low E^O)
Mn	Oxygen Production	phenolate imidazole?	Open sided	O_2 evolved (high E^O)

and substrate. The external reducing equivalents are used to partially reduce the O_2 so activating it, Fig. 2. The best example of control is that operating in cytochrome P-450. There is little doubt that the metal enzymes have attained a high degree of evolutionary perfection and that detailed knowledge of them will lead to advances in chemistry.

The idea of the perfection of a catalyst for a reaction must take into account not only the special geometries which may be of the greatest suitability for particular steps but it must also include an appropriate set of dynamic possibilities. These dynamic features have been recognised in the movements of iron in haem-enzymes both on redox change and on oxygen uptake.

The Inherent Dangers of Oxidation Reactions

Since all organic materials are unstable in an oxygen atmosphere there is the ever present problem of electrophilic attack by a vast range of oxidation products from oxygen, nitrogen, sulphur, halogen, and even carbon compounds. It appears that these dangers are inherent in the chemistry of life and that there can be no escape from oxidative degradation except by repair. This subject is so large that we leave it for a separate article.

Conclusion

This article shows the role of the higher oxidation states of inorganic elements. There is special concentration upon detoxification and protection devices rather than on energy-production and general metabolism which are already very familiar. It is quite striking how oxidative attack has evolved in biological systems so as to assist elimination of unwanted chemicals and to destroy parasites. The bulk chemicals used in this attack are oxygen, chlorine, bromine and iodine. The catalyst and carrier centres use selenium, sulphur and iron. The major hazard of this attack is the production of carcinogens, which probably act through mutagenic attack on DNA. Biology does not provide fail-safe systems.

TABLE XI

Dangers of High Redox Potential Chemicals

Reagent	Chemical Attack	Diseases Caused (?)
N/O compounds	Electrophilic (DNA)	Cancer
Halogenated Compounds	Alkylating Agents (DNA)	Cancer
Superoxide	Electrophilic (Proteins)	Inflamation
Oxygen	Weak	General slow degradation

Protection appears to involve in part the reducing ability of carbon compounds catalysed by transfer through sulphur and selenium compounds.

It is also of great interest that quite separate roles have been found for iron, manganese, copper, molybdenum, cobalt, chromium, nickel and zinc. We are just beginning to see the selective chemical pressures which have led to this functional discrimination. Perhaps this understanding will improve our own chemistry.

References

Bossur, F.P., Chellapa, K.L. and Margerum, D.W. (1977). *J. Am. Chem. Soc.* **99**, 2195.

Gunsalus, I.C., Pedersen, T.C. and Sliger, S.G. (1975). *Ann. Rev. Biochem.* **377**.

Hayaishi, O. (1962). 'Oxygenases", Academic Press, New York.

Howie, J.K. and Sawyer, D.T. (1976). *J. Amer. Chem. Soc.* **98**, 6698.

Hrycay, E.G., Gustafson, J., Ingelman, M., Sandberg, B. and Ernster, L. (1976). *Eur. J. Biochem.* **61**, 43.

Koppenol, W.H., Van Buuren, K.J.H., Butler, J. and Braams, R. (1976). *Biochim. Biophys. Acta.* **449**, 157-168.

Lontie, R. and Witters, R. (1972). *In* 'Inorganic Biochemistry' (Ed. G.I. Eichorn), Chap. 11. Elsevier, New York.

Vanneste, W.H. and Suberbuhler, A. (1974). *In* 'Molecular Mechanisms of Oxygen Activation' (Ed. O. Hayaishi), p. 371. Academic Press, New York.

Wechsler, C.J., Hofman, B.M. and Basolo, F. (1975). *J. Am. Chem. Soc.* **97**, 5278.

Williams, R.J.P. (1973). *Biochem. Soc. Transaction* **1**, 1.

6 THE SUPEROXIDE ION AND THE TOXICITY OF MOLECULAR OXYGEN

H. Allen O. Hill

Inorganic Chemistry Laboratory, South Parks Road, Oxford, England.

When, some 3×10^9 years ago, the blue-green algae made their appearance (Schopf and Barghoovn, 1967), their ability to effect photosynthesis using water as the source of electrons had an inestimable consequence: molecular oxygen. Oxygen, whilst abundant, had hitherto been found in oxides and though some, such as sulphate, nitrate and carbonate, acted, and still do act (Doelle, 1975) as oxidants in biological processes, the creation of a ubiquitous gaseous oxidant must have led to a selection for those organisms which could use molecular oxygen (or dioxygen as it is more properly called) as an electron acceptor. They had simultaneously to control and couple more oxidations and presumably protect themselves against side reactions. Those which could not became extinct or could, and can, only survive in the greatly decreased number of anaerobic environments.

We are concerned here with the relationship in biological systems between dioxygen and superoxide, how these systems protect themselves against the consequences of a temporal or spatial misplacement of dioxygen, superoxide and products therefrom, and the manifestation of inadequate protection against them.

H.A.O. HILL

Dioxygen and its reduction products

   Though dioxygen contains a strong bond (Brix and Herzberg, 1954), its bonds to many other elements are even
stronger. Most elements and the majority of compounds are
thermodynamically unstable with respect to oxidation by
dioxygen. This pertains to most constituent compounds
of living organisms and consequently living organisms
themselves are thermodynamically unstable, in the presence of dioxygen, with respect to carbon dioxide and
water. Fortunately, these reactions are not realised
since there exists a significant kinetic constraint without which the evolution of the Earth, both geologically
and biologically, would have been very, very different.
Wherein lies this kinetic impediment?

   In spite of its simplicity a one-electron molecular
orbital description provides us with a convenient and
informative introduction to the properties of dioxygen
(Figure 1). The electronic configuration is

$$(\sigma_g 1s)^2 (\sigma_u 1s)^2 (\sigma_g 2s)^2 (\sigma_u 2s)^2 (\sigma_g 2p)^2 (\pi_u 2p)^4 (\pi_g 2p)^2 \qquad [1]$$

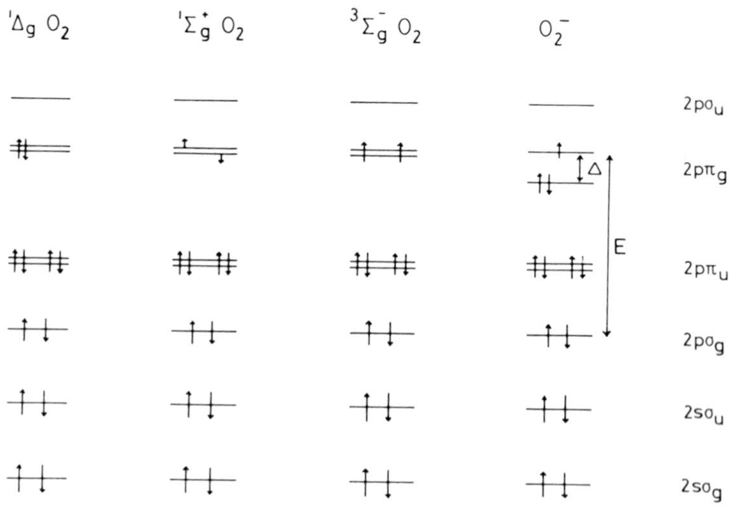

Fig. 1. A schematic one-electron molecular orbital diagram for dioxygen and related species. The parameters, Δ and E, used in the
description of the O_2^- ions, are referred to in the text.

giving rise to a $^3\Sigma_g^-$ ground state, with two unpaired electrons, as required by Hund's first rule. The excited
states, $^1\Delta_g$, $^1\Sigma_g^+$, lie respectively at 94.7 kJmol^{-1} and
157.8 kJmol^{-1} above the ground state giving some idea of
the electrostatic forces involved in determining the
energies of states with different charge and spin dis-
tributions. The kinetic constraint referred to above lies
simply in the fact that dioxygen has two unpaired electrons and most other compounds have none. This provides
a fundamental impediment to the reaction of dioxygen with
most molecules for as the two reactants approach there is
a spin restriction associated with the Pauli Principle
which makes reaction improbable unless there is a change
in spin state. However, not only does this require energy
it also requires time. The time for a change in spin
state, 10^{-8}s, is long compared to the lifetime of many
collision complexes, 10^{-12}s, hence the chance of it taking
place within the lifetime of the collision complex is low.
These constraints do not apply to reactions with single
electrons, hydrogen atoms, or other atoms or molecules
containing unpaired electrons. The chemistry of dioxygen
is therefore mostly a chemistry of one-electron transfer
reactions, catalysis or the formation by e.g. radiation,
of high-energy reactants (Taube, 1965).

The successive addition of electrons to dioxygen
generates the intermediates shown in Fig. 2. The relationship between all these molecules is succinctly encapsula-
ted in the oxidation state diagram shown in Fig. 3. This
provides a convenient method of displaying the oxidation
reduction potentials measured relative to the element in
its standard state. The gradient of the line joining two
points is equal to the oxidation reduction potential of
the couple formed by the two species which the points
represent. The more positive the gradient the stronger
the couple is as an oxidant. Thus it can be observed that
dioxygen, O_2^-, HO_2, H_2O_2 and HO_2^- are all oxidants rela-

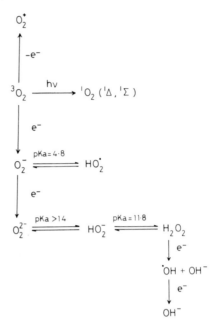

Fig. 2. The relationship between dioxygen and its oxidation and reduction products.

tive to water. Most powerful of all is the hydroxyl radical. Another feature emerges on inspection of Fig. 3. If the point representing a species lies above the line joining two adjacent species then thermodynamically the 'central' species can disproportionate to the other two species. Thus at all pH values, H_2O_2 is unstable with respect to O_2 and H_2O; likewise the superoxide ion is thermodynamically unstable with respect to dioxygen and hydrogen peroxide. However the pH dependence of this tendency to disproportionate (Czapski, 1971) is such that it is much less marked at higher pH. Indeed when the redox potentials in aprotic solution are considered, this tendency is reversed in that the peroxide ion O_2^{2-} should react with dioxygen to give the superoxide ion. All anions are stabilised by protons or indeed cations. The electron affinity of dioxygen is only slightly favourable (Table I) and decidedly unfavourable for the

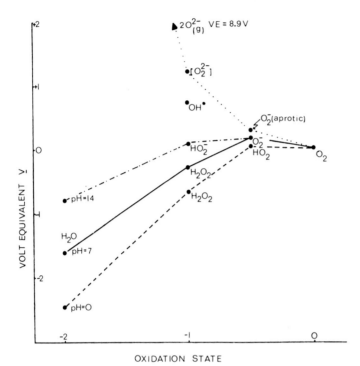

Fig. 3. The oxidation state diagram for dioxygen. The Volt Equivalent is the standard electrode potential at 25°C (*versus* the normal hydrogen electrode), multiplied by the oxidation state (the *formal* charge per atom of element *in a compound or ion*, assuming that each oxygen atom has a charge of -2 and each hydrogen atom a charge of +1); the oxidation state of the elemental form is zero). The potentials shown are therefore those of the cell:

$$Pt,H_2 \left| a_{H^+} = 1 \right| \left|\left| \begin{array}{l} \text{Standard couple} \\ pO_2 = 1 \text{ atm} \\ pH, \text{ solvent, as described} \end{array} \right. \right| Pt.$$

The data are taken from Fee and Valentine, 1977.

process $O_{2(g)} + 2e \rightarrow O_{2(g)}^{2-}$. Therefore the peroxide ion can exist only when stabilised by bonding, be it to protons, cations or some organic moiety. Hence the marked de-stabilisation relative to O_2^- as the medium becomes depleted in protons and/or metal ions.

The reduction products of dioxygen do not suffer from the same kinetic constraints to which dioxygen is subject. Consequently the effective use of dioxygen as an oxidant

TABLE I

Some Properties of Dioxygen

	Ionisation Potential (kJ mol^{-1})	Electron Affinity (kJ mol^{-1})	Dissociation Energy (kJ mol^{-1})	Bond Length (pm)	Stretching Frequency (cm^{-1})	Reference
O_2^+				112	1860	Cotton and Wilkinson, 1972
$O_2(^3\Sigma_g^-)$	1164	41 $-646-$ ($O_2(g) \rightarrow O_2^{2-}(g)$)	493	121	1556	Cotton and Wilkinson, 1972; Watanabe, 1957; Pack and Phelps, 1966; Wood and D'Orazio, 1965
O_2^-			268($HO_2^{\cdot}$)	133	1145	Cotton and Wilkinson, 1972; Benson and Shaw, 1970; Vaska, 1976
O_2^{2-}			213(H_2O_2)	149	$\sim$ 770	Cotton and Wilkinson, 1972; Benson and Shaw, 1970; Abrahams et al., 1951

may require concomitant reduction. This apparent paradox
is present in many chemical and biochemical reactions in
that the use of dioxygen as an oxidant requires the
simultaneous presence of a reductant as, for example, with
monooxygenases (Massey and Hemmerich, 1975). It is in
such situations that we may expect to find evidence for
the presence of superoxide and peroxide. Whether we suc-
ceed in finding them also depends on whether they remain
bound to the enzyme generating or utilising them and of
course on their lifetimes. It is most striking that cytochrome oxidase (Caughey *et al.*, 1976), which is required
for the reaction $O_2 + 4H^+ + 4e \rightarrow 2H_2O$, contains four
prosthetic groups, two hemes and two copper ions, presumably capable of discharging four electrons, albeit
one-by-one, to dioxygen without the release of any intermediate reduction products. Reductions of dioxygen to
products other than superoxide are discussed in chapter
5.

The Chemistry of the Superoxide Ion

Just as the removal of one electron from the anti-
bonding $(\pi_g 2p)^2$ orbitals to give O_2^+ leads to a shortening
of the O-O distance, Table 1, so the addition of one
electron lengthens it. The configuration $(\pi_g 2p)^3$ gives
rise to a paramagnetic $^2\Pi$ ground state with one unpaired
electron. The magnetic behaviour is deceptively simple
since, due to spin-orbit coupling two states are derived
from the interaction of the spin- and orbital angular
momentum to give $^2\Pi_{3/2}$ and $^2\Pi_{\frac{1}{2}}$ of which the former has
the lower energy. This description provides (Kanzig and
Cohen, 1959) an adequate explanation of the properties
of the *free* ion. In any real chemical environment e.g. in
a lattice, in solution, or on a surface the consequent reduction in symmetry leads to a removal of the degeneracy
of the ground state with a concomitant quenching of the
orbital contribution. The magnetic properties and,

a fortiori, the e.p.r. spectroscopy of the O_2^- ion will be particularly sensitive. The g-values of the *free* ion, $g_{11} = 4$; $g_\perp = 0$, will change to $4 > g_z > 2$ and $2 > g_x$, $g_y > 0$ (Fig. 4). The e.p.r. spectra of the superoxide ion generally exhibit (Bennett *et al.*, 1968) axial or near-axial symmetry, and a theoretical treatment gives (Kanzig and Cohen, 1959), as useful approximations:

$$g_x = g_e [\Delta^2/(\lambda^2+\Delta^2)]^{\frac{1}{2}} - \lambda/E\{[-\lambda^2/(\lambda^2+\Delta^2)]^{\frac{1}{2}} - \Delta(\lambda^2+\Delta^2)^{\frac{1}{2}}+1\} \qquad [2]$$

$$g_y = g_e [\Delta^2/(\lambda^2+\Delta^2)]^{\frac{1}{2}} - \lambda/E\{[\lambda^2/(\lambda^2+\Delta^2)]^{\frac{1}{2}} - \Delta(\lambda^2+\Delta^2)^{\frac{1}{2}}-1\} \qquad [3]$$

$$g_z = g_{11} = g_e + 2\lambda(\lambda^2+\Delta^2)^{-\frac{1}{2}} \qquad [4]$$

where $g_e = 2.0023$, Δ is the splitting due to the environment, λ is the spin-orbit coupling constant and E is the energy indicated in Fig. 1. Spectra have been obtained of O_2^- in a wide variety of environments including surfaces, ionic lattices, aqueous and non-aqueous solutions, and in metal complexes. The variation in g_{11} is indeed found and has led to considerable confusion though it can be fitted very well to the theoretical expression (Fig. 4).

The most important species to be considered in biological systems are those of O_2^-, HO_2 (Czapski, 1971), $O_2^-(H_2O)_n$ (Green and Hill, unpublished results) and ion pairs containing O_2^- (Bray *et al.*, 1977). Though $g_\perp \sim 2.0$, g_{11} is sensitive to the environments encountered in biological systems.

The superoxide ion can be prepared in a number of ways of which the most common are the direct reaction of dioxygen with electropositive metals, electrolysis (Maricle and Hodgson, 1965), and pulse radiolysis (Behar *et al.*, 1970). The oxidation of H_2O_2 to give the superoxide ion is difficult but does result from reaction with cerium(IV) (Saito and bielski, 1961) or periodate

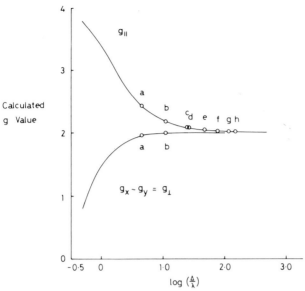

Fig. 4. A plot of calculated $g_\perp$ and g_{11} values against $\log(\Delta/\lambda)$ using equations given in Kanzig and Cohen (1959) and based on a diagram given by Knowles *et al.* (1969). A value of 350 was taken (Lunsford and Jayne, 1966) for E/λ (see Fig. 1). Experimental values of g_{11} are indicated: a, O_2^- in KCl (Lunsford, 1973); b, in NaO_2 (Lunsford, 1973); c, in H_2O (Knowles *et al.*, 1969); d, on MgO (Lunsford, 1973); e, in N,N-dimethylformamide (Maricle and Hodgson, 1965); f, on SnO_2 (Lunsford, 1973). The experimental values of $g_\perp$ are given for examples a and b. In all other examples $g_\perp \sim 2.00$.

IO_4^- (Knowles *et al.*, 1969).

For some time the chemistry of the superoxide ion was distinguished only by its tendency to disproportionate (Czapski, 1971). This is the most obvious reaction of the superoxide ion in protic media. It has been thoroughly investigated and the following data are available (Bielski, 1977).

$$HO_2 + HO_2 \rightarrow H_2O_2 + O_2 \qquad k_1 = 7.6 \times 10^5 \text{ Mol}^{-1}\text{s}^{-1} \qquad [5]$$

$$HO_2 + O_2^- \rightarrow HO_2^- + O_2 \qquad k_2 = 8.9 \times 10^7 \text{ Mol}^{-1}\text{s}^{-1} \qquad [6]$$

$$O_2^- + O_2^- \rightarrow O_2^{2-} + O_2 \qquad k_3 = <0.3 \text{ Mol}^{-1}\text{s}^{-1} \qquad [7]$$

The overall second-order rate constant is greater at that
pH corresponding to the pK_a of HO_2 consistent with reaction [6] having the fastest rate. Reaction [7] is very
slow which is not surprising in view of the comments
above on the relative stabilities of the two ions. The
stability of O_2^- depends not only on proton concentration
but also on the metal ion, small ions such as Li^+ decreasing the life-time in solution (Hill and Turner,
unpublished), larger ions such as Ba^{2+} increasing it
(Bray et al., 1977) though it is not known if this is a
disproportionation reaction. The latter is catalysed by
iron and copper complexes though the mechanism has not
been unravelled (Bielski and Gebicki (1970).

Reactions of the Superoxide Ion

O_2^- has a number of options open to it: it can act as
a reductant, an oxidant or a nucleophil and though the
exact role it undertakes has not always been defined,
this provides a convenient method of discussing them.

Some of the reactions in which O_2^- acts as a reductant are given in Table II. Apart from those of value as
assay methods (McCord et al., 1977b) the most interesting
are the reactions with quinones and peroxides. With the
former the equilibria are dependent on the redox potential of the quinone. For example for the reaction (Patel
and Willson, 1973) in water of vitamin K with O_2^-

$$\text{Quinone} + O_2^- \underset{k_{-1}}{\overset{k_1}{\rightleftharpoons}} \text{semiquinone} + O_2 \qquad [8]$$

k_1 is $<0.0002 \times 10^9$ $Mol^{-1}sec^{-1}$; k_{-1} is 0.2×10^9 $Mol^{-1}s^{-1}$,
whereas for benzoquinone k_1 is 1.0×10^9 $Mol^{-1}s^{-1}$ and
k_{-1} is $<0.01 \times 10^9$ $Mol^{-1}s^{-1}$. In aprotic media the O_2^- is
a more effective reductant and many quinones, including
Vitamin K are reduced to the semiquinone. Alkylhydro-
peroxides react (Peters and Foote, 1976) with O_2^- in
acetonitrile to give products which can reasonably be

considered to derive from radical intermediates e.g.,

$$(CH_3)_3CCOOH \xrightarrow{O_2^-} CH_3\overset{O}{\overset{\|}{C}}CH_3 + (CH_3)_3COH \qquad [9]$$

Dialkylperoxides react much more slowly.

Given the redox potentials illustrated by Fig. 3, it would be expected that O_2^- would be a much better oxidant in protic media. This is consistent with the observations to date and indeed of the reactions so far reported, Table 2, a number are of interest. For example the oxidation (Misra and Fridovich, 1972b) of epinephrine used as a convenient, though complex assay gives readily adrenochrome. With quinols the reaction is complex and summarised in Fig. 5. It is not surprising that the details of the reaction with a given quinol are difficult to elucidate! Both ascorbic acid and α-tocopherol have been considered as 'radical' traps. It is therefore no surprise to find that O_2^- has been shown to react with the former (Wong et al., 1975) and with a model (Nishikimi and Machlin, 1975) of the latter. In a very interesting study it was shown (Bielski and Chan, 1976) that the reactivity of NADH towards O_2^- is markedly enhanced by binding to lactate dehydrogenase. This serves as a warning that it can be dangerous to translate data obtained in simple systems to the complex situations which can arise in biology. Finally we have in the reaction of cyclohexenone with superoxide a nice illustration that the generation of one reactive material would lead to the formation of another, in the epoxide.

The charge of the superoxide ion obviously confers on it the characteristics of a nucleophil. Such qualities like those of many nucleophils will be less apparent in protic solvents where the reactivity of species such as $O_2^-(H_2O)_n$ is markedly less than the parent anion, than in aprotic media. Consequently few reactions have been

TABLE II

Some reactions of superoxide in aqueous and aprotic media

(a) Superoxide as a reducing agent

Compound	Products	Reference
$C(NO_2)_4$	$\longrightarrow$ $C(NO_2)_3^- + NO_2 + O_2$	Rabani et al (1965)
Fe (III) cytochrome c	$\longrightarrow$ Fe (II) cytochrome c $+ O_2$	McCord et al (1968)
[benzoquinone structure, O at top and bottom]	$\rightleftharpoons$ *[semiquinone structure, O^- top, $O^{\bullet}$ bottom]* $+ O_2$	Bors et al (1974) Patel et al (1973)
[nitrobenzene, NO_2]	$\longrightarrow$ *[nitrobenzene radical anion, NO_2, ⊖]* $+ O_2$	LeBerre et al (1966)
H_2O_2	$\xrightarrow{M^{n+}}$ $OH^- + OH^{\bullet} + O_2$	Haber et al (1934) Cohen (1977) Kellogg et al (1977)
Cu (II) (o-phenanthroline)$_2$	$\longrightarrow$ Cu (I) (o-phenanthroline)$_2 + O_2$	Valentine et al (1975)
Mn (III) tetraphenylporphrin	$\longrightarrow$ Mn (II) tetraphenylporphrin $+ O_2$	Valentine et al (1976)

(b) Superoxide as an oxidant

Compound	Products	Reference
[adrenaline structure: OH, HO, HO, NH, CH_3]	$\longrightarrow$ *[adrenochrome structure: O, OH, ^-O, N^+, CH_3]*	Misra et al (1972)
[catechol disulfonate: HO, HO, SO_3^-, SO_3^-]	$\longrightarrow$ *[radical: ^-O, $^{\bullet}O$, SO_3^-, SO_3^-]* $+ H_2O_2$	Millow (1970)
Lactate dehydrogenase-NADH	Lactate dehydrogenase NAD$^{\bullet}$ $+ H_2O_2$	Bielski et al (1976)
[catechol: OH, OH]	$\longrightarrow$ *[o-benzosemiquinone: $O^{\bullet}$, O^-]* $+ H_2O_2$	Lee-Ruff (1977)
RCHO	$\longrightarrow$ $RCO_2^- + O_2 + OH^-$	LeBerre et al (1966)
[cyclohexenone]	$\longrightarrow$ *[epoxy cyclohexanone]*	Dietz et al (1970)

(c) <u>Superoxide as a nucleophil</u> TABLE II cont.

Compound	Products	Reference
Fe(III) protoporphrin dimethyl ester perchlorate	O_2 Fe(II) protoporphrin dimethyl ester + ClO_4^-	Hill et al (1974)

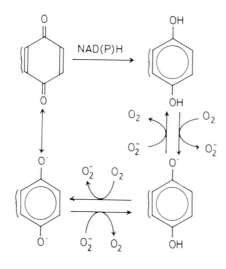

		Corey et al (1975)
2RX $\longrightarrow$	$RO_2R + O_2 + 2X^-$	Dietz et al (1970)
RX $\longrightarrow$	ROH	Corey et al (1975) San Filippo (1975)
AquoCo(III) cobalamin $\longrightarrow$	Superoxo Co(III) cobalamin	Ellis et al (1973)
		Lee-Ruff (1977)
		Harbour et al (1975)

Fig. 5. The reactions of dioxygen and superoxide with quinols, semiquinones and quinones. The position of any given equilibrium depends on the redox potentials of the quinol/semiquinone/quinone couples and on the solvent.

discovered to date in protic media about which it could
be stated with confidence that the superoxide ion acts
solely as a nucleophil. In solvents such as dimethylsulphoxide it is a different story in that the usual
electrophilic reagents undergo attack by O_2^-. The superoxide ion and its conjugate acid both react with metal
ions or complexes in some instances with concomitant
reduction (Valentine and Quinn, 1976) of the metal ion,
in other instances forming a stable salt or metal complex
(Ellis et $al.$, 1973). If the latter includes a metal ion
with an accessible lower oxidation state then it is sometimes difficult to acquire information on which oxidation
state of the metal ion is the more appropriate. In, for
example, the reaction of O_2^- with a derivative of iron
(III) protoporphyrin IX, the product has a spectrum almost
identical with that of oxymyoglobin (Hill et $al.$, 1974).
Whether the description of the latter as a superoxyiron
(III) complex or a dioxyiron(II) complex is more appropriate is a matter of lively debate.

It is well-known that the superoxide can act as a
bridging ligand in metal complexes and such complexes may
be of relevance in the autoxidation (Chin, D.-H., et $al.$,
1977) of other metal complexes such as iron(II) porphyrins. Of recent work most interesting are the reactions
with those complexes of manganese (Valentine and Quinn,
1976) and copper (Green and Hill, unpublished; Valentine
and Curtis, 1975), both of which may be of relevance to
the superoxide dismutases and the former is also of
interest in connection with the protein assembly involved
in the evolution of dioxygen (Heath, 1973) of chloro-
plasts.

It should be made clear that in many of the reactions
discussed above, the mechanisms are far from understood.
In particular the evidence is usually lacking that the
species attacking the organic or inorganic moiety is in
fact O_2^- whether as its conjugate acid, solvate or ion

pair. Furthermore few of these reactions are such as to make one fearful of the consequence of the generation and release of O_2^- in a biological environment. The same cannot be said for the sinister results of the possible incestuous reactions between O_2^- and the other reduction products of dioxygen. For example it is conceivable that dioxygen generated by either the dismutation (Khan, 1976) of O_2^- or by the reaction (Kellogg and Fridovich, 1975) of O_2^- or H_2O_2 is in one of the $^1O_2$ states. Opinions differ greatly as to whether these have been demonstrated; if formed they would be capable of wreaking havoc in a bio- logical environment (Foote and Wexler, 1964). The same is true, only more so, for the hydroxyl radical (Draganic and Draganic, 1971). The so-called Haber-Weiss reaction $O_2^- + H_2O_2 \rightarrow O_2 + OH^- + OH\cdot$ (Haber, and Weiss, 1934) is thought (Halliwell, 1976) to proceed at a very slow rate. Nevertheless it is conceivable (Cohen, 1977) that, just as most chemical environments are contaminated with trace metals, biological environments may contain suffi- cient for the catalysed Haber-Weiss reaction to be im- portant (Kellogg and Fridovich, 1977), e.g.:

$$Fe^{III} + O_2^- \rightarrow Fe^{II} + O_2 \qquad [10]$$

$$Fe^{II} + H_2O_2 \rightarrow Fe^{III} + OH^- + OH\cdot. \qquad [11]$$

There is some evidence that such reactions do indeed occur though the interpretation of e.g. combined inhibition by catalase and superoxide dismutase, is be- devilled by the lack of truly specific reagents for $OH\cdot$, O_2^- and $^1O_2$. It is an attractive postulate that, to avoid the generation of a species as destructive as $OH\cdot$, two efficient enzymes were evolved to prevent the formation of the precursors, H_2O_2 and O_2^-. These same species are just those which result from the irradiation of oxygena- ted aqueous solution (Czapski, 1971) and it is intriguing

to consider that, as well as having to contend with the
inadvertent chemical generation of such species as a
consequence of making use of dioxygen as the terminal
electron acceptor, early primitive organisms, and as we
shall see, not so primitive late ones, may have been required to evolve a protection against the ravages of radiation.
Of course most biological fluids contain other dissolved
small molecules. For example, though at first sight,
dissolved carbon dioxide might seem innocuous, consider
the following set of reactions:

$$O_2^- + CO_2 \rightleftharpoons CO_2^- + O_2 \qquad\qquad [12]$$

$$O_2^- + CO_2 \rightleftharpoons CO_4^- \qquad\qquad [13]$$

$$CO_2 + HO_2^- \rightleftharpoons HCO_4^- \qquad\qquad [14]$$

$$2CO_2 + 2O_2^- \rightleftharpoons C_2O_6^{2-} + O_2 \qquad\qquad [15]$$

Evidence exists for each of these though not necessarily
under the same conditions but the reactivity of the products remains to be explored (Michelson and Durosay, 1977;
Spangler and Collins, 1975; Firsova et $al.$, 1963;
Mel'nikov et $al.$, 1962).

Superoxide Dismutase

We have referred above to superoxide dismutases
(Fridovich, 1974, 1975; Malmström et $al.$, 1975; Michelson
et $al.$, 1977a) as scavengers of O_2^-. The identification of
the cupreins, as they were previously known, as catalysts
for the disproportionation reaction is a story of happen-
stance, serendipity, skill and imagination (McCord and
Fridovich, 1977). The cupreins had languished for close
to thirty years as the Cinderellas of copper proteins.
Their chance came, when they were invited to meet the
various derivatives of dioxygen and the oxyanions; the

superoxide 'shoe' fitted. All aerobic organisms so far
investigated have been shown to contain at least one
superoxide dismutase (Fridovich, 1974). For eukaryotes,
the cytoplasm contains a protein MW ~ 32,000 daltons,
composed of two subunits each containing one zinc and one
copper ion. They have been isolated from fungi (Misra and
Fridovich, 1972a; Rapp *et al.*, 1973), yeast (Weser *et al.*,
1972b), plants (Sawada *et al.*, 1972; Asada *et al.*, 1972;
Beauchamp and Fridovich, 1973), and vertebrates (Bannister *et al.*, 1971; Bannister and Wood, 1970). The mitochondrial enzyme contains manganese (Weisiger and Fridovich, 1973; Ravindranath and Fridovich, 1975) (in some
species this protein is also found in the cytoplasm
(McCord *et al.*, 1977a)) as interestingly, in view of the
symbiotic theory of mitochondrial evolution, do many
prokaryotes (Table III). Some of the latter also contain
an iron superoxide dismutase. Thus all three readily
available and accessible transition metals having suitable redox couples have been employed. Comparatively
little is known about the manganese and iron enzymes
though it has been possible to effect some metal sub-
stitutions but with concomitant loss of activity (Ose and
Fridovich, 1976).

The copper-zinc protein is well-studied. The crystal
structure reveals (Richardson *et al.*, 1975), as spectroscopic methods had predicted (Fee and Gaber, 1972), that
the copper and zinc ions are in close proximity (Fig. 6).
The copper is coordinated to three histidine ligands, the
zinc ion to a glutamate and three histidines, one of
which (His 61) is interposed between the two metal ions
and may act as a bridging ligand. Recent work has shown
(Cass *et al.*, 1978) that the zinc ion is required to pre-
form the copper binding site and is responsive to changes
at the copper site; e.g. change in oxidation state or
ligand binding (Cass and Hill, unpublished). By doing so
it may influence the course of the reaction which may

TABLE III

Manganese and Iron-containing Superoxide Dismutase

Species	Relative Molecular Mass (daltons)	Subunits	Metal g-atom/mole	Reference
Animal:				
Chicken liver mitochondria	80,000	4 × 20,000	2.3 Mn	Weisiger and Fridovich, 1973
Human liver	85,000	4 × 21,000	3.9 Mn	McCord *et al.*, 1977
Fungi:				
Saccharomyces cerevisiae mitochondria	100,000	4 × 25,000	3.8 Mn	Ravindranath and Fridovich, 1975
Pleurotus olearius (Basidiomycetes)	76,000	4 × 20,000 (two types)	2.1 Mn	Lavelle *et al.*, 1974
	78,000	4 × 20,000	1.9 Mn	Lavelle and Michelson, 1975
Eukaryotic Algae:				
Porphyridium cruentum (Rhodophyta)	40,000	2 × 20,000	1.0 Mn	Misra and Fridovich, 1977
Prokaryotic Algae (Blue-Green)				
Plectonema boryanum	37,000	2 × 19,000	0.94 Fe	Misra and Keele, 1975
	42,000	2 × 21,000	2.0 Fe	Asada *et al.*, 1975
Spirulina platensis	37,000	2 × 18,400	1.0 Fe	Lumsden and Hall, 1974; Lumsden *et al.*, 1976
Bacteria:				
Streptococcus mutans	42,000	2 × 19,000	1.2 Mn	Weisiger and Fridovich, 1973; Vance *et al.*, 1972
	-	n × 20,000	Mn	Vance *et al.*, 1972
Bacillus stearothermophilus	40,000	2 × 20,000	1 Mn	Bridgen *et al.*, 1976
Thermus aquaticus	80,000	4 × 21,000	2.1 Mn	Sato and Harris, 1977

Mycobacterium phlei.	80,000	4 X 20,000	1.7 Mn + 1.2 Fe + 0.8 Zn	Chikata *et al.*, 1975
Mycobacterium sp. strain Takeo	62,000	3 X 21,000	1.7 Mn	Kusunose *et al.*, 1976
Mycobacterium lepraemurium	45,000	2 X 22,000	1.3 Mn	Schihara *et al.*, 1977
Rhodopseudomonas sphaeroides	37,000	2 X 18,000	1.1 Mn	Lumsden *et al.*, 1976
Escherichia coli	40,000	2 X 20,000	1.2 Mn	Weisiger and Fridovich, 1973
				Keele *et al.*, 1970
	39,000	2 X 18,000	1.0 - 1.8 Fe	Yost and Fridovich, 1973;
				Slykhouse and Fee, 1976
Bacillus megaterium	40,000	2 X 20,000	1.0 Fe	Anastasi *et al.*, 1976
Pseudomonas ovalis	40,000	2 X 20,000	1.4 Fe*	Yamakura, 1976
Photobacterium leiognathi	40,000	2 X 20,000	1.6 Fe	Puget and Michelson, 1974
Photobacterium sepia	40,000	2 X 20,000	1.4 Fe	Puget and Michelson, 1974
Desulfovibrio desulfuricans	43,000	2 X 21,500	1.6 Fe	Lumsden and Hall, 1975

*After dialysis and crystallization, iron content dropped to 1.1 g-atom/mole with proportional drops in catalygic activity. Authors conclude native enzyme contains 2 catalytically active Fe, one more tightly bound than the other.

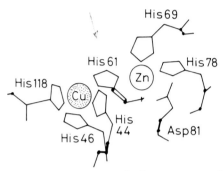

Fig. 6. The immediate environment of the copper and zinc ions in bovine superoxide dismutase adapted from Richardson *et al.*, 1975.

proceed as follows (Hodgson and Fridovich, 1975):

$$E\ Cu(II)[His^-][HisH]_3 + O_2^- \xrightarrow{H^+} E\ Cu(I)(HisH)_4 + O_2$$
$$[16]$$

$$E\ Cu(I)[HisH]_4 + O_2^- \xrightarrow{H^+} E\ Cu(II)[His^-][HisH]_3 + H_2O_2$$
$$[17]$$

in which His 61 acts as a ligand to copper(II) as the imidazole anion. It is unlikely to do so to copper(I) hence it could be considered to act as a proton sink. No detailed mechanism has yet been proposed which accounts satisfactorily for the relationship between proton and electron transfer. It might be thought strange to require a catalyst for a reaction which proceeds quite quickly on its own. Alas all things are relative. For example the rate of the catalysed reaction at pH 7.4 (Fielden *et al.*, 1974) is close to diffusion controlled $2.3 \times 10^9\ Mol^{-1}s^{-1}$. Not only that, but, as has been pointed out, if the concentration of superoxide dismutase in cells is $\sim 10^{-5}$ M and the concentration of O_2^- is very much less than that, then the effective rate in the presence of the enzyme might be as much as 10^9 times faster than the uncatalysed reaction. It is tempting to suggest that the close similarity between the copper-zinc enzymes derived from dif-

ferent species points to an important role for the enzyme;
once having found a successful catalyst it was left well
alone. A comparison of the ^{1}H n.m.r. spectra of the dismutase from a number of species suggests (Cass and Hill,
unpublished) that their structures in solution do not
differ significantly from that of the bovine enzyme. This
would be consistent with the low antigenicity (Metzger
et al., 1968) of the enzyme and the significant degree of
cross-reaction (Tegelström, 1975) of the protein from one
species with the antibody raised against the protein of
another.

   To date the evidence is that the sole biological substrate for these proteins is O_2^- though the hunt continues
for others. For example it is possible that the protein
could act as a general radical dismutase perhaps for small
oxygen, sulphur or nitrogen-containing radicals generated
either by oxidation or radiolysis. The role of these
proteins as superoxide dismutases has sometimes been
queried (Weser, 1973) because simple metal ions or complexes can act (Joester et al., 1972; Weser et al., 1972a;
Brigelius et al., 1975) as catalysts for the disproportionation with the implication that enzymes should always
be inherently superior as catalysts to all simple complexes. This may be true for reactions in which the substrate are large or the reaction is complex but it probably represents our poor understanding of reaction mechanisms and/or our inadequate synthetic abilities. Even
so, metalloproteins have to have properties, such as
solubility, sequestering ability, or membrane permeability,
in addition to their functions as catalysts, compatible
with their operation in biological surroundings.

   Identification of superoxide dismutase has been of
great value in the detection of the superoxide ion in many
reactions. Though superoxide dismutase also reacts (Bray
et al., 1974) with H_2O_2 and a number of inhibitors including (Rotilio et al., 1972) CN^- it is safe to assume

that, if the presence of SOD perturbs a reaction and other proteins or denatured SOD have no effect, O_2^- is present in the system. A negative result does *not* mean that O_2^- is absent because the concentration of SOD *at the site* of generation of O_2^- may be zero. For example if O_2^- is generated inside a cell then, if the superoxide dismutase is added extracellularly no perturbation will be observed. Even with reactions *in vitro* the O_2^- may remain bound to the enzyme or due to higher local concentrations if other possible reactants fail to reach the dismutase. It is always worthwhile checking the effect of small O_2^- scavengers (Misra and Fridovich, 1972a) even though these are not as specific.

The Superoxide ion in Biological Systems

Table IV gives a selection of those systems in which O_2^- has been detected or implicated. Most of these are concerned with the reduction of dioxygen, there being many fewer examples of O_2^- being clearly derived from other sources. We can conveniently divide them into two categories which we can describe as (a) the inadvertent generation of the superoxide ion and (b) the required generation of O_2^-.

(a) Many of these are known as autoxidation reactions. For example the autoxidation of heme proteins has been shown (Misra and Fridovich, 1972a) to generate O_2^-. If we write, oxyhemoglobin as $HbFe^{III}O_2^-$ then one would expect that the superoxide ion might simply be displaced by a competing nucleophile. This does indeed seem to be the case. In addition the auto-oxidation has been found (Rotilio *et al.*, 1977) to be inhibited by ethylenediamminetetracetic acid and promoted by adrenaline suggesting that metal ions or adrenaline can scavenge the O_2^- hence displacing the equilibrium. In a similar category we can place the autoxidation of flavins (Ballou *et al.*, 1969) quinols (Misra and Fridovich, 1972a; Cadenas *et al.*, 1977)

TABLE IV

In vivo *sources of superoxide*

Enzyme	Reaction	Co-factors	References
Xanthine Oxidase	Xanthine → Urate	FAD, Mo, Fe/S	Edmondson *et al.*, 1972
Aldehyde Oxidase	R CHO → R COOH	FAD, Mo, Fe/S	Rajagopalan and Handler, 1964
Dihydro-orotate hydrogenase	Dihydro-orotate → Orotate	Flavoprotein	Miller and Massey, 1965
Indoleamine-2,3 -dioxygenase	L-Tryptophan → L-Formylkynurenine	Heme-Fe	Ohnishi *et al.*, 1977
Ferredoxin-NADPH reductase	NADPH + ox.Ferredoxin NADP$^+$ + red.Ferredoxin	Flavoprotein	Nakamura, 1972
Microsomal hydroxylases	RH → ROH	Wide variety of enzymes and substrates e.g. Cu: Dopamine-β-hydroxylase; Heme Fe: P450 dependent hydroxylases	Coon *et al.*, 1973
Galactose oxidase	o-Galactose → D-galacto-hexodialdose	Cu	Hamilton *et al.*, 1973
Photosystem II		Complex membranous system found in chloroplasts Cu, Mn	Allen and Hall, 1973

(Fig. 5) and ferredoxins (Orme-Johnson and Beinert, 1969; Nilsson *et al.*, 1969; Misra and Fridovich, 1971).

Turning to category (b), there is now considerable evidence that many enzymatic reactions generated O_2^-, either as the oxidant or as a precursor of the oxidant e.g. H_2O_2, $OH^\cdot$, $'\Delta O_2$ and their complexed forms. An interesting variation on this theme is the enzyme indoleamine-2,3-dioxygenase which appears (Hirata and Hayaishi, 1977) to require O_2^- as a substrate. The reaction (Fig. 7) has O_2^- as a preformed oxidant to give the $Fe^{III}O_2^-$ complex which is then further reduced to $Fe^{II}O_2^-$ or $Fe^{III}O_2^{2-}$ which the evidence suggests is, or forms, a most reactive material. Similar complexes are probably present at some

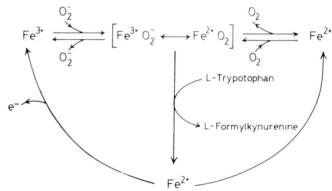

Fig. 7. A representation of the mechanism of the reaction catalysed by indoleamine-2,3-dioxygenase. Fe^{3+} and Fe^{2+} represent the two oxidation states of the heme protein.

stage in the reactions of cytochrome P.450. One of these enzymes has been the subject of some elegant investigations and indeed it has been observed (Strobel and Coon, 1971) that in the presence of high Cl^- concentrations, superoxide dismutase inhibits effectively.

The synthesis of prothrombin, a key constituent of the blood clotting cascade, has been shown (Esnouf *et al.*, 1978) to involve O_2^-, though whether as reactant or precursor of the reactant is not yet clear. The reaction is (Stenflo and Suttie (1977) an oxidative carboxylation

of some glutamic acid residues in the precursor of prothrombin to yield the γ-carboxyglutamate residues in prothrombin. It takes place in or on liver microsomes and
requires Vitamin K, an NADH-dependent reductast, oxygen,
carbon dioxide and a carboxylase. It has been shown
(Esnouf *et al.*, 1978) that SOD inhibits both the carboxylation and the formation of the Vitamin K epoxide. It is possible that the carboxylation step involves the formation of
species such as $C_2O_6^{2-}$, CO_4^{2-}, CO_2^- or CO_3^- directly from
CO_2 and O_2^- as described earlier. It is interesting that
similar species have been proposed (Michelson and Durosay,
1977) to account for the stimulation of photo-induced
hemolysis by carbonate or bicarbonate.

A most fascinating example of the use of dioxygen and
the generation of O_2^- by a biological system lies in the
act of phagocytosis. Contact with foreign materials
prompts certain cells, e.g. neutrophils, which specialise
in attacking bacteria, to take up (Carnutte *et al.*, 1975)
dioxygen which, *via* an NADPH-linked reductase gives O_2^-,
approx. 70% of the dioxygen being used (Weening *et al.*,
1975) for this purpose. In turn the same percentage of
H_2O_2 is formed (Root and Metcalf, 1975) from O_2.
Hydrogen peroxide, in the presence of myeloperoxidase
and Cl^-, generates (Harrison and Schultz, 1976) hypo-
chlorite. It is still a matter of opinion whether the O_2^-
leads to the production of $OH^\cdot$ and/or $^1O_2$. So this
microscopic version (Fig. 8) of chemical warfare leads,
or could lead, to extremely reactive and destructive entities. What of the chemical consequences for any organism
which harbours, indeed relies on, this justified agressor?
The reliance is well-illustrated by patients with chronic
granulomatous disease, who have an inborn defect in phago-
cyte metabolism. No uptake of dioxygen leading to O_2^- and
H_2O_2 production is observed with the consequence that such
patients are (Carnutte *et al.*, 1975) subject to recurrent
and severe bacterial infections. On the other hand if

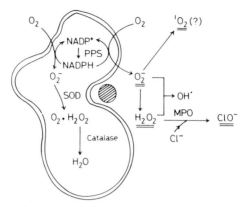

Fig. 8. A schematic illustration of the use of dioxygen by a phago-
cyte during the act of phagocytosis. The foreign material is represented by that hatched circle. Reactive (toxic) entities are shown
underlined; SOD is superoxide dismutase; MPO is myeloperoxidase;
PPS represents the pentose phosphate shunt for the regeneration of
NADPH.

somehow the phagocytes were induced to discharge in all
directions by a situation which, as far as the host organism is concerned, did not really warrant it, then considerable damage could occur to other host tissues. It is
with the consequences of a surfeit or deficiency of either
O_2^- or superoxide dismutase with which we now concern
ourselves.

Physiological Consequences

Whether or not 'life is an incurable disease' (Cowley,
1656), the problems for aerobic organisms in handling dioxygen are ever present. This is most easily seen by
introducing higher animals to an atmosphere of pure dioxygen. The respiratory system is affected first and
breathing becomes laboured and death is due to pulminary
oedema. Considerable cell damage is found, many biological
materials having been oxidised. In an impressive series
of experiments Fridovich and his colleagues showed (McCord
et al., 1971; Gregory and Fridovich, 1973 a and b; McCord
et al., 1973; Hassan and Fridovich, 1977) that the ability

of various microorganisms to survive high dioxygen concentrations depends on the levels of superoxide dismutase.
Furthermore it was possible to show that the synthesis
of superoxide dismutase could be induced by hyperbaric
dioxygen, to the extent that the resultant strain of
bacteria could survive 20 atmospheres of dioxygen. Though
the evidence for mammals is not as clear-cut, a similar
conclusion can be reached (Crapo and Tierny, 1973;
Stevens and Autor, 1977). Superoxide dismutase does protect.

There should therefore be a distinction between aerobic and anaerobic organisms. This is true (Table V) to
the extent that *all* aerobic organisms contain superoxide
dismutase be it Cu/Zn, Mn and/or Fe-containing. At first
it was thought (McCord *et al.*, 1971) that superoxide dismutases are not present in strict anaerobes. Aero-tolerant
organisms are intermediate in type in that although they
may contain a dismutase they lack catalase. Recent work
has shown (Hewitt and Morris, 1975; Hatchikian *et al.*,
1977), however, that some strict anaerobes do contain a
dismutase. There are a number of explanations forthcoming: that O_2^- is not the only substrate for the enzyme; that anaerobes still have had to endure dioxygen
since it could have arisen by e.g. radiolysis of water
(Lumsden and Hall, 1975); that they acquired (Hatchikian
et al., 1977) the ability to synthesise the enzyme as a
result of gene transfer. The last of these explanations
is obviously amenable to investigation. The second is
interesting since attention was drawn above to the relationship between the products of the radiation of oxygenated water and those which could result from the reduction of dioxygen. That hardy organism *Micrococcus
radiodurans* contains (McCord *et al.*, 1971) significantly
more superoxide dismutase than comparable organisms and
considerably more catalase. Experiments on mammals have
shown that superoxide dismutase, as well as other radical

TABLE V

Superoxide dismutase and catalase contents
of a variety of micro-organisms

Species	Superoxide Dismutase Units/mg	Catalase Units/mg	Reference
Aerobes			
Escherichia coli	1.8	6.1	a
Salmonella typhimurium	1.4	2.4	a
Halobacterium salinarium	2.1	3.4	a
Rhizobium japonicum	2.6	0.7	a
Micrococcus radiodurans	7.0	289	a
Saccharomyces cerevisiae	3.7	13.5	a
Mycobacterium sp.	2.9	2.7	a
Pseudomonas sp.	2.0	22.5	a
Aerotolerant anaerobes			
Butyribacterium rettgeri	1.6	0	a
Streptococcus faecalis	0.8	0	a
Streptococcus mutans	0.5	0	a
Streptococcus bovis	0.3	0	a
Streptococcus mitis	0.2	0	a
Streptococcus lactis	1.4	0	a
Zrymobacterium oroticum	0.6	0	a
Lactobacillus plantarum	0	0	a
Strict anaerobes			
Veillonella alcalescens	0	0	a
Clostridium pasteurianum	0	0	a
Clostridium stricklandii	0	0	a
Clostridium lentoputrescens	0	0	a
Clostridium cellobioparum	0	0	a
Butyrivibrio fibrisolvens	0	0.1	a
Chlorobium thiosulfatophilum NCIB 8346	+++	NT*	b
Chromatium sp. NCIB 8348	+	NT	b
Desulfotomaculum nigrificans NCIB 8395	++	NT	b
Desulfovibrio desulfuricans NCIB 8307	+	NT	b
Clostridium perfringens NCIB 11105	+++	NT	b
Desulfovibrio vulgaris NCIB 8303	+	+	c
Desulfovibrio gigas NCIB 9332	+	+	c
Desulfovibria desulfuricans NCIB 8310	+	+	c
NCIB 8307	-	-	c
Desulfovibrio salexigens NCIB 8403	-	-	c

*NT - not tested; a, McCord *et al.*, 1972; b, Hewitt and Morris, 1975; c, Hatchikian *et al.*, 1977.

scavengers e.g. α-tocopherol does offer (Petkau *et al.*, 1976) some protection against radiation.

Some antibiotics require (White and White, 1968) dioxygen for maximum potency. Bacteria which have high dismutase content (Gregory and Fridovich, 1973a) are more resistant. Paraquat, the dimethyl 4,4-bipyridyl cation, is thought (Autor, 1974) to generate superoxide ions in the presence of O_2. The issue of whether superoxide dismutase protects against the onslaught of this widely employed herbicide is undecided.

So far we have referred to the effects of the inadvertent production. Phagocytosis is an example in which the production of O_2^- is required. What if the use of O_2^- were to be prevented by an excess of superoxide dismutase? Similarly what if those biosynthetic reactions or metabolic routes which require O_2^- at some stage were to be impeded? It has even been suggested (Michelson *et al.*, 1977b) that O_2^- serves a useful purpose by scavenging $OH^{\cdot}$ by the reaction $OH^{\cdot} + O_2^- \rightarrow OH^- + O_2$. Whether the local concentration of O_2^- is ever sufficient to make it an effective scavenger is open to question. The gene controlling the synthesis of the Cu/Zn SOD in humans is (Tan *et al.*, 1973) located on chromosome 21. It is this chromosome which is doubled in Downs syndrome and those with this genetic defect have been shown to have more superoxide dismutase than normal. Few physiological consequences thereof have been identified. However, though the amount of O_2^- produced by the phagocytes during the resting period of such patients is normal, the amount of O_2^- detected when the phagocytes are activated is reduced (Corberand *et al.*, 1974).

Of the other clinical problems so far reported to be affected by superoxide dismutase, those which have attracted the most attention (McCord, 1974) have associated problems of inflammation. Of these, various arthritic conditions have been singled out (Menander-

Huber and Huber, 1977) for investigation and extensive
trials have taken place. It has been suggested that if an
auto-immune reaction is the cause of this complaint the
phagocytes reacting with e.g. immunoglobulin (Johnston
et al., 1976), generate the destructive entities referred
to above.

Finally, and most speculatively, the long-standing
'radical theory of ageing' may not be unconnected (Willson,
1977) with a number of the topics discussed in this
article. If the oxidation of membrane lipids is related
to cellular ageing it is interesting that superoxide
dismutase prevents (Kellogg and Fridovich, 1977) lipid
peroxidation. It has been suggested (Fridovich *et al.*,
1977) that a constant low level of irreparable damage due
to imperfect scavenging of superoxide might be a factor
in cellular ageing though this does not seem to be so
for human erythrocytes (Michelson *et al.*, 1977b). The
investigation of the age-related levels of superoxide
dismutase, particularly the Mn-containing enzyme in other
tissues would be well worth investigation. If the defence
mechanisms of an organism become less effective with age
then the inability to deal with the toxic effects of that
otherwise obligatory element, oxygen, might have profound
consequences. The paradox presented by the two opposing
features of the role of dioxygen in biological processes
(Fig. 9) has proved, and will continue to prove, a most
seductive problem for all manner of men.

Acknowledgements

I would like to thank the Science Research Council,
through the Oxford Enzyme Group, and the Medical Research
Council for financial support and my collaborators,
Professor W.H. Bannister, Dr. J.V. Bannister, Dr. M.P.
Esnouf and Messrs. A. McEuan, A.E.G. Cass and M.R. Green
for their help.

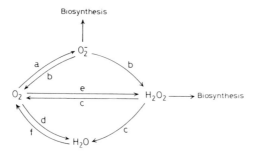

Fig. 9. A schematic representation of the biological oxygen cycle:
(a) various enzymes, see Table IV and text; (b) superoxide dismutase; (c) catalase; (d) cytochrome oxidase; (e) various enzymes
e.g. amino-acid oxidases; (f) photosynthesis.

References

Abrahams, S.C., Collin, R.L. and Lipscomb, W.N. (1951). *Acta Cryst.*
 4, 15.
Allen, J.F. and Hall, D.O. (1973). *Biochem. Biophys. Res. Comm.* **52**,
 856.
Anastasi, A., Bannister, J.V. and Bannister, W.H. (1976). *Int. J.
 Biochem.* **7**, 541.
Asada, K., Urano, M. and Takehashi, M. (1972). *Seikagaku* **44**, 624.
Asada, K., Yoshikawa, K., Takahashi, H., Masda, Y. and Enmanji, K.
 (1975). *J. Biol. Chem.* **250**, 2801.
Autor, A.P. (1974). *Life Sci.* **14**, 1309.
Ballou, D., Palmer, G. and Massey, V. (1969). *Biochem. Biophys. Res.
 Comm.* **36**, 898.
Bannister, J.V., Bannister, W.H. and Wood, E. J. (1971). *Eur. J.
 Biochem.* **18**, 178, 187.
Bannister, W.H. and Wood, E.J. (1970). *Life Sci.* Section 2, **9**, 229.
Beauchamp, C.O. and Fridovich, I. (1973). *Biochim. Biophys. Acta*
 317, 50.
Behar, D., Czapski, G., Dorfman, L.M., Rabini, J. and Schwarz, H.A.
 (1970). *J. Phys. Chem.* **74**, 3209.
Bennett, J.E., Mile, B. and Thomas, A. (1968). *Trans. Faraday Soc.*
 64, 3200.
Benson, S.W. and Shaw, R. (1970) in *Organic Peroxides* (Swern, D.,
 ed.) Wiley-Interscience, New York.
Bielski, B.H.J. and Gebicki, J.M. (1970). *Adv. Rad. Chem.* **2**, 177.
Bielski, B.H.J. and Chan, P.C. (1976). *J. Biol. Chem.* **251**, 3841.
Bielski, B.H.J., communciated at the 'International Conference on
 Singlet Oxygen and Related Species in Chemistry and Biology',
 Pinawa, Manitoba, Canada. 1977, see Bielski, G.H.J. and Allen, A.O.
 (1977). *J. Phys. Chem.* **81**, 1048-1050.
Bors, W., Saran, M., Lengfelder, E., Spottl, R. and Michel, C.
 (1974). *Curr. Top. Radiat. Res.* **9**, 247.
Bray, R.C., Cockle, S.A., Fielden, E.M., Roberts, P.B., Rotilio, G.
 and Calabrese, L. (1974). *Biochem. J.* **139**, 43.

Bray, R.C., Mautner, G.W., Fielden, E.M. and Carle, C.I. (1977). *In* 'Superoxide and Superoxide Dismutases' (Michelson, A.M., McCord, J.M. and Fridovich, I., eds.), Academic Press, London, pp. 61-75.

Bridgen, J., Harris, J.I. and Kolb, E. (1976). *J. Mol. Biol.* **105**, 333.

Brigelius, R., Hartmann, H.-J., Bors, W., Saran, M., Lengfelder, E. and Weser, U. (1975). *Hoppe-Seylers Zeit.* **356**, 739.

Brix, P. and Herzberg, G. (1954). *Can. J. Phys.* **32**, 110.

Cadenas, E., Boveris, A., Ragan, C.I. and Stoppani, A.O.M. (1977). *Arch. Biochem. Biophys.* **180**, 248.

Carnutte, J.T., Kipnes, R.S. and Babior, B.M. (1975). *New Engl. J. Med.* **293**, 629.

Cass, A.E.G., Hill, H.A.O., Bannister, J.V. and Bannister, W.H. (1978). *Biochem. J.* in press.

Cass, A.E.G. and Hill, H.A.O., unpublished results.

Caughey, W.S., Wallace, W.J., Volpe, J.A. and Yoshikawa, S. (1976). *In* 'The Enzymes', 3rd edn., Vol. XIII (Boyer, P.D., ed.), Academic Press, London, p. 299.

Chikata, Y., Kusunose, E., Ichihara, K. and Kusunose, M. (1975). *Osaka City Med. J.* **21**, 127.

Chin Der-Hang, Gaudio, J.D., La Mar, G.N. and Balch, A.L. (1977). *J. Amer. Chem. Soc.* **99**, 5486 and references therein.

Cohen, G. (1977). *In* 'Superoxide and Superoxide Dismutases (Michelson, A.M., McCord, J.M. and Fridovich, I., eds.). Academic Press, London, p. 317.

Coon, M.J., Strobel, H.W. and Boyer, R.F. (1973). *Drug Metabol. Disposit.* **1**, 92.

Corberand, J., de Larrard, B., Pris, J., Colombies, P. (1974). *Nouvelle Revue Français d'Hematologie* **14**, 298.

Corey, E.J., Nicolaou, K.C., Shibasaki, M., Machida, Y. and Shiner, C.S. (1975). *Tet. Lett.* 3183.

Cotton, F.A. and Wilkinson, G. (1972). 'Advanced Inorganic Chemistry', 3rd edn., Interscience, New York.

Cowley, A. (1656). A Letter to Dr. Scarborough, Humphrey Moseley, London.

Crapo, J. and Tierny, D. (1973). *Clin. Res.* **21**, 222.

Czapski, G. (1971). *Ann. Rev. Phys. Chem.* **22**, 171.

Dietz, R., Forno, A.E.J., Larcombe, B.E. and Peover, M.E. (1970). *J. Chem. Soc.* (**B**) 816.

Doelle, H.W. (1975). 'Bacterial Metabolism', 2nd edn., Academic Press, London.

Draganic, I.G. and Draganic, Z.D. (1971). *In* 'The Radiation Chemistry of Water', Academic Press.

Edmondson, D.E., Massey, V., Palmer, G., Beacham, L.M. and Elion, G.B. (1972). *J. Biol. Chem.* 247, 1597.

Ellis, J., Pratt, J.M. and Green, M. (1973). *J. Chem. Soc. Chem. Comm.* 781.

Esnouf, M.P., Green, M.R., Hill, H.A.O., Irvine, G.B. and Walter, S.J. (1978). *Biochem. J.* in press.

Fee, J.A. and Gaber, B.P. (1972). *J. Biol. Chem.* **247**, 60.

Fee, J.A. and Valentine, J.S. (1977). *In* 'Superoxide and Superoxide Dismutases', Academic Press, London.

Fielden, E.M., Roberts, P.B., Bray, R.C., Lowe, D.J., Mautner,
    G.N., Rotilio, G. and Calabrese, L. (1974). *Biochem. J.* **139**, 49.
Firsova, T.P., Molodkina, A.N., Morozova, T.G., and Astenova (1963).
    Russ. J. Inorg. Chem. **8**, 140.
Foote, C.S. and Wexler, S. (1964). *J. Amer. Chem. Soc.* **86**, 3879.
Fridovich, I. (1974). *Advan. Enzymol.* **41**, 35.
Fridovich, I. (1975). *Ann. Rev. Biochem.* **44**, 147-159.
Fridovich, I., McCord, J.M., Michelson, A.M. (1977). *In* 'Superoxide
    and Superoxide Dismutases' (Michelson, A.M., McCord, J.M. and
    Fridovich, I., eds.), Academic Press, London, p. 551.
Goscin, S.A. and Fridovich, I. (1973). *Biochim. ? iophys. Acta.* **289**,
    276.
Green, M.R. and Hill, H.A.O., unpublished results.
Gregory, E.M. and Fridovich, I. (1973a). *J. Bacteriol.* **114**, 543.
Gregory, E.M. and Fridovich, I. (1973b). *J. Bacteriol.* **114**, 1193.
Haber, F. and Weiss, J. (1934). *Proc. Roy. Soc. Ser. A.* **147**, 332.
Halliwell, B. (1976). *FEBS Lett.* **72**, 8.
Hamilton, G.A., Libby, R.D. and Hartzell, C.R. (1973). *Biochem.
    Biophys. Res. Comm.* **55**, 333.
Harbour, J.R. and Bolton, J.R. (1975). *Biochem. Biophys. Res. Comm.*
    64, 803.
Harrison, J.E. and Schultz, J. (1976). *J. Biol. Chem.* **251**, 1371.
Hassan, H.M. and Fridovich, I. (1977). *J. Biol. Chem.* **252**, 7667.
Hatchikian, C.E., Le Gall, J. and Bell, G.R. (1977). *In* 'Superoxide
    and Superoxide Dismutases' (Michelson, A.M., McCord, J.M. and
    Fridovich, I., eds.). Academic Press, London, p. 159.
Heath, R.L. (1973). *Int. Rev. Cytol.* **34**, 49.
Hewitt, J. and Morris, J.G. (1975). *FEBS Lett.* **50**, 315.
Hill, H.A.O., Turner, D.R. and Pellizer, G. (1974). *Biochem. Biophys.
    Res. Comm.* **56**, 739.
Hill, H.A.O. and Turner, D.R., unpublished results.
Hirala, F. and Hayaishi, O. (1977). *In* 'Superoxide and Superoxide
    Dismutases' (Michelson, A.M., McCord, J.M. and Fridovich, I.,
    eds.), Academic Press, London, pp. 395-406.
Hodgson, E.K. and Fridovich, I. (1975). *Biochem.* **14**, 5294.
Joester, K.-E., Jung, G., Weber, U. and Weser, U. (1972). *FEBS Lett.*
    25, 25.
Johnson, R.A. and Nidy, E.G. (1975). *J. Org. Chem.* **40**, 1680.
Johnston, R.B., Lehmeyer, J.E. and Guthrie, L.A. (1976). *J. Exp. Med.*
    143, 1551.
Kanzig, W. and Cohen, M.H. (1959). *Phys. Rev. Letts.* **3**, 509.
Keele, B.B., McCord, J.M. and Fridovich, I. (1970). *J. Biol. Chem.*
    245, 6176.
Kellogg, E.W.; III, and Fridovich, I. (1975). *J. Biol. Chem.* **250**,
    8812.
Kellogg, E.W., III, and Fridovich, I. (1977). *J. Biol. Chem.* **252**,
    6721.
Khan, A.U. (1976). *J. Phys. Chem.* **80**, 2219.
Knowles, P.F., Gibson, J.F., Pick, F.M. and Bray, R.C. (1969).
    Biochem. J. **111**, 53.
Kusunose, M., Noda, Y., Schihara, K. and Kusunose, E. (1976). *Arch.
    Microbiol.* **108**, 65.
Lavelle, F., Durosay, P., Michelson, A.M. (1974). *Biochimie* **56**, 451.

LeBerre, A. and Berguer, Y. (1966). *Bull. Soc. Chim. France*, p. 2368.

Lee-Ruff, E. (1977). *Chem. Soc. Rev.* **6**, 195-214.

Lumsden, J. and Hall, D.O. (1974). *Biochem. Biophys. Res. Comm.* **58**, 35.

Lumsden, J. and Hall, D.O. (1975). *Nature* **257**, 670.

Lumsden, J., Cammack, R., Hall, D.O. (1976). *Biochim. Biophys. Acta* **438**, 380.

Lunsford, J.H. and Jayne, J.P. (1966). *J. Chem. Phys.* **44**, 1487.

Lunsford, J.H. (1973). *Catalysis Rev.* **8**, 135.

Malmström, B.G., Andreasson, L.-E. and Reinhammar, B. (1975). *In* 'The Enzymes', 3rd edn., vol. XII (Boyer, P.D., ed.), Academic Press, London, p. 507.

Maricle, D.L. and Hodgson, W.G. (1965). *Anal. Chem.* **37**, 1562.

Massey, V. and Hemmerich, P. (1975). *In* 'The Enzymes', 3rd edn., Vol. XII (Boyer, P.D., ed.), Academic Press, London, p. 191; Ullrick, V. and Duppel, W. (1975). ibid., p. 253.

Mel'nikov, A. Kh., Firsova, T.P. and Molodkina, A.N. (1962). *Russ. J. Inorg. Chem.* **7**, 633.

Metzger, H., Shapiro, M., Mosimann, J. and Vinton, J. (1968). *Nature* **219**, 1166.

McCord, J.M. and Fridovich, I. (1968). *J. Biol. Chem.* **243**, 5753.

McCord, J.M., Keele, B.B., Jr. and Fridovich, I. (1971). *Proc. Natl. Acad. Sci. U.S.A.* **68**, 1024.

McCord, J.M., Beauchamp, C.O., Goscin, S., Misra, H.P. and Fridovich, I. (1973). *In* 'Oxidases and Related Redox Systems', Proc. 2nd Inst. Symp. (King, T.E., Mason, H.S. and Morrison, M., eds.). University Park Press, Baltimore, p. 51.

McCord, J.M. (1974). *Science* **185**, 529.

McCord, J.M., Boyle, J.A., Day, E.D., Jr., Rizzolo, L.J. and Salin, M.L. (1977a). *In* 'Superoxide and Superoxide Dismutases' (Michelson, A.M., McCord, J.M. and Fridovich, I., eds.), Academic Press, London, p. 129.

McCord, J.M., Crapo, J.D. and Fridovich, I. (1977b). *In* 'Superoxide and Superoxide Dismutases' (Michelson, A.M., McCord, J.M. and Fridovich, I., eds.), Academic Press, London, p. 11.

McCord, J.M. and Fridovich, I. (1977). *In* 'Superoxide and Superoxide Dismutases' (Michelson, A.M., McCord, J.M. and Fridovich, I., eds.), Academic Press, London, 1977, pp. 1-10.

Menander-Huber, K.B. and Huber, W. (1977). *In* 'Superoxide and Superoxide Dismutases' (Michelson, A.M., McCord, J.M. and Fridovich, I., eds.), Academic Press, London, p. 537.

Michelson, A.M. and Durosay, P. (1977). *Photochem. and Photobiol.* **25**, 55.

Michelson, A.M., McCord, J.M. and Fridovich, I., eds. (1977a). 'Superoxide and Superoxide Dismutases', Academic Press, London.

Michelson, A.M., Puget, K., Durosay, P. and Bonneau, J.C. (1977b). *In* 'Superoxide and Superoxide Dismutases' (Michelson, A.M., McCord, J.M. and Fridovich, I., eds.), Academic Press, London, p. 467.

Miller, R.W. and Massey, V. (1965). *J. Biol. Chem.* **240**, 1466.

Millow, R.W. (1970). *Can. J. Biochem.* **48**, 935.

Misra, H.P. and Fridovich, I. (1971). *J. Biol. Chem.* **246**, 6886.

Misra, H.P. and Fridovich, I. (1972a). *J. Biol. Chem.* **247**, 6960.

Misra, H.P. and Fridovich, I. (1972b). *J. Biol. Chem.* **247**, 3170.

Misra, H.P. and Fridovich, I. (1972). *J. Biol. Chem.* **247**, 3410.
Misra, H.P. and Keele, B.B. (1975). *Biochim. Biophys. Acta* **379**, 418.
Misra, H.P. and Fridovich, I. (1977). *J. Biol. Chem.* **252**, 6421.
Nakamura, S. (1972). *Biochem. Biophys. Res. Comm.* **48**, 1215.
Nilsson, R., Pick, F.M. and Bray, R.C. (1969). *Biochim. Biophys. Acta.* **192**, 145.
Nishikimi, M. and Machlin, L.J. (1975). *Arch. Biochem. Biophys.* **170**, 684.
Ohnishi, T., Hirata, F. and Hayaishi, O. (1977). *J. Biol. Chem.* **252**, 4643.
Orme-Johnson, W.H. and Beinert, H. (1969). *Biochem. Biophys. Res. Comm.* **36**, 905.
Ose, D.E. and Fridovich, I. (1976). *J. Biol. Chem.* **251**, 1217.
Pack, J.L. and Phelphs, A.V. (1966). *J. Chem. Phys.* **44**, 1870.
Patel, K.B. and Willson, R.L. (1973). *J.C.S. Faraday I* **69**, 814.
Peters, J.W. and Foote, C.S. (1976). *J. Amer. Chem. Soc.* **98**, 873.
Petkau, A., Chelack, W.S. and Pleskach, S.D. (1976). *Int. J. Radiat. Biol.* **29**, 297.
Poupko, R. and Rosenthal, I. (1973). *J. Phys. Chem.* **77**, 1722.
Puget, K. and Michelson, A.M. (1974). *Biochimie* **56**, 1255.
Rabani, J., Mulac, W.A. and Matheson, M.S. (1965). *J. Phys. Chem.* **69**, 53.
Rajagopalan, K.V. and Handler, P. (1964). *J. Biol. Chem.* **239**, 2022.
Rapp, U., Adams, W.C. and Miller, R.W. (1973). *Can. J. Biochem.* **51**, 158.
Ravindranath, S.O. and Fridovich, I. (1975). *J. Biol. Chem.* **250**, 6107.
Richardson, J.S., Thomas, K.A., Rubin, B.H. and Richardson, D.C. (1975). *Proc. Nat. Acad. Sci. U.S.A.* **72**, 1349; Richardson, D.C. (1977). *In* 'Superoxide and Superoxide Dismutases' (Michelson, A.M., McCord, J.M. and Fridovich, I, eds.), Academic Press, London, p. 217.
Root, R.K. and Metcalf, J. (1975). *Clin. Res.* **23**, 311A.
Rotilio, G., Fioretti, E., Falcioni, G. and Brunori, M. (1977). *In* 'Superoxide and Superoxide Dismutases' (Michelson, A.M., McCord, J.M. and Fridovich, I., eds.), Academic Press, London, p. 239.
Rotilio, G., Morpurgo, L., Giovagnoli, C., Calabrese, L. and Mondovi, B. (1972). *Biochemistry (USA)* **11**, 2187.
Saito, E. and Bielski, B.H.J. (1961). *J. Amer. Chem. Soc.* **83**, 4467.
San Filippo, J. Jr., Chern, C.I. and Valentine, J.S. (1975). *J. Org. Chem.* **40**, 1077.
Sato, S. and Harris, J.I. (1977). *Eur. J. Biochem.* **73**, 373.
Sawada, Y., Ohyama, T. and Yamakazi, I. (1972). *Biochim. Biophys. Acta.* **268**, 305.
Schihara, K., Kusunose, E., Kusunose, M. and Mori, T. (1977). *J. Biochem. (Tokyo)* **81**, 1427.
Schopf, J.W. and Barghoorn, E.S. (1967). *Science* **156**, 508.
Slykhouse, T.O. and Fee, J.A. (1976). *J. Biol. Chem.* **251**, 5472.
Spangler, G.E. and Collins, C.I. (1975). *Anal. Chem.* **47**, 393-402.
Stenflo, J. and Suttie, J.W. (1977). *Ann. Rev. Biochem.* **46**, 157.
Stevens, J.B. and Autor, A.P. (1977). *Lab. Invest.* **37**, 470.
Strobel, H.W. and Coon, M.J. (1971). *J. xiol. Chem.* **246**, 7826.
Tan, Y.H., Tischfield, J., Epstein, C.J., Ruddle, F.M. (1973). *J. Exp. Med.* **137**, 317.

Taube, H. (1965). 'Oxygen: Chemistry, Structure and Excited States', Little, Brown, Boston.
Tegelström, H. (1975). *Hereditas* **81**, 185.
Valentine, J.S. and Curtis, A.B. (1975). *J. Am. Chem. Soc.* **97**, 224.
Valentine, J.S. and Quinn, A.E. (1976). *Inorg. Chem.* **15**, 1997.
Vance, P.G., Keele, B.B. and Rajagopalan, K.U. (1972). *J. Biol. Chem.* **247**, 4782.
Vaska, L. (1976). *Accounts Chem. Res.* **9**, 175.
Watanabe, K. (1957). *J. Chem. Phys.* **26**, 542.
Weening, R.S., Wever, R. and Roos, D. (1975). *J. Lab. Clin. Med.* **85**, 245.
Weisiger, R.A. and Fridovich, I. (1973). *J. Biol. Chem.* **248**, 3582.
Weser, U., Joester, K.-E., Paschen, W. and Jung, G. (1972a). *Z. Physiol. Chem.* **353**, 1576.
Weser, U., Pring, R., Schallies, A., Fretzdorff, A., Kraus, P., Völter, W. and Voetsch, W. (1972b). *Hoppe Seylers Zeit.* **353**, 1821.
Weser, U. (1973). *Struct. Bonding* **17**, 1.
White, H.L. and White, J.R. (1968). *Mol. Pharmacol.* **4**, 549.
Willson, R.L. (1977). *Chem. Ind.* **183**.
Wong, K., Morgan, A.R. and Paranchyck (1975). *C.J. Biochem.* **52**, 950.
Wood, R.H. and D'Orazio, L.A. (1965). *J. Phys. Chem.* **69**, 2558.
Yamakura, F. (1976). *Biochim. Biophys. Acta* **422**, 280.
Yost, F.J. and Fridovich, I. (1973). *J. Biol. Chem.* **248**, 4905.

7 SILICON IN THE BIOSPHERE

J.D. Birchall

*Imperial Chemical Industries Limited, Mond Division
Runcorn, Cheshire, U.K.*

In this review an attempt is made to trace the pathway of
silicon from the rocks and soil minerals of the earth's
crust to the small (but not insignificant) amounts found
in living organisms, drawing attention to what is known
of the element within the complex structure of living organisms and seeking throughout for a pattern in its movement and binding to organic molecules.

Any enquiry as to the biological significance of an
element requires detailed knowledge of its concentration
and location and real progress in unravelling function
can be made only when techniques are available capable of
detecting of the order of 10^{-14}g in 10^{-9}g of material
(Williams, 1977). Whilst the presence of silicon in living
organisms and its possible role have received attention
for many years, it is only comparatively recently that
analytical techniques with appropriate sensitivity and
accuracy have become available; the time may be right for
a focussed attack on the problem.

Silicon constitutes about 20 atomic per cent of the
earth's crust (Table I) (Deevey, 1970). It is invariably
associated with oxygen as the dioxide, SiO_2, and in the
silicates of rocks, minerals and soil constituents. The
silicates are based on $\{SiO_4\}^{4-}$ tetrahedra either isolated
(as in the ortho-silicates) or as polymers in which tetra-

TABLE I

Abundance of some elements in lithosphere and biosphere

Element	Abundance — Atomic per cent	
	Lithosphere	Biosphere
H	2.92	49.8
O	60.4	24.9
C	0.16	24.9
N	–	0.27
Ca	1.88	0.073
K	1.37	0.046
Si	20.5	0.033
Mg	1.77	0.031
P	0.08	0.030
S	0.04	0.17
Al	6.2	0.016

hedra, joined at corners via oxygen, are arranged in
chains or sheets or in three-dimensional networks. The
varied structures of the silicate minerals result from
such arrangements of tetrahedra, from the ability of
$\{AlO_4\}^{4-}$ to replace $\{SiO_4\}^{4-}$ and from the need for additional cations in the structures so formed to maintain
charge neutrality. The transport of silicon and its intimate association with living organisms must involve
solubilisation of the massive forms of silica and silicate, a process requiring their depolymerisation. Broadly,
the transport to be considered is illustrated in Fig. 1.

The Scale of the Transfer

When silica and silicates are in contact with water,
there is invariably some hydrolysis of Si-O-Si bonds and
silicic acid is liberated in very small quantities into
the aqueous phase. At near-neutral pH, the solubility of
silicic acid is of the order of 100 ppm and at concentrations below this it exists as the monomer, $Si(OH)_4$ (Iler,
1955). At concentrations much above 100 ppm monomeric
silicic acid tends to polymerise, eventually forming
colloidal silica. Above pH9, silicic acid begins to

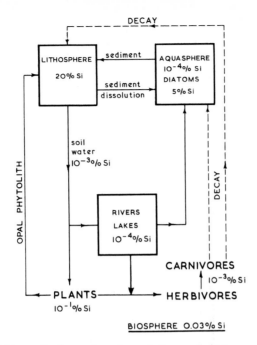

Fig. 1. The Silicon Cycle and order of Concentration.

ionise and its solubility rises sharply. The solubility
of silicic acid is much reduced in the presence of
cations such as calcium, aluminium and iron and can be
increased in the presence of dissolved organic materials.
Whilst the solubility of silicic acid in pure water is of
the order of 100–140 ppm within the pH range 1–9 (Alexander, Heston and Iler, 1954), much lower values are
found in the interstitial water of soils and in sea water.
In soil waters, values in the range 13–60 ppm are reported and in sea water and lakes the concentration of dissolved silica is reported as being about 5 ppm (Jones and
Hendrick, 1963; McKeaque and Cline, 1963). The low value
in soil water is thought to be due to the presence of
cations and to the adsorption of silicic acid from solution
by iron and aluminium oxides (Jones and Handreck, 1963).
    In spite of the low solubility, the movement of sili-
con from the land mass to the world ocean is of the order

of 10^8 tonne/year (Livingstone, 1963) and the dissolution of opaline silica (the exoskeletons of diatoms etc.) adds some 10^{12} tonne/year dissolved silica to the world ocean. Biospheric carbon incorporation has been estimated to be of the order of 10^{10} tonne/year (Deevey, 1970), a figure that puts into perspective the massive movement of silicon.

The Transfer $Si_{(lithosphere)}$ to $Si_{(plants)}$

The rate of plant production is about 10^{11} tonne/year some 40% of this being marine phytoplankton. Terrestrial vegetation contains on average 0.15% by weight of Si (Table II) derived from soil water.

TABLE II

The Average Percentage by Weight of Elements in Terrestrial Vegetation

Element	%	Element	%	Element	%
O	70	Fe	2×10^{-2}	Br	1×10^{-4}
C	18	Mn	7×10^{-3}	Mo	5×10^{-5}
H	10.5	F	3×10^{-3}	Y	4×10^{-5}
Ca	0.5	Ba	3×10^{-3}	Ni	2×10^{-5}
N	0.3	Al	2×10^{-3}	V	2×10^{-5}
K	0.3	Sr	2×10^{-3}	Pb	2×10^{-5}
Si	0.15	B	1×10^{-3}	Li	1×10^{-5}
Mg	7×10^{-2}	Zn	3×10^{-4}	U	1×10^{-5}
P	7×10^{-2}	Rb	2×10^{-4}	Ga	3×10^{-6}
S	5×10^{-2}	Cs	2×10^{-4}	Co	2×10^{-6}
Cl	4×10^{-2}	Ti	1×10^{-4}	I	1×10^{-6}
Na	2×10^{-2}	Cu	1×10^{-4}	Ra	2×10^{-12}

The level of silicon varies widely in plants of different species, being high in Graminacae (the ash of which can contain 30–60% SiO_2) and low in legumes and dicotyledons (Jones and Handreck, 1967). Equisetum (horsetail) is particularly rich in SiO_2 and the ash of this primitive plant can contain up to 90% SiO_2. Much of the silicic acid entering plants via the transpiration stream is deposited in the extremities as opal (amorphous, hydrated silica, $SiO_2 \cdot nH_2O$), although α-quartz has been reported (Lanning *et al.*, 1958). These opaline bodies, or phytoliths, often have characteristic shapes resulting from their formation within certain cells or other struc-tures of the plant. The presence of silica in plants, and especially of opal phytoliths, is not without consequence and, for example, cattle grazing on high-silica grasses may develop urolithiasis, a disorder in which siliceous calculi (uroliths) form in the bladder and enter the urethra, when obstruction may result in fatal bladder rupture (Connell *et al.*, 1959; Forman *et al.*, 1959). Again the presence of silica in forage is thought to re-duce digestibility and hence nutritional value (Van Soest *et al.*, 1968). It appears that silica reduces the avail-ability of cell wall carbohydrate, either by mechanically protecting the organic constituent (as does lignin) or by inhibiting cellulolytic enzymes.

The opal phytoliths of pasture plants are harder than the teeth of grazing animals and, since the amounts in-gested are large and continuous (say 14 Kg SiO_2 per year for sheep) it has been suggested that opal is a major factor in tooth wear (Baker *et al.*, 1959).

The intake and presence of silica in plants may not be without advantage. The presence of silicic acid in soil water or plant fluids, or the presence of opaline silica in plant tissue, may reduce the toxic effect of metals such as manganese (Okuda *et al.*, 1962). It has been sug-gested that the presence of solid silica within plant tissue may influence the susceptibility of cereals to fungal attack and, for instance, an increase in the re-sistance of rice to brown spot, stem rot and blast dis-ease has been claimed to result from a high silica

content, particularly in the leaves (Izawa and Kume, 1959).
The mechanism of protection is not known but since most
parasitic fungi penetrate their hosts by boring through
the epidermal cell walls (Butler and Jones, 1949), the
silicification of cell walls may provide a mechanical
barrier or reduce the accessibility of carbohydrate to
the lytic enzymes introduced by parasitic fungi.

The deposition of solid silica within plant structures
may, by mechanically strengthening leaves or stem, be
advantageous. For example, the disposition of leaves may
be affected and it has been claimed that the leaves of
the rice plant may show a drooping habit when their silica content is low but are more erect when the supply of
silicic acid is high (Mitsui and Takatoh, 1963). When
cereals are provided with high levels of nitrogen resulting in rapid growth, longer and weaker stems may be
formed. These bend easily and 'lodging' under wind and
rain stress occurs with significant loss on harvesting.
A high degree of silicification may, by stiffening the
stem, reduce this tendency to 'lodge' (Jones and Handreck,
1967).

The concentration of silica varies widely among the
different parts of a plant. In mature oat plants, for
example, the distribution of silica (total Si as SiO_2) is
reported as being typically (Jones, 1963) —

Husk	10.9%	SiO_2 on dry substance			
Nodules	10.3%	"	"	"	"
Internodes	1.4%	"	"	"	"
Leaf	5.3%	"	"	"	"
Leaf sheath	4.5%	"	"	"	"
Root	1.8%	"	"	"	"
Grain	0.12%	"	"	"	"

The availability of silicon, its transportation and binding

The hydrolysis of Si-O-Si bonds in silica and in the
silicate minerals will eventually release silicic acid
into solution and it is in this form that silicon enters
the biosphere. The chemistry of silicic acid (and its

polymers) is dominated by the following characteristics
(Iler, 1955).

(i) the ability of monomeric silicic acid to condense
to form progressively, dimer, trimer etc. and eventually
polysilicic acid as hydrated, amorphous silica. The rate
of condensation is a maximum at about pH 5.5 and is
strongly influenced by the presence of organic polar
molecules.

(ii) the capacity of silicic acid and its polymers to
associate with organic polar molecules by hydrogen bonding.

(iii) the ability of silanol groups (e.g. on silicic
acid and its polymers and on colloidal silica) to react
with basic metal hydroxides. This is the mechanism suggested for the strong adsorption of silicic acid by iron
and aluminium oxides. Reactions such as

$$:SiOH + Fe^{III}(OH)_2^+ \rightarrow (:SiOFe^{III}O) + H^+ + H_2O$$

have been proposed (Iler, 1955).

(iv) the reactivity of silanol groups with hydroxy-
compounds to form esters

$$:SiOH + ROH \rightarrow :SiOR + H_2O,$$

and the ease with which ester exchange can occur.

These characteristics dictate the fate of the silicic
acid released by mineral degradation.

In addition to purely inorganic weathering, there is
evidence that the degradation of rocks and soil minerals
is influenced by the presence of plants and soil micro-
organisms. The organic residues of plant decomposition
such as fulvic and humic acid contribute to the degradation of soil minerals by reacting with polyvalent cations
(especially aluminium) to form complexes or chelates. A
silica-rich residue remains from which silicic acid is

likely to be more readily leached than from the parent
alumino-silicate. Organic acids such as aspartic acid,
acetic acid, tartaric acid, citric acid and tannic acid
have been shown to cause a significant increase in the
dissolution of the framework cations (Al, Fe) of minerals
such as kaolinite, flint clays, montmorillonite etc. and,
for example, in the presence of strong complexing agents,
the concentration of Si was found to be 50-fold that in
pure water (Huang and Keller, 1972). This increase
appears to result from the removal of aluminium rather
than from any direct complexation of silicon itself.
However, the complexing ability of polyphenolic compounds
(and especially the o-diphenols) has been studied and
their significance in soil discussed (Bloomfield, 1966).
These materials are of special interest in view of the
existence and ease of formation of complexes such as:

For example, diatomite has been found to react in a
solution of catechol at pH 8.2 to the extent of 30% in
ten days, forming the tricatechol silicic acid ester
(Weiss *et al.*, 1961). Fulvic and humic acid contain pheno-
lic groups (or their progenitors) and, apart from the
ability of such groups to complex metals, there is the
possibility of direct attack on silica itself, especially
on the siliceous residues remaining following the extraction of framework aluminium. There is evidence, too, that
plant exudates and microbial products attack mineral
structure. For example, 2-ketogluconic acid, produced by
pseudomonas, has been shown to assist the dissolution of
silicates (Henderson and Duff, 1963). Compounds present
in lichen such as evernic acid, atranorin, salazinic
acid and roccellic acid, which may account for up to 8%

of the dry weight of lichen, have been shown to extract
Al, Fe, Ca and Mg from minerals such as biotite, granite
and basalt and thus to increase the potential availability
of silicon (Iskandar and Syers, 1972).

Evernic acid

Atranorin

Salazinic acid   Roccellic acid

notably Equisetum, which accumulate silica in significant
quantities, do not depend on silica dissolved in soil
water for their supply but are capable of directly promoting the release of silicic acid from soil minerals
(Lovering and Engel, 1967). Such is the ability of Equisetum to accumulate silica that it has been estimated that
all the silica in 1 acre-foot of basalt (about 2000 tons)
would have passed through the plants (covering 1 acre) in
5000 years! Plants such as Equisetum have a high concentration of silicon in their sap − 150 ppm Si being common. In experiments in which Equisetum was grown in a
nutrient medium containing 11.5 μg Si/ml (and hence present as monomeric silicic acid) the concentration of Si in
solution fell to almost zero after 15 days whereas in a
control solution, not in contact with the plants, the
concentration did not change significantly. Monomeric
silicic acid can be distinguished from polymeric forms by
its rapid reactivity with molybdate (molybdate blue reaction) and, if at this stage in the experiment a solution
containing polymeric silicic acid was added to the nutrient solution, then, within 12 hours, 75% of this added

polymer appeared as molybdate-reactive (monomeric) silicic
acid (Lauwers and Heinen, 1974). This suggests that Equisetum is able to depolymerise silica and to take up the
monosilicic acid formed. These experiments were not con-
ducted under sterile conditions so that the effect of
micro-organisms cannot be excluded. However, the solutions employed contained few bacteria so that root exudates as active agent of mineralysis seem probable.

Bacteria such as Proteus Mirabilis take up monomeric
silicic acid in an energy-dependent process, the silicon
content of an incubation medium (containing also glucose
and succinate) falling from 30 μg Si/ml to 26 μg Si/ml
within 26 days and falling further with the replenishment
of succinate (Lauwers and Heinen, 1974). If, in such experiments, only polymeric silicic acid is supplied, depolymerisation appears to be brought about by the intact
cells. Such depolymerisation may explain the observation
that some bacteria attack a smooth glass surface, leaving
grooves shown in electron micrographs to have dimensions
matching those of the bacteria (Oberlies and Pohlmann,
1958; Barber and Shone, 1966).

Present knowledge suggests, then, that in order to
become available for assimilation by plants, the silica
and silicates of soil must be depolymerised and solubil-
ised to silicic acid of low molecular weight. This is
brought about by inorganic weathering but the process is
aided very considerably by organic compounds resulting
from plant decomposition, by root exudates and by microbial intervention. It seems likely that in the case of
many plants the uptake of silicic acid is 'passive',
silicic acid moving across the root and into the transpiration stream by diffusion and mass flow. Its concentration by water loss then results in polymerisation and
the deposition of opaline silica. Hence a linear relationship will exist between uptake and concentration in the
growth medium. Such a relationship is often found and is

exemplified for oat plants in Table III.

TABLE III
(Jones and Hendreck, 1965)

*Uptake of Silica by Oats in relation to the Level of Silica
in the Soil Solution and the Amount of Water Transpired*

Soil (pH 5.6)	SiO_2 in solution (ppm)	Dry matter (g/plant)	SiO_2 in plants %	SiO_2 in plants mg/plant	Water transpired (kg/plant)	SiO_2 expected (mg/plant)
Wollongbar krasnozem	7	7.07	0.40	28.3	3.86	27.0
University sandy loam	54	6.40	2.77	177	3.27	176
Penola black clay	67	6.92	3.96	274	3.90	261

Certain plants appear to be able to *exclude* silicic
acid and this may account for the fact that dicotyledons
contain 10–20 times less silicon than graminacae. Crimson
clover (representing dicotyledons) and oats have been
grown in identical media, when the former plants contained only 5–10% of the silica found in the oats although
the plants had similar transpiration ratios (Handreck and
Jones, 1967). Again, the silicic acid uptake of legumin-
ous and graminaceous plants grown in the same soil has
been compared; barley and rye grass contained 1.95 and
1.58 per cent SiO_2 respectively, whereas red clover and
blue lupin contained only 0.12 and 0.24 per cent SiO_2.
The mechanism of silicic acid exclusions is unknown
(Grass-Brauckmann, 1953).

Rice appears to have a special ability to accumulate
silicic acid, apparently absorbing silicic acid at a
greater rate than it absorbs water (Okuda and Takahashi,
1964). For example, the concentration of monosilicic acid
in a culture solution fell from 100 ppm to 10 ppm SiO_2 in
37 hours following the introduction of intact plants. The
concentration of silicic acid in xylem sap was always

greater than in the external solution. Plants introduced to a solution containing 100 ppm SiO_2 had 650 ppm SiO_2 in the xylem sap after 37 hours (Okuda and Takahashi, 1964). There is evidence that the uptake of silicic acid is associated with metabolic processes in the roots and that although root uptake is largely independent of transpiration rate, the subsequent translocation of the silicic acid towards the leaves is related to transpiration rate (Baba *et al.*, 1956).

It thus seems that plants may be classified into three groups:

(a) those in which silicic acid uptake is directly related to the concentration in the supply and the transpiration rate

(b) those in which there is active uptake

(c) those in which there is active exclusion.

As regards active uptake, the observation that the silification of *Oryza sativa L* was increased by the presence of poly-2-vinylpyridine-1-oxide (PVPNO) in the nutrient solution — which contained 100 ppm SiO_2 — is a strong indication of the possible role of root exudates (Wynn-Parry, 1975). Root exudates may reduce the tendency for silicic acid to polymerise (a characteristic of PVPNO); the absence of such exudates may allow a degree of polymerisation such that the polysilicic acid formed cannot pass through root membranes.

It has been mentioned that rice sap can contain several hundred ppm of SiO_2 and that the sap of Equisetum is rich in silicic acid, although monomeric silicic acid exists in water below pH 9 only at levels less than about 100 ppm. This suggests that, in sap, silicic acid is stabilised against polymerisation and deposition in the lower parts of the transpiration stream and, indeed, there has been much debate as to the nature of the silicic acid in plant sap, whether it is free or is bound to some organic moiety. An examination of the I.R. and U.V. spectra of sap

from Equisetum (which contained about 300 ppm SiO_2) suggested the presence of Si-C bonds with phenyl groups as
the organic moiety (Lovering and Engel, 1967). Confusion
is possible because the absorption peaks for $Si-C_6H_5$ and
for epsomite (an important sap constituent) are close.
However, the presence of phenyl groups has led to the
suggestion that the silicon in Equisetum sap is associated with organic moieties resembling fragments of the
complex molecular structures of humic and fulvic acid.
It seems more likely that, rather than Si-C bonds, the
Si-OH groups of silicic acid become associated with aromatic structures through reaction with o-diphenols perhaps originating from humus and lignin decay.

    The stability of silicic acid in plant sap need not
involve compound formation since silicic acid is capable
of associating with organic molecules (having − O −,
NH_2, COOH, OH etc. groups) by hydrogen bonding. It is
well known that the polymerisation of silicic acid is
retarded in the presence of, for example, polyethers,
alcohols, amides, etc. (Iler, 1955) and there are many
related molecules in plant sap. As an example of the
stabilisation of silanol groups against condensation, the
attachment of a short polyether chain (4 ether oxygen
atoms) to a silane triol via an Si-C bond has been shown
to result in marked stability, such that a 50% w/w
aqueous solution contained 75% monomeric triol as indicated by ^{29}Si N.M.R. studies (Birchall *et al.*, 1977).
The fate of most of the silicon entering a plant is to
be deposited as opaline silica. However, there is evidence that a minor proportion becomes bound to the structural polysaccharides of cell walls. Studies (Engel,
1953) in which rye straw was subject to extraction with
water and water/methanol, yielded an ether-soluble, Si-
containing material as a small proportion of the total Si
present. The organic component was identified as galactose, present in the ratio of 1:2 with silicic acid.

Free galactose could be obtained only by saponification
with 2% NaOH. The extraction of growing rye straw with
1:3 methanol/benzene was found to give an ether-soluble
Si-galactose complex in which the Si to galactose ratio
was 1:1. Structures such as those shown below were suggested.

H_2COH $H_2C-O-Si\lessgtr$

$H_2C-O-Si-O-$

NHac

The organic binding of Si in Proteus Mirabilis is indicated and claims have been made that the incubation of
Proteus Mirabilis in nutrient containing silicon (as
silicate) results in the binding of silicon to organic
moieties and that complexes can be liberated by sonifica-
tion of the cells and detected in the cell-free extract
(Heinen, 1968, 1965a and b). The suggestion is made that
silicic acid carbohydrate esters are formed from which
ortho-silicic acid and carbohydrate (e.g. hexoses) are
released by hydrolysis.

Studies to distinguish 'bound' and 'free' silicon are
fraught with difficulty. 'Total' silicon is readily
determined by the molybdate blue method following alkali
fusion. 'Bound' and 'free' Si are often distinguished by
reactivity to molybdate, it being assumed that compounds
containing Si—C and Si—O—C bonds will not react. This
assumes that the bonds remain intact under the conditions
of the test. Confusion may arise too, as a result of the
ability of silicic acid to associate through hydrogen
bonding with a variety of polar molecules and hence to
display a much reduced tendency to polymerise. This can
be mistakenly attributed to compound formation. Thus,
studies (Holzapfel, 1951) in which the solubility of
quartz in water was found to be enhanced in the presence
of carbohydrates (glucose, galactose, lactose, etc.) may

reflect the ability of such compounds to retard polymerisation rather than true compound formation. However,
the totality of evidence appears to favour strong association between silicic acid and carbohydrates, although
there have been no convincing attempts to isolate and
characterise the materials.

The Diatom

The one species for which silicon is well-established
to be essential is the diatom, a species of fundamental
importance in the food chain since marine phytoplankton
account for approaching half the earth's primary plant
production and the fixation of about 10^{10} tonne/year of
carbon (Strickland, 1972; Golley, 1972). The diatom has
the ability to take in silicic acid from the very low concentration in its environment (a few parts per million)
and to construct a siliceous ($SiO_2 \cdot nH_2O$) shell or frustule
of complex morphology. On the death of the organism, the
siliceous remains (10^{-5} mg SiO_2 per cell) accumulate as
sediment and some 10^6 Km^2 of ocean floor is covered by
diatomaceous ooze from which there is some dissolution
and recycling of silicon (Fairbridge, 1966).

The concept of silicic acid uptake, transport and
polycondensation to form the frustule has proved far too
simple a view and it is increasingly considered that
silicic acid is intimately involved in complex metabolic
processes. Diatom cell division ceases with the depletion
of silicic acid from the environment (Lewin and Chen,
1968) and the production of DNA stops (Darley and Volcani,
1969). Germanium acts as a specific inhibitor for cell
growth by blocking incorporation of silicon (Werner, 1966).

The complex events following the transfer of growing
cells of *Cyclotella Cryptica* to a silicic acid-free environment have been studied in some detail (Werner, 1966,
1967, 1968a and b, 1970) as have the effects of added
silicic acid on *Navicula Pelliculosa* cells previously

starved of silicon (Coombs, 1967 and 1968). In the case
of cells transferred to an Si-depleted environment, cell
division was inhibited within one hour and within four
hours there was inhibition of net protein production and
DNA synthesis. This was followed by the inhibition of
chlorophyll synthesis (5 hours) and by a doubling in
fatty acid synthesis (6–9 hours). After 12 hours, photosynthesis was reduced by 80%. The recovery of activity in
Si-starved *Navicula Pelliculosa* cells following the addition of silicic acid was manifested by the increased incorporation of label from $^{14}CO_2$ into aspartate, glutamate,
citrate and 2-oxoglutarate within thirty minutes; chlorophyll synthesis resumed after one hour; lipid synthesis
decreased throughout and RNA and net protein synthesis
recovered so that, within seven hours, new cell division
had started.

The gradual decay and final cessation of major metabolic activity in the absence of silicon and the demonstration of renewed metabolic activity in Si-starved
cells exposed to silicic acid, indicate the involvement
of silicon in major metabolic pathways (Werner, 1977),
i.e. in the citric acid cycle between acetyl CoA (fatty
acid synthesis) and 2-oxoglutarate; in the synthesis of
special proteins in cell organelles; in the regulation of
respiration and chrysolaminaran utilisation, and in the
regulation of chlorophyll synthesis, all suggesting a
special silicic acid metabolism in the organelles of
diatoms.

A heat stable factor can be washed from cells with the
result that silicic acid uptake is reduced. Treatment with
aspartic or glutamic acid fully restores uptake (Lewin
and Chen, 1968). Reduced sulphur compounds restore silicic acid uptake in washed cells.

The formation of the siliceous frustule occurs in
association with a membrane, the silicalemma, and may
result from specific interaction between protein and

silicic acid with the protein acting as a template.
During silificication a marked increase in protein concentration has been noted (Coombs and Volcani, 1968).
The silicalemma is proteinaceous and rich in serine and
threonine (Reimann *et al.*, 1966). This protein may act as
a template by presenting a layer of hydroxyl groups
(attached to serine and threonine and perhaps spaced by
other amino acids) able to react with silicic acid to
produce an ordered first layer of the frustule (Hecky *et
al.*, 1973). It may be that morphological features of the
frustule and species variation result from different
sequencing of the template protein. Certainly, the fact
that the complex morphology of the frustule is repeated
faithfully over many cell divisions suggests strongly
that it is laid down on a template, itself genetically
determined. That a protein can act as a template for
silicic acid should occasion no surprise, since the adsorption of amino acids, peptides etc. on quartz surfaces
is recorded (Siefert, N., 1959; Seifert, H., 1961) and
the interaction between biopolymer monolayers and silicic
acid has been studied in some detail (Holt and Went,
1959; Holt and Clark, 1955; Minones *et al.*, 1973).

The frustule is associated with carbohydrate although
carbohydrate incorporation is significant mostly after
silicification, probably as structural polysaccharides
such as mannan, xylan, sulphated polysaccharide etc.
(Hecky *et al.*, 1973). Such polysaccharides may act as an
outer buffer coat, a function of which could be to prevent
dissolution of the frustule in an environment under-
saturated with respect to silicic acid. In this respect
the incorporation of a small proportion of aluminium or
iron would be expected to depress solubility greatly and
this may explain the increased rate of dissolution produced by washing cells with acid (Werner, 1977; Iler,
1955).

Silicon in Higher Animals

Herbivores will, of course, ingest the silicon associated with vegetation and much of this will be in the
form of opaline phytolith. A sheep ingesting 1 kg dry
matter per day will take in between 40 g and 2 g per day
of SiO_2 (Jones and Handreck, 1967) depending on the
nature of the feed. Just over 2% of this will be in the
form of dissolved silicic acid. It is suggested that
only a very small proportion of the total silicon ingested is retained in tissue. If a figure of 70 ppm SiO_2
(dry basis) is taken as the tissue level in sheep, then
it is calculated that only 0.016% of the ingested SiO_2
remains in the tissue of a three year old animal of 25 kg
dry weight having ingested 40 g SiO_2 per day (Jones and
Handreck, 1967).

Attempts to obtain a balance between the ingestion,
excretion and retention of silicon in animals have proved
difficult and confusing. Usually, sodium silicate, silica
sol or ethyl silicate have been used as the source of
silicon and the evidence is that most of this is excreted in the urine in inorganic form, perhaps at a rate
determined by the solubility of silicic acid (Sauer *et
al.*, 1959). Such experiments take no account of the strong
probability that a proportion of the silicon in natural
materials is organically bound. From such experiments,
however, has emerged the suggestion that silicon exists
in animal tissue in inorganic form (silicic acid and its
polymers) and in organically bound form, the latter being
retained in tissue and the former excreted rapidly.

The analysis of whole organs for silicon has been
carried out by numerous workers for many years and the
levels reported show a fall as analytical methods (almost always a variation of the molybdate blue method)
and precautions against contamination, became more rig-
orous (King, 1933 and 1939; Holt, 1950; Spector, 1956).
Whole tissue levels of the order of a few mg SiO_2 per

100 g dry weight are commonly reported. The normal human
blood level is less than 10 ppm SiO_2 and the daily human
urinary output is about 20 mg daily (Goldwater, 1936).

The analysis of whole organ is less important and
less informative than the identification of the binding
site of silicon at the gross level of tissue type and,
ultimately, at the molecular level. The first identification of connective tissue as being particularly rich in
silicon was made as long ago as 1902 (Kahle, 1914) when
'Wharton's jelly' (embryonic connective tissue) was concluded to be the tissue having, at 0.024% SiO_2 on dry
matter, the highest level of silicon of all the tissues
then examined. Connective (or 'collagenous') tissue may
be regarded as a gel of ground substance in which are
embedded cells (major type the fibroblast) and fibres.
The ground substance is composed mainly of mucopolysaccharides and the proportion and type of fibrous and
ground substance material present determine the characteristic properties of connective tissue — the flexibility and toughness of cartilage, the elasticity of the
aorta etc. In 1932, the isolation of a jelly-like material
containing silicic acid in association with a carbohydrate
was reported to result from the electrodialysis of ox
tendon and other material (Johlin, 1932), perhaps the first
indication that the polysaccharides of connective tissue
are peculiarly rich in silicon.

The early work appears to have been particularly per-
ceptive in the light of the most recent findings. Sophisticated experiments to examine the effect of silicon
insufficiency on the development of young rats and chicks
directed attention to changes in connective tissue and
bone development and hence to the analysis for silicon
of the collagenous and polysaccharidic components of connective tissue (Schwarz, 1973).

The criteria for an essential element are: the development of a deficiency state on diets depleted in that

element but otherwise satisfactory for health, and a return to normal growth and health when the diet is supplemented with that element. It has become possible to isolate animals from access to metals, glass, dust and other
sources of trace element contamination by the use of all-
plastic apparatus, then to feed a chemically defined diet
(Schwarz, 1967 and 1970). A significant reduction in the
growth of rats was produced by a silicon-deficient diet
using such a regime and the addition of 50 mg Si (as sodium metasilicate) per 100 g of diet caused an increase
in the rate of growth (Schwarz, 1972) (Table IV).

TABLE IV

*Growth Effects of Dietary Silicon on Rats Maintained in a
Trace Element Controlled Environment on Low Silicon Diets*

	Number of animals	Average daily weight gain (g)	% increase over controls	p value
Basal diet A				
control	15*	1.51 ± 0.11[†]		
50 mg% Si	11	2.02 ± 0.08	33.8	< 0.005
Basal diet B				
control	12*	1.19 ± 0.06		
50 mg% Si	11	1.49 ± 0.06	25.2	< 0.005

*Summary of two experiments
[†]Average ± standard error of the mean (S.E.M.)

The structure of the skull and pigmentation of the
incisors were affected by silicon deficiency. The results
of silicon-deficiency experiments on chicks were reported
almost simultaneously (Carlisle, 1972). The chick was
chosen because of a high rate of skeletal development,
because of the absence of a parental contribution via
milk and because, by deutectomy, any contribution via
the yolk inside the chick at the time of hatching could
be prevented. In some chicks, the diet was supplemented
by the addition of sodium metasilicate at the level of

100 ppm Si in the diet. The addition of silicon again
promoted a significant increase in growth rate over that
of the deficient animals.

TABLE V

Growth Response of Chicks to Silicon Supplementation

Study No.	Chicks (No.)	Average daily weight gain in 23 days (g) (mean ± S.E.M.)		Difference (%)	p
		Unsupplemented group	Supplemented group		
1	36	2.37 ± 0.11	3.10 ± 0.10	30.0	< .01
2	30	3.25 ± 0.09	4.20 ± 0.09	30.0	< .02
3	48	2.57 ± 0.09	3.85 ± 0.11	49.8	< .01

There were notable differences between the chicks on
the basal and silicon-supplemented diets, the former being
stunted with shorter leg bones having smaller circumfer-
ence and thinner cortex; relatively flexible metatarsal
bones and fragile femur and tibia were noted. In the deficient birds, the beak was flexible and the wattles absent, the comb being poorly developed.

These findings support an earlier observation associating silicon with the process of calcification in young
bone (Carlisle, 1970). In this work, quantitative electron probe microanalysis for Ca, P and Si was performed
on specimens of normal tibia from young mice and rats.
The spatial distribution of silicon and calcium were
found by traversing across young tibia, hence examining
the periosteum, osteoid layer and bone trabecular regions.
In the periosteal regions, both calcium and silicon levels
were low, whereas in the adjacent osteoid layer silicon-
rich regions were found containing of the order of 0.5%
Si. The silicon content of fully mineralised bone is reported to be low. Traverses across the metaphyseal region
of a longitudinal section of tibia revealed silicon levels

rising to 0.01–0.06% Si in the pre-osseous border and
reaching 0.12% at the edge of a trabecule, then falling
as the calcium content approached that of fully mineralised bone. Hence, it appears that silicon is localised in
those regions in which there is low to moderate calcium
content and active calcification is in progress. Furthermore, the silicon in such zones is not distributed uniformly but is concentrated in sites corresponding to bony
spicules in the process of formation and, within these,
there may be up to 1% Si. The processes initiating the
mineralisation of tissue are not by any means fully
known and these findings strongly suggest the involvement
of silicon.

A more detailed examination of the bones of chicks fed
on Si-deficient and Si-supplemented diets has shown notable differences in composition, the most significant
being a reduced water content in the tibia and femur of
Si-deficient animals. These could contain as much as 35%
less water (Carlisle, 1974 and 1976). The fact that a
major water-binding component of cartilage is mucopolysaccharide prompted a study of articular cartilage composition. Articular cartilage was removed from the tibia
of experimental animals and was found to be present in
larger quantities in Si-supplemented animals and also
contained a greater proportion of hexosamine (0.359% as
against 0.296% wet tissue), supporting a view that the
greater water content of the bones of Si-supplemented
chicks is related to a higher content of mucopolysaccharide in cartilage.

The cockerel comb is regarded as 'target tissue',
being largely a polysaccharidic structure. Here again,
the amount of connective tissue and the hexosamine content were found to be significantly larger in the Si-
supplemented birds. There was also a difference in the
silicon content of the tissue from the two groups
(Carlisle, 1974) (Table VI).

TABLE VI

Effect of Silicon Intake on Comb Composition

Diet	Tissue (wet wt) mg	Total Hexosamine (wet wt) mg	Per cent Hexosamine (wet) %	Silicon (dry wt) ppm
Low Si	90.30 ± 4.99	0.085 ± 0.012	0.094 ± 0.003	11.4 ± 0.36
Supplemented	134.80 ± 10.20	0.175 ± 0.020	0.130 ± 0.009	21.2 ± 3.02

Studies have also revealed abnormalities in the formation of the skull of chicks fed on Si-deficient diets,
with gross changes in architecture, the overall appearance being narrower and shorter. X-ray and histological
examination indicated less trabeculae, reduced calcification and less collagen (Carlisle, 1977).

The most recent work, then, indicates that silicon is
intimately involved in the process of bone formation
being present at relatively high levels in the regions of
active calcification. There are indications too that
silicon is distributed within the sub-cellular organelles
of the osteoblast, the active bone-forming cell (Carlisle,
1975) and there are strong indications that silicon is
involved in mucopolysaccharide synthesis.

An examination of the components of tissue such as
cartilage, bovine nasal septum, umbilical cord etc. has
shown silicon to be bound within the structure of polysaccharidic biopolymers such as hyaluronic acid, chon-
droitin-4-sulphate, dermatan sulphate and heparan sulphate (Schwarz, 1973) in a form not reactive with molybdate, stable to 8 M urea (and hence unlikely to be associated by hydrogen bonds) and not liberated either by
autoclaving or by treatment with weak acid or alkali. In
the case of hyaluronic acid, treatment with hyaluronidase
did not liberate silicic acid. In work of this type,
total silicon is determined colorimetrically following
carbonate fusion; 'free', unbound silicic acid is deter-

TABLE VII

Free, Total, and Bound Si in Glycosaminoglycans
Polyuronides, and Some Glycans

Substance and Source	Si (µg/g)		
	Free	Total	Bound (total minus free)
Glycosaminoglycans			
Hyaluronic acid			
(a) Human umbilical cord	25	354	329
(b) Human umbilical cord	1533	1892	359
(c) Bovine vitreous humor	980	949	-
Chondroitin 4-sulfate			
(d) Notocord of rock sturgeon	44	598	554
(e) Rat costal cartilage	30	361	331
Chondroitin 6-sulfate			
(f) Human umbilical cord	45	123	78
(g) Human cartilage	36	227	191
(h) Sturgeon cartilage	64	121	57
Dermatan sulfate			
(i) Hog Mucosal tissue	46	548	502
Heparan sulfate			
(j) Beef lung	39	466	427
Heparin			
(k) Hog Mucosal tissue	33	175	142
Keratan sulfate-1			
(l) Bovine cornea	31	37	-
Keratan sulfate-2			
(m) Human costal cartilage	37	105	68
Polyuronides			
Pectin			
(n) Citrus fruit	5	2586	2581
Alginic acid			
(o) Horsetail kelp	-	43	-
(p) Horsetail kelp	5	456	451
Polyglycans			
Glycogen			
(q) Rabbit liver	8	34	26
Starch			
(r) Corn	-	22	-
Dextran			
(s) Leuconostoc mesenteroides	19	22	-
Inulin			
(t) Dahlia tubers	15	29	14

mined by colorimetric analysis and 'bound' silicon is
taken as the difference between the two, supported by the
fact that the release of silicon requires drastic treatment. The results obtained for a variety of glycosaminoglycans, polyuronides etc. are given in Table VII
(Schwarz, 1973).

In view of the interest in the possible binding of
silicon in plants, the finding of a high level of bound
silicon in pectin is of particular note. This level corresponds to one silicon atom per 10,000 molecular weight
units. In the mucopolysaccharides of animal origin, the
bound silicon level corresponds to one atom per 130–180
repeating units.

Bound silicon has been found in collagen from various
sources, the level depending on origin and method of
isolation (Schwarz, 1974). Typical values are reported
below:

Salt soluble from rat skin	547	μg Si/g dry wt			
Acid soluble from rat skin	479	"	"	"	"
Acid soluble from mouse skin	1938	"	"	"	"
Acid soluble from calf skin	3296	"	"	"	"
Bovine Achilles tendon	90–496	"	"	"	"
Bovine articular cartilage	307–979	"	"	"	"
Lathyritic pig skin	999	"	"	"	"
Glomerular membrane	825	"	"	"	"

These levels suggest the presence of 3–6 atoms of Si to
each α-protein chain in the collagen molecule.

It must be recognised that the components of connective tissue isolated by proteolytic digestion, alkali
extraction etc. are artefacts in that the very isolation
process alters the characteristics that exist *in vivo*.
The products are essentially fragments of the polysacchar-
ide/protein complexes existing in the living organism.
However, the finding of bound silicon in the isolated
collagenous and polysaccharidic components, possibly
bound via an ester linkage,

$$\cdot Si - O - C - polymer$$

has led to the suggestion that silicon acts as a labile
cross-linking site, allowing inter- and intra-chain
associations between protein and polysaccharide (or between two polysaccharides) and hence contributing to the
structural integrity of connective tissue (Schwarz, 1973;
Carlisle, 1974). Cross linking could be of the type,

$$polymer - C - O - \overset{\displaystyle |}{\underset{\displaystyle |}{Si}} - O - C - polymer$$

in which lability is provided by switching of the Si—O
bond (Schwarz, 1974). The presence of Si bound at intervals to minority sugars in a polymeric glycan chain could
have important effects determining polymer conformation.
Hyaluronate, for example, is reported to exist as a
double helical structure (Atkins, 1973; Dea *et al.*, 1973)
and viscosity studies, gel filtration behaviour etc. suggest that the two anti-parallel chains may come apart on
appropriate chemical treatment. The presence of some 100
ppm of bound silicon in bovine vitreous humor hyaluronate
has led to the suggestion that silicon may bridge the two
anti-parallel chains through neutral sugars or through
the hydroxyl group of the uronic acid moiety (Varma,
1974).

Present knowledge, then, suggests that silicon has a
structural role in connective tissue; that it is involved
in mucopolysaccharide synthesis and that it is implicated
in the mineralisation of the pre-osseous matrix.

It is difficult to resist the view that there are certain features in common between a suggested 'organisa-
tional' role for silicon in the protein/polysaccharide
complexes of connective tissue and the suggested protein-
template mediated silicification of diatoms.

Silicon, Disease and Ageing

Changes in connective tissue are found in certain disease conditions and in the process of ageing. Embryonic

connective tissue is particularly rich in mucopolysaccharides and since these appear to be a major binding site for silicon, it is perhaps not surprising that the level of silicon in certain tissue has been reported as falling with age. In the rabbit, for example, the silicon content of the aorta is reported to fall by 84% between 12 weeks and 18–24 months, there being also a fall in the level in skin (83%) and thymus (96%) within that period (Carlisle, 1974). The silicon content of the human aorta has been reported to fall with age as indicated in Table VIII (Loeper, 1967).

TABLE VIII

Silicon Content of Human Aorta with Age

Age — Years	Si content of aorta μg Si/100 mg N
infants	205 ± 44
10–20	160 ± 43
20–30	125 ± 30
30–40	86 ± 16

In this study, an attempt was made to identify the aortal tissue component binding silicon by extracting mucopolysaccharide, elastin and collagen; the silicon content of the fractions were:

mucopolysaccharide	up to 1270 μg Si/g dry wt.
elastin	170–500 " " " "
collagen	ca. 220 " " " "

The level of Si in human plasma is reported to be of the order of 1 μg Si/g and hence 10–20 times less than the level in systemic organs (Austin, 1973). A tendency was noted for human plasma Si level to increase with age and for the ratio bound/unbound silicon in plasma to rise.

The silicon content of rat skin shows a 60% decrease between 5 weeks and 30 months (Leslie *et al.*, 1962). In studies of the variation with age in the plasma and intestinal tissue levels of silicon in male and female rats, significant sex-linked differences have been noted

(Charnot, 1971). In both sexes, the Si level in plasma
rises with age (note the finding in human plasma (Austin,
1973)) and there is a decrease in the level of Si bound
in intestinal tissue, although there is a greater loss of
silicon from tissue (to plasma) in the senescent female
(Table IX).

TABLE IX

*Silicon Levels in Plasma and Intestinal Tissue
in Rats with Sex and Age*

Rat	Plasma Si mg/l		Intestinal tissue Si μg/g	
	male	female	male	female
normal	6.19 ± 1.1	6.59 ± 1.09	15.8 ± 1.68	34.06 ± 5.4
senescent	9.1 ± 1.45	14.35 ± 2.75	11.66 ± 2.93	6.58 ± 0.69

The decrease in tissue silicon in the senescent female
animal is of particular interest in view of the claim that
in sterilised or post-menopausal woman, a decrease in
silicon level is observed (Charnot, 1959).

Few examinations have been reported of the silicon content of connective tissue in disease conditions, although
there is some evidence that the silicon level of the arterial wall falls with increasing sclerotic damage (Loeper,
1967) (Table X).

An extraordinary concentration of silicon has been
noted in the rims and cores of the cerebral plaques in
Alzheimer's disease, a condition resulting in presenile
dementia (Austin, 1973). Silicon levels between 1600 and
14,600 μg/g were found in the plaques, the level in normal human cerebrum being of the order of 7 μg/g. The
raised silicon level is local to the plaque, there being
no general increase in areas of cerebrum remote from the
plaque. The cause of this curious localisation is unknown.

TABLE X

*Silicon Content of Human Aortal Tissue having
Various Degrees of Sclerotic Damage*

Condition of Aorta	Si content µg Si/100 mgN
Normal	180 ± 21
Moderately altered: light subendothelial lipid deposits: rarefaction of elastin fibre; meta- chromasia of ground substance	105 ± 12
Greatly altered; lipid deposits unite media and intima; rarefaction of elastin fibres; calcium excess	63 ± 8

Silicon at the Cell Level

The fact that silicon has been found bound within the structure of mucopolysaccharides suggests that the cells responsible for the synthesis of these materials have the ability to insert silicon after the macromolecule has been formed or, alternatively, incorporate pre-formed mono- or di-saccharide Si-derivatives as minority components of the macromolecule. The process may be related to the attachment of silicon to polyuronides such as pectin in plants and to the suggested formation of carbohydrate/silicic acid esters by Proteus Mirabilis. The metabolic requirement for silicon in the diatom has already been mentioned as has a sub-cellular distribution of silicon in the osteoblast.

Electron probe microanalysis (E.P.M.) in its latest refinements (energy dispersive X-ray analysis in the scanning transmission mode) is a powerful tool for the examination of tissue for silicon since it has the potential to detect amounts of the order of 10^{-18} g and, indeed, silicon has been detected in animal cells by this method (Weavers, 1973; Mehard and Volcani, 1975). There are problems associated with the preparation of specimens for examination, notably the possibility of a redistribu-

tion of the elements during the process of section preparation as a result of momentary thawing of the frozen
material as the knife passes through it, and contamination is an ever present problem. Studies of the distribution of sodium (as NaCl) in frozen carboxymethyl cellulose
gels has indicated that little, if any, redistribution
occurs due to thawing in section preparation (Appleton,
1974) and various methods have been developed to avoid
redistribution and to retain sub-cellular structure, such
as the infiltration of tissue with glycerol to avoid ice
damage and embedment in low viscosity epoxy resin (Appleton, 1974; Mehard, 1975).

In rat liver slices, silicon has been detected by
E.P.M. in mitochondria, rough endoplastic reticulum and
nucleus (Mehard, 1976) Table XI shows the counts for Si
and P obtained in the study.

TABLE XI

E.P.M. for Si and P in Rat Liver Cells

Organelle	Counts	
	Si	P
Nucleus	49.6 ± 12.3	199.0 ± 15.9
Nucleolus	48.8 ± 8.7	65.0 ± 3.7
Mitochondria	26.2 ± 6.9	23.7 ± 5.2
Endoplastic Reticulum	65.0 ± 31.0	22.2 ± 4.5
Embedment	37.0 ± 26.0	5.0 ± 5.0

To support this finding, cell organelles were isolated
and analysed for silicon by the silicomolybdate method
(adapted for the elimination of interference by phosphorus) with the results shown below (Table XII).

Electron probe microanalysis of the spinosum layer of
rat skin has indicated a remarkable localisation of silicon (together with Mg and P) in desmosomal areas, none
being detected in cytoplasmic areas only a few microns
distant (Appleton, 1974). It seems doubtful that con-

TABLE XII

Si and P in Isolated Rat Liver Cell Organelles

Cell Fraction	Concentration µg/mg protein	
	Si	P
Nuclei	3.0	165.4
Mitochondria	0.44	8.8
Microsomes	0.48	13.6

tamination could result in such a pattern of distribution. The significance of the presence of silicon in cell junction zones is obscure.

The Toxicity of Particulate Silica

This paper has so far been concerned with the presence of silicon in biological systems at the near-atomic level of dispersion but no review would be complete without reference to the toxicity of particulate silica and some silicates. The inhalation or injection of particulate silica (especially in its crystalline modifications) results in a fibrogenic response. It is generally agreed that the initial step in the process is the phagocytosis of silica particles by macrophages, followed by the death of the cells and the release of the particles which are then available to other macrophage cells. The accumulation of dead macrophage cells stimulates collagen synthesis by neighbouring fibroblasts. There are two questions: what is the mechanism by which cells are killed by ingested silica? How does cell death stimulate fibroblast activity? The first question has been answered with some certainty and it is considered that silica particles ingested by a cell cause damage to lysosomal membranes, increasing permeability and releasing lysosomal enzymes. Membrane damage is though to result from the strong interaction, by hydrogen bonding, between membrane constituents such as proteins and phospholipids and the surface

silanol groups of the silica (Allison, 1968). Such interactions have been demonstrated in model experiments involving protein monolayers (Minones *et al.*, 1973), the
production of turbidity in protein solutions in the
presence of polymerised silicic acid (Bergman and Nelson,
1962) and in studies of the effect of silicic acid on the
permeability of black lecithin membranes (Körösy and
Taboch, 1973). There is evidence indicating that the
changes in the electrical properties of lecithin membranes
produced by neutralised alkali silicate solutions are
time dependent, suggesting that the growth of a silicic
acid polymer to a critical size is necessary for significant membrane damage. This is consistent with a view
that there is 'adsorption' of membrane constituents by a
siliceous particle and hence a distortion of membrane
structure; mono- and oligo-silicic acid would have little
effect, being too small to produce distortion. The finding that poly-2-vinylpyridine-1-oxide, which hydrogen-
bonds strongly to a silica surface, is able to reduce the
toxicity of silica to macrophages, strongly supports the
hydrogen bonding mechanism of membrane damage (Schlipköter *et al.*, 1963).

Events following silica-induced macrophage death are
relatively obscure and the mechanism of fibroblast stimulation is not known with any certainty. Silica itself
causes no stimulation of fibroblast cells. There are
sound indications that macrophages cultured with silica
particles produce a soluble factor which, when added to
fibroblast cultures as a cell-free extract, stimulates
collagen synthesis (Heppleston and Styles, 1967). The
factor also stimulates the synthesis of collagen and other
proteins in experimental granulation tissue and can be obtained by treating certain sub-cellular particles (from
intact macrophages) with silica (Aalto *et al.*, 1976).

The Possible Significance of Silicon

It is perhaps appropriate at this point to attempt a synthesis of ideas and to review areas of speculation. The fate of the majority of silicon entering the biosphere is to form loose and transient associations with organic molecules (mainly by hydrogen bonding) and to be excreted largely in inorganic form or repolymerised as hydrated silica. Polymerisation may be mediated by a biopolymer to determine morphology as in the diatom. The full significance of silicon must reside in that small proportion of the total passing through the biosphere which becomes strongly bound to the organic molecules and macromolecules of living systems. The chemistry throughout appears to be that of silicic acid and the $Si - O - C$ bond, there being little, if any, evidence for the formation of $Si - C$ bonds in nature. What is the role and function of this bound silicon, especially in higher animals? According to one author, the human body contains about seven grams of silicon (Monceaux, 1960).

Silicon presents unusual problems to the investigator. Its very ubiquity makes it difficult to avoid adventitious contamination and analytical methods of sufficient sensitivity have only relatively recently been applied to the problem of location and concentration. Much of the work carried out has relied on the 'molybdate blue' reaction perfected to the extent that about 1 µg SiO_2 can be detected and interference by phosphorus eliminated (Engel and Holzapfel, 1960). Atomic adsorption spectroscopy has pushed the limit to about 10^{-2} µg SiO_2 (Werner, 1970) and mass spectroscopy to about 10^{-6} µg SiO_2 (Goering *et al.*, 1973). Tracer techniques have been applied but there are difficulties, notably the short half-life of radio-isotopes (e.g. ^{31}Si 2.6 hours) or the need to differentiate parent and daughter nuclide activity (e.g. $^{32}Si \rightarrow ^{32}P$).

In spite of these difficulties it now seems to be

J.D. BIRCHALL

reasonably well established that silicon is bound in the
biopolymers of mesenchymal origin in animals and in plant
polyuronides. Perhaps through its association with glyco-
proteins etc. silicon is apparently implicated in the
process of mineralisation of pre-osseous tissue. Hence,
in plants and animals silicon is associated with structural components.

Silicic acid (and its polymers) can associate with
organic molecules in three ways. Being a weak acid, dis-
sociated and undissociated hydroxyls coexist, so that both
reaction with bases and hydrogen bonding between undissociated hydroxyls and polar molecules are possible.
Esters are readily formed and the present evidence favours
this as the mode of attachment to minority sugars in the
polysaccharide component of connective tissue biopolymers
and in polyuronides. The suggestion has been made (Schwarz
(Schwarz, 1973) that labile cross-links between polysaccharides and between polysaccharide and protein could
result from bridging of the type

$$R_1 - O - \overset{|}{\underset{|}{Si}} - O - R_2$$

and

$$R_1 - O - \overset{|}{\underset{|}{Si}} - O - \overset{|}{\underset{|}{Si}} - O - R_2,$$

the lability of the bridge being due to cleavage (and
oxygen switching) of the $Si - O$ bond in the ester linkage.
Such a labile mode of inter- or intra-macromolecular
association could contribute importantly to structure and
could result in an effectively high molecular weight *in
vivo*, although this might readily be degraded in the extraction processes used to isolate tissue components.
Such associations could contribute to the functions of
ground substance (control of the transport and exchange
of electrolytes and the diffusion of various molecules

and macro-molecules; water binding etc.) and of connective tissue itself (resilience, permeability etc.).

The suggestion of a labile mode of association involving Si − O − C bonds together with indications that silicon is located in cell desmosomal areas may, if the latter is confirmed, provide a clue to cell adhesion. Is silicon bonded to sugar residues at junction areas?

The association of silicon with the mesenchymal function has naturally led to speculation as to the possible significance of silicon in the changes in connective tissue that occur in ageing and in disease conditions such as rheumatoid disease, atherosclerosis etc. and most attention has been given to the latter. The reported decrease in the level of silicon in tissue with age may be directly related to the decrease with age in the level of mucopolysaccharide in connective tissue. For example, the incorporation of sulphate (as $^{35}SO_4$) into chondroitin sulphate in the human aorta has been reported as falling with age (Hauss and Junge-Hülsing, 1961) and such a change could account for a falling aortal silicon level, since bound silicon has been found in chondroitin sulphate. On the other hand, there is evidence that mesenchymal activity is raised above the age-dependent norm in arteriosclerotic tissue, yet silicon levels are reported to be lower (Loeper et al., 1967). If, as is suggested by several workers, silicon has a role in the integrity of connective tissue in vivo, in the filtration characteristics of ground substance (and perhaps in the integrity of cell junctions in the endothelia), then it is possible to speculate that reduced silicon levels in arterial tissue might result in greater susceptibility to haemodynamic damage and enhanced lipid infiltration. There is as yet no firm evidence to support such speculation.

In cholestrol-fed rabbits, treatment with derivatives of dimethyl silanediol or methyl silane triol has been claimed to reduce lipid infiltration into the aortal wall

and to reduce the tissue response to lipid, in particular
the degeneration of elastin fibres, and this has led to
the suggestion that silicon is somehow associated with
the inhibition of elastase activity (Gendre, 1972). One
major problem in such studies is that the silicon derivatives used have been unstable, hydrogen bonded complexes
and their fate, following administration is unknown
(Fourtillan and Dupin, 1973; Fourtillan *et al.*, 1972).
For example, one compound used in experiments on
cholesterol-fed rabbits has been the glycerol ester of
dimethyl silanediol. In water, this hydrolyses to produce
the free diol and glycerol with subsequent polymerisation of the diol to linear and cyclic products (Fourtillan
et al., 1972). There is no experimental evidence that
silicon, introduced in such a form, reaches the artery
wall or, indeed, what direction it takes. Clearly, tracer
studies are needed to resolve the problem and perhaps *in
vitro* studies on elastin degradation would be revealing.

A major problem is that virtually nothing is known of
the processes by which silicon becomes bound within the
biopolymers of connective tissue, nor is much known of
the form of silicon most acceptable to the body, although
experiments with silicon-deficient animals have apparently
indicated that organic derivatives are more effective in
reversing the effects of deficiency than inorganic silicates (Charnot, 1959; Schwarz, 1977). The evidence is
that in nature, the most durable organic binding of silicon is with sugars and yet the interactions of sugars and
silicic acid is a neglected area of chemistry — contrast
the situation with boron and phosphorus.

If the major dietary source of organically bound silicon is indeed in structural components of animal and
vegetable tissue, then it seems probable that there has
been a considerable decline in intake over past decades
and this has prompted speculation as to the consequences
of such a dietary change. A major source of bound silicon

will be cereal fibre and, in connection with the association of cereal fibre intake, colonic cancer, diverticu-
lar disease etc., it has been remarked that the amount of
fibre in Western diet has declined by about 3000 per cent
over the past ninety years (Burkitt, 1971). Some authori-
ties have proposed an inverse relationship between dietary
fibre intake and serum cholesterol levels (Trowell, 1972a
and b; Burkitt *et al.*, 1974) and there are reports that
fibre-rich diets protect against hypercholesterolemia
(Mathur *et al.*, 1968). This has led to the hypothesis
that a lack of 'biologically available' silicon in modern
diets plays a part in the aetiology of atherosclerosis
(Schwarz, 1977). It is pointed out that 'dietary fibre'
is poorly defined chemically, the term being used to
cover food constituents resistant to digestion — cellulose, pectins, lignin etc. It is suggested that only certain types of fibre are effective in lowering serum
cholesterol levels and a relationship is proposed between
the silicon content of fibre and serum cholesterol reduction. An analysis of a variety of fibre sources revealed
marked differences in silicon content between raw and
refined materials (Schwarz, 1977). For example, cellulose
contained only 6 ppm Si but sugar beet pulp contained
23,110 ppm and sugar cane pulp 11,270 ppm. The hulls and
straw of rice, oat and wheat contained levels of silicon
of this order whereas wheat bran contained an order of
magnitude less and wheat flour (65% extraction) contained
only 21 ppm. It is suggested that such differences could
account for inconsistencies in the reported effect of
fibre on serum cholesterol levels.

A number of mechanisms for the beneficial effects of
fibre-silicon on lipid levels have been proposed. The
silicon-containing species could be the site of bile-acid
binding in the digestive tract, enhancing the elimination
of cholesterol metabolic end-products. The direct binding
of cholesterol could also play a part. Again, organo-

silicates may absorb and function as constituents of connective tissue, so influencing the integrity of the arterial wall. Finally, it has been suggested that an active
silicon derivative could directly participate in the
intermediary metabolism of steroids and bile acids
(Schwarz, 1977).

In the context of a role for silicon in atherosclerosis, a study of two cohorts of men in Finland, having a
more than twofold difference in the frequency of death
from coronary disease, was followed by an analysis of the
water supply to the separate groups. In the high risk
group, the water supply contained significantly less
silicon than obtained in the low risk group (Schwarz *et
al.*, 1977). A tenuous association between calcium and
silicon has been a feature of this review and it is interesting to speculate on any association between this
observation and the known relationship between deaths
from coronary disease and water hardness. The subject remains one of tantalising uncertainty.

The importance of connective tissue integrity in
tumour invasion (and the destruction of connective tissue
by proteases released by tumours) has been emphasised by
several workers (Tarin, 1972) so that the claim that certain organo-silicic acid derivatives can promote a local
mesenchymal response, resulting in the formation of a
collagen capsule at the periphery of a tumour, is of considerable interest. In studies of implanted Walker's
carcinoma in rats, treatment with an organo-silicic acid
derivative has been claimed to promote such a local response resulting in regression (Voronkov, 1971). Interest
has been mainly in the silatranes, in which toxicity is
related to the nature of the substituent, R. When R is
aryl, high toxicity to mammals results, whereas when R is
alkyl or alkoxy the compounds have low toxicity (Voronkov
and Lukevics, 1969). Silatranes hydrolyse at a rate depending on water solubility (and hence on the nature of

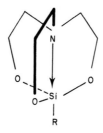

R) and the silane triol so formed will polymerise.

   The observation of a specific mesenchymal response
following the use of a silicic acid derivative has proved
difficult to reproduce (Birchall and Longstaff, unpublished) and the structure of the compound with which the
original observation was made has not been revealed in
detail. However, the possibility that silicic acid derivatives may directly stimulate fibroblast activity and
protein synthesis cannot be dismissed, since the nature
of the soluble fibroblast stimulant released from silica-
treated macrophages remains unknown. A soluble factor
capable of stimulating protein synthesis is produced by
the treatment with silica of sub-cellular material from
intact macrophages (Aalto *et al.*, 1976). In this context,
it is worth recalling that the earliest (1910–1914) proposed therapeutic use of silicic acid derivatives was in
the promotion of collagen encapsulation of pulmonary
tubercular lesions and rapid cicatrization, and such
effects were claimed to be produced in experimentally
infected animals in response to treatment with the product
of transesterification of tetraethoxysilane and glycerol
(Knorr, 1916; Kahle, 1914; Rössle, 1914; Flaskämper,
1924). However, the concept of direct fibroblast stimulation by a silicic acid derivative remains unproven.

   This review is concerned with the 'natural' chemistry
of silicon and hence with the chemistry of the Si — O — C
bond, whereas most attempts to synthesise therapeutic
compounds have involved the Si — C bond (foreign to and
rarely, if ever, cleaved in living systems) and it is

inappropriate to review this area in any detail. The
Si — C bond is used for its very inertness in contact
with living tissue, in the silicones. Few organo-silicon
compounds have shown activity greater than in the carbon
equivalent. The trimethyl silylation of testosterone has
been shown to increase activity by altering lipid solubility and hence transport across lipid membranes (Chang
and Jain, 1966) and a number of compounds have been reported to lower blood-pressure (Garson and Kirchner,
1971). A compound of particular interest is 2,6-cis-
diphenylhexamethylcyclosiloxane.

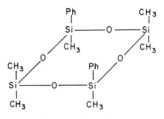

    The biological activity of this compound has been
studied in some depth; it exhibits oestrogenic and anti-
gonadotropic activity and has been used in clinical
trials against protrate cancer (Bennet and Aberg, 1975).
For further information on the biolocial activity of
organo-silicon compounds, the reader should refer to
various reviews (Voronkov and Lukevics, 1969; Fessenden,
R.J. and Fessenden, J.S., 1967).

    Progress in the therapeutic use of silicon will re-
main largely a matter of chance until much more is known
of the pathway, location and binding of the element in
living systems and of the silicic acid chemistry involved.

    Does a theme emerge from this review? Certainly, there
appears to be throughout an association between silicon
(as silicic acid) and carbohydrates (perhaps with calcium
as a participant), in consequence of which there may result subtle effects on organisation and structure, mineral
binding, cohesion and adhesion. Much remains to be done,
but such knowledge as now exists suggests that the rewards

might be considerable.

References

Aalto, M., Potila, M. and Kulonen, E. (1976). *Expl. Cell. Res.* **97**, 193-202.

Alexander, G.B., Heston, W.M. and Iler, R.K. (1954). *J. Phys. Chem.* **58**, 483.

Allison, A.C. (1968). *Proc. Royal Soc.* B 171, 19-30.

Appleton, T.C. (1974). *In* 'Electron Microscopy and Cytochemistry', p. 236 (Ed. E. Wisse *et al.*), North Holland/American Elsevier.

Atkins, E.D.T. and Sheehan, J.K. (1973). *Science* **179**, 562-564.

Austin, J.H. (1973). *Prog. Brain Research* **40**, 485-95.

Baba, I., Iwata, I., Takahashi, Y. and Kittaka, A. (1956). *Nippon Sakumotsu Gakkai Kiji* 24, 169-172.

Baker, G., Jones, L.H.P. and Wardrop, I.D. (1959). *Nature* **184**, 1583.

Barber, D. and Shone, M. (1966). *J. Exp. Biol.* **17**, 569-578.

Bennet, D.R. and Aberg, B. (1975). *Acta Pharma et Toxicol.* **36**, Supplement III, 1-147.

Bergman, I. and Nelson, Sybil E. (1962). *J. Coll. Sci.* **17**, 823-837.

Birchall, J.D., Howard, A.J. and Carey, J.G. (1977). *Nature* **266**, No. 5598, 154, March 10.

Birchall, J.D. and Longstaff, E. Unpublished data.

Bloomfield, C. (1966). *J. Sci. Food and Agric.* **17**, 39-43.

Burkitt, D.P. (1971). Third David Kissen Memorial Lecture. Cancer Priorities (Brit. Cancer Council) 3rd Symp. Edinburgh, 124-138.

Burkitt, D.P., Walker, A.R.P. and Painter, N.S. (1974). *J. Amer. Med. Assoc.* **229**, No. 8, 1068, August 19.

Butler, E.J. and Jones, S.C. (1949). 'Plant Pathology', Macmillan.

Carlisle, Edith (1972). *Science* **178**, 619-21. 10 November.

Carlisle, Edith (1970). *Science* **167**, 279. 16 January.

Carlisle, Edith (1974). *Trace Elem. Metab. Animals Proc. Int. Symp.* 2, 407-23.

Carlisle, Edith (1976). *J. Nutrition* **106** (4), 478-84.

Carlisle, Edith (1977). *Fed. Proc.* **36** (3), 1123. March.

Carlisle, Edith (1975). *Fed. Proc.* **34** (3), 927. March.

Carlisle, Edith (1974). *Fed. Proc.* **33** (6), 1758. June.

Chang, E. and Jain, V.K. (1966). *J. Med. Chem.* **9**, 433-435.

Charnot, A. (1959). *Produits Pharm.* **3**, 126-130.

Charnot, Y. (1971). *Ann. Endocrin.* **32** (3), 397-402.

Connell, R., Whiting, F. and Forman, S.A. (1959). *Can. J. Comparitive Med.* Vol. XXIII, No. 2. February.

Coombs, J. *et al.* (1967). *Expl. Cell. Res.* **47**, 315-28.

Coombs, J. and Volcani, B.E. (1968). *Planta (Berl.)* **80**, 264-279.

Darley, W.M. and Volcani, B.E. (1969). *Expl. Cell. Res.* **58**, 334-42.

Dea, I.C.M., Moorhouse, R., Rees, D.A., Arnott, S., Guss, J.M. and Balazs, E.A. (1973). *Science* **179**, 560-562.

Deevey, E.S. (1970). *Scientific American* **223**, No. 3, 148-158. Sept.

Engel, W. (1953). *Planta* **41**, 358-390.

Engel, W. and Holzapfel, L. (1960). *Beitr. Silikose-Forsch* 4, 67-71.

Fairbridge, R.W. (1966). *In* 'The Encyclopedia of Oceanography', pp. 469-74, Reinhold, New York, 1966.

Fessenden, R.J. and Fessenden, J.S. (1967). *Advances in Drug Research*

4, 95-132.
Flaskämper, A. (1924). *Z. Tuberk.* **39**, 257-60.
Forman, S.A., Whiting, F. and Connell, R. (1959). *Can. J. Comparative Med.* Vol. XXIII, No. 5, May.
Fourtillan, J.B., Dupin, J.P. and Lemoigne, B. (1972). *Bull. Soc. Pharm. Bordeaux* III, 200-210.
Fourtillan, J.-B. and Dupin, J.-P. (1973).*Chimie Therapeutique* 2, 207-214, March/April.
Garson, L.R. and Kirchner, L.K. (1971). *J. Pharm. Soc.* **60**, 1113-1127.
Gendre, P. (1972). *Bull. Pharm. Soc. Bordeaux* III, 3-13.
Goering, J.J. *et al.* (1973). *Deep Sea Res.* **20**, 777-789.
Goldwater, L.J. (1936). *J. Ind. Hyg. Toxicol.* **18**, 163.
Golley, F.B. (1972). *In* 'Ecosystem structure and function', (Ed. J.A. Wiens), 69-88. Oregon State University Press, Corvallis.
Grass-Brauckmann, E. (1953). *Z. Pflanzenernachr. Dueng Bodenk* **62**, 19-23.
Handreck, K.A. and Jones, L.H.P. (1967). *Australian J. Biol. Sci.* **20**, 483-485.
Hauss, W.H. and Junge-Hülsing, G. (1961). *Dtsch med. Wsch.* **86**, 763.
Hecky, R.E., Mopper, K. and Degeus, E.T. (1973). *Mar. Biol.* **19**, 323-331.
Henderson, M.K. and Duff, R.B. (1963). *J. Soil Sci.* **14**, 236-4.
Heinen, W. (1965a). *Arch. für Mikrobiol.* **52**, 69-79.
Heinen, W. (1965b). *Arch. für Mikrobiol.* **52**, 49-68.
Heinen, W. (1968). *Acta Bot. Neerl.* **17**(2), 105. April.
Heppleston, A.W. and Styles, J.A. (1967). *Nature* **214**, 521-522.
Holt, P.F. (1950). *Brit. J. Ind. Med.* **7**, 12.
Holt, P.F. and Clark, S.G. (1955). *Proc. Biochem. Soc.* **60**, XXX.
Holt, P.F. and Went, C.W. (1959). *Trans. Far Soc.* **55**, 1435-50.
Holzapfel, L. (1951). *Zeit für Electrochem.* **55** (6).
Huang, W.H. and Keller, W.D. (1972). *Nature* **239**, 149. 30 October.
Iler, R.K. (1955). 'The Colloid Chemistry of Silica and Silicates', Cornell University Press.
Iskandar, I.K. and Syers, J.K. (1972). *J. Soil Sci.* **23** (3), 255-265.
Izawa, G. and Kume I. (1959). *Hyogo Noka. Daigaku Kenko Hokoku* **4**, 13-17.
Johlin, J.M. (1932). *Proc. Soc. Expl. Med.* **29**, 760-761.
Jones, L.H.P. and Handreck, K.A. (1963). *Nature* **198**, No. 4883, p. 852, 1 June.
Jones, L.H.P. (1963). *Plant and Soil* **18**, 358.
Jones, L.H.P. and Handreck, K.A. (1965). *Plant Soil* **23**, 79-96.
Jones, L.H.P. and Handreck, K.A. (1967). *Adv. Agron.* **19**, 107-149.
Kahle, H. (1914). *Münch. Med. Woch.* **14**, 752-756.
King, E.J. and Stantial, H. (1933). *Biochem. J.* **27**, 990.
King, E.J. (1939). *Biochem. J.* **33**, 949.
Knorr, L. (1916). U.S. Pat. 1,178,731.
Körösy, Fe de and Taboch, M.F. (1973). *Biochemica et Biophysica Acta* **291**, 608-611.
Lanning, F.C., Ponnaiya, B.W.X. and Crumpton, C.F. (1958). *Plant Physiol.* **33**, 339-343.
Lauwers, A.M. and Heinen, W. (1974). *Arch. Mikrobiol.* **95**, 67-78.
Leslie, J.G., Kung-Ying, T.K. and McGavack, T.H. (1962). *Proc. Soc. Exp. Biol. Med.* **110**, 218.
Lewin, J.C. and Chen, C.H. (1968). *J. Phycol.* **4**, 161-6.

Livingstone, D. (1963). *U.S. Geol. Survey Prof. Paper 440*, **G63**, 1963.
Loeper, J. *et al.* (1967). *Giornale di Clinica Med.* **47**, 7, 596-605.
Lovering, T.S. and Engel, C. (1967). *Geol. Survey, Prof. Paper 594B*, U.S. Govrn. Printing Office.
McKeaque, J.A. and Cline, M.G. (1963). *Adv. Agron.* **V**, 15, 339-396.
Mathur, K.S., Khan, M.A. and Sharma, R.D. (1968). *Br. Med. J.* **1**, 30-31.
Mehard, C.W. and Volcani, B.E. (1975). *J. Histochem. Cytochem.* **23**, 348-358.
Mehard, C.W. and Volcani, B.E. (1976). *Cell Tissue Res.* **166**, 255-263.
Minones, J. *et al.* (1973). *J. Colloid and Interface Science* **42**, No. 3, 503-515. March.
Mitsui, S. and Takatoh, H. (1963). *Soil Sci. Plant Nutr. (Tokyo)* **9**, 49-53.
Monceaux, R.H. (1960). *Produits Pharm.* **15**, No. 3, 99-109. March.
Oberlies, F. and Pohlmann, G. (1958). *Naturwissenschaften* **45**, 487.
Okuda, A. and Takahashi, E. (1962). *Nippon Dojo Hiryogaku Zasshi* **33**, 1-8.
Okuda, A. and Takahashi, E. (1964). *In* 'The Mineral Nutrition of the Rice Plant', Symp. Intern. Rice Res, pp. 123-146. John Hopkins Press, Baltimore.
Reimann, B.E.F., Lewin, J.C. and Volcani, B.E. (1966). *J. Phycol.* **2**, 74-84.
Rössle, R. (1914). *Münch. Med. Woch.* **14**, 756-757.
Sauer, D.F., Laughland, D.H. and Davidson, W.M. (1959). *Canadian J. Biochem. & Physiol.* **37**, 1173-1181.
Schlipköter, N.W. *et al.* (1963). *Germ. Med. Mon.* **12**, 509-514.
H. Schultz (attributed) (1914). Reported by Kahle, H. in *Münch. Med. Woch* **14**, 725-56.
Schwarz, K. and Smith, J.C. (1967). *Nutr.* **93**, 182.
Schwarz, K. (1970). *In* 'Trace Element Metabolism in Animals' (ed. C.F. Mills), Livingston, Edinburgh.
Schwarz, K. and Milne, D.B. (1972). *Nature* **239**, 333, October.
Schwarz, K. (1973). *Proc. Nat. Sci. U.S.A.* Vol. 70, No. 5, 1608-1612. May.
Schwarz, K. and Chen, S.C. (1974). *Fed. Proc.* **33** (3), 704.
Schwarz, K. (1974). *Fed. Proc.* **33** (6), 1748, June.
Schwarz, K., Ricci, B.A., Punsar, S. and Karvonen, M.J. (1977). *The Lancet*, 5 March, p. 538.
Schwarz, K. (1977). *The Lancet*, 26 February, p. 454.
Siefert, N. (1959). *Naturwissenschaften* **46**, 261.
Seifert, H. (1061). *Naturwissenschaften* **48**, 713-714.
Spector, W.S. (1956). 'Handbook of Biological Data', W.B. Saunders Co. Philadelphia.
Strickland, J.D. (1972). *Oceanogr. mar. biol. Ann. Rev.* **10**, 349-414.
Tarin, D. (Ed.) (1972). 'Tissue Interactions in Carcinogenesis', Academic Press, 1972.
Trowell, H. (1972). *Amer. J. Clin. Nutr.* **25**, 926-932.
Trowell, H. (1972). *Rev. Evr. Etuel Clin. Biol.* **17**, 345-349.
Underwood, E.J. (1971). 'Trace elements in Human and Animal Nutrition', Academic Press, New York.
Van Soest, P.J. and Jones, L.H.P. (1968). *J. Dairy Sci.* **51**, 1644-8.
Varma, R. *et al.* (1974). *Biochimica et Biophysica Acta* **263**, 548-588.
Voronkov, M.G. (1971). *Acad. Nauk. SSR* **200** No. 4, 967-969.

Voronkov, M.G. and Lukevics, E. (1969). *Russ. Chem. Rev.* **38**, 12 975-986.
Weavers, B.A. (1973). *J. Microscopy* **97**, 331-341.
Weiss, A., Reiff, G. and Weiss, A. (1961). *Z. Anorg. allg. Chem.* **311**, 151-79.
Werner, D. (1966). *Arch. Mikrobiol.* **55**, 278-308.
Werner, D. (1967). *Arch. Mikrobiol.* **57**, 51-60.
Werner, D. (1968a). *Z. Naturforsch.* **23**, 268-72.
Werner, D. (1968b). *Ber. dt. bot. Ges.* **81**, 425-9.
Werner, D. (1970). *Helgoländer wiss Meeresunters* **20**, 97-103.
Werner, D. (1970). *Hoppe-Seylers Z. physiol. Chem.* **351**, 134-5.
Werner, D. (Ed.) (1977). 'The Biology of Diatoms', pp. 110-142. Blackwell Scientific Publications, 1977.
Williams, R.J.P. (1977). *Chemistry in Britain* **13**, No. 7, p. 267. July.
Wynn-Parry, D. (1975). *Ann. Bot.* **39**, 815-8.

8 A SHORT NOTE ON SELENIUM BIOCHEMISTRY

R.J.P. Williams

Inorganic Chemistry Laboratory, South Parks Road, Oxford OX1 3QR.

It was intended to have a full chapter of this book devoted to the involvement of Selenium in biology. Prof. K. Schwarz had agreed to write the chapter and his untimely death leaves a gap in the book which none of the other contributors could possibly fill. All of us wish to express our admiration of the persistence and skill which he showed in chasing down general trace element requirements, not just for selenium, in animal biology.

Selenium is not an obvious element for biology to have selected as a catalyst, Table VII of Chapter V, yet it is intimately connected in as yet an unknown and very important way with the reactions of peroxides and their possible decomposition products, hydroxyl radicals, activated oxygen, and superoxide. Again the roles of selenium and vitamin E appear to be linked as antioxidants. The involvement with glutathione is also very puzzling, see below. The biochemistry of selenium is also related to the susceptibility of animals to disease, see references. An especially intriguing possibility is the link between low selenium levels and cancer which has been revived recently by the work of Schrauzer (1976). Here this point will not be strongly stressed. Rather we shall survey the chemistry of selenium and then speculate upon its possible biological role. We act within the belief that a trace element is used by biological systems because the chemistry of the element has some outstanding peculiarity, which is turned into functional value.

Selenium Chemistry

Oxidation States and Reactivity

This is one of a few non-metal elements which can show
variable valence within the redox range open to biological
systems, Fig. 3 Chapter 5. Thus it differs strikingly from
P, Si, N, Cl, and B. Moreover its oxidation states have
somewhat similar relative stabilities differentiating the
character of the element from that of oxygen. The residual
non-metals of consequence are carbon and sulphur and several of the redox potentials of these elements are not too
dissimilar from those of selenium. The higher oxidation
states of selenium are however not very stable even in
organic compounds when we compare them with the sulphur
oxidation states. Selenium oxides are quite strong oxidising agents.

The ability of selenium to form compounds of coordination number greater than four distinguishes it strongly
from carbon and marginally from sulphur. As in all groups
of non-metals the heavier the non-metal the better it acts
as a leaving group – selenium is a better leaving group
than sulphur or carbon, a fact related to the ease of increase of coordination number. Putting the kinetic and
thermodynamic facts together we can say that selenium–
oxygen bonds in the lower positive oxidation states are
less thermodynamically stable and more kinetically reactive than the corresponding sulphur compounds. This pins
one special advantage of selenium to oxygen atom transfer
reactions such as

$$SeO + X \rightarrow Se + XO$$

and $$SeO_2 + X \rightarrow SeO + XO$$

This use of selenium is well-known in organic chemistry.
It must be remembered however that these oxides of selenium are too oxidising to be anything but intermediates in

biological systems.

At the same time that we observed a decrease in oxide stability relative to the elemental state on going down a non-metal group of the Periodic Table we also observe a loss of stability of the hydride, i.e. of the X–H bond e.g. C–H > Si–H; N–H > P–H; O–H > S–H > Se–H. Overall then the elemental state, i.e. here the Se–Se bond, is more stable relative to reduction or oxidation than is the S–S bond. By way of contrast at the top of the Group di-oxygen, O_2, is very unstable with respect to H_2O.

The stability of carbon–selenium bonds follows the usual trend that heavier non-metal to carbon bonds are not particularly stable. For example the free energy of forma-tion of $(CH_3)_2X$ where X is oxygen, sulphur, selenium or tellurium falls in the sequence $O \gg S > Se > Te$.

When an organo–selenium compound is exposed to oxygen the reaction

$$O + (CH_3)Se \rightarrow HCHO + CH_3SeH$$

occurs. Compare

$$O + (CH_3)_2S \rightarrow (CH_3)_2SO$$

Similarly metal atoms can strip selenium from organo–selenium compounds

$$M + (CH_3)_2Se \rightarrow MSe + C_2H_6.$$

This stripping is very effective with the heavier metals of the transition metal series and of the B-Group elements, e.g. copper or mercury.

The above chemistry suggests and indeed it is found that selenium is most likely to appear in biological sys-tems as a selenol, RSe–H, or as a selenium ether, R_2Se. Corresponding forms of sulphur are cysteine and methionine.

Selenols and Seleno-ethers

Apart from their redox reactions, selenols have the
peculiar property that they are not very weak acids.
Thus H_2Se has a pK_a of 3.8 (H_2S, 8.0) and the estimated
pK_a for RSeH (Seleno-cysteine) is 6.0. Thus selenols in
enzymes will be found in the *anionic* form. Selenols are
then very good nucleophiles and are also very good leaving
groups. Here selenium has a distinct advantage over sulphur as a reactive centre.

The ether grouping R_2Se is able to react to give the
higher state of methylation $R_2Se^+CH_3$ — a reaction it has
in common with thio-ethers. However this reaction has
little advantage in catalysis over the thio-ether methyl-
transfers. The author does not sense any particular peculiarity in seleno-ethers in contrast with the special
advantage of selenols.

It is also interesting to consider the reaction of
RSe^- with acids other than the proton. Of course RSe^- is
an extremely strong binding agent for heavy metals such
as Hg^{2+}. Selenium compounds are used in detoxification.
However it will attack much poorer acids such as sulphur
compounds to give polymers $RSe(S)_n^-$. This reaction parallels the formation of poly-sulphides in water, but of
course the reaction will go at a much lower pH with selen-
ides.

Selenium in Biology

This simple account of the chemistry of selenium reveals some features which could be of great importance in
the selection of selenium as a biological catalyst. To
see these in their proper perspective we shall assume that
selenium can occur in biology in two forms $-CH_2-Se-CH_3$,
selenomethionine, and $-CH_2SeH$, selenocysteine. We shall
also assume following the chemistry of sulphur that there
is the possibility of the formation of Se—Se links, diselenide bridges, compare disulphide bridges, and that

these bridges would be rather stable. Referring to the above discussion of redox chemistry and Fig. 3 of Chapter 5 we can estimate the redox potentials of these compounds at pH = 0, i.e.

$$2-SeH \rightarrow -Se-Se- \qquad -0.3 \text{ volts}$$

$$SeO \quad \rightarrow \quad Se + H_2O \qquad +0.4 \text{ volts}$$

(At pH = 7 the first reaction will be less reducing — say -0.2 volts). We shall now look a little more closely at the chemistry of these four compounds of selenium in relation to the known biological importance of the element in association with a few proteins.

Glutathione Peroxidase

Glutathione peroxidase carries out the catalysis of the reaction

$$2RSH + H_2O_2 \rightarrow RS-SR + 2H_2O$$

It can also use organic peroxides. The fact that a selenium compound is involved in the catalysis means that the first three elements of Group VI of the Periodic Table are reacting together and possibly the redox equivalents go in the sequence

when the role of the selenium is in the transfer of oxidising equivalents.

The structure of this enzyme, Ladenstein *et al*. (1977), shows that two selenium atoms are not so close together, < 20 Å apart although on separate subunits. This suggests the following mechanism for reaction is more likely

$$H_2O_2 + ESe^- \quad + \quad RSH \xrightarrow{\ H^+\ } ESe - SR + 2H_2O$$

$$RS^- + ESe - SR \longrightarrow ESe^- + RS-SR$$

where the intermediate E–Se–OOH is possible.

It will be interesting to see if this type of reaction sequence does occur with the enzyme.

Note again that there may be several other requirements for selenium in enzymes presently unknown. In particular T. Stadtman (1977) has drawn attention to two oxido-reduction enzymes, a reductase and a hydrogenase, which require selenium, formate dehydrogenase and glycinate reductase. Glycine reductase involves glutathione as a cofactor and the reaction is a reductive deamination. The usual route for removal of ammonia is via oxidation using pyridoxal phosphate. At first site the function of selenium which has been shown to be present as seleno-cysteine by Stadtman could possibly be through formation of a C–Se bond as in $Se-CH_2-CO_2H$. The energetics of the reaction are extremely interesting since this inter-mediate would be reminiscent of the reactions of activa-ted thiols. The final product is acetate.

The formate dehydrogenase gives carbon dioxide. A plausible reaction path is the attack of the $(RSe)_2$ group on the formic acid.

$$(RSe-SR) \text{ or } (RSe)_2 + H-C{\overset{\displaystyle /\!/\,O}{\underset{\displaystyle \backslash OH}{}}} \rightarrow RSe - C{\overset{\displaystyle /\!/\,O}{\underset{\displaystyle \backslash OH}{}}} + RSe^- \ (+RS^-) + H^+$$

$$\rightarrow RSe - C{\overset{\displaystyle /\!/\,O}{\underset{\displaystyle \backslash OH}{}}} \rightarrow RSe^- + CO_2 + H^+$$

The selenide could be in a mixed Se–S bridge rather than a diselenide.

A final note of interest is the finding of a small

selenium protein in muscle. The link between muscle degeneration and selenium may be related to the functions
of this protein.

Selenium and Other Trace Chemical Requirements

The relationship between selenium and sulphur is obvious and to some extent one element must be able to
substitute for the other. This type of substitution has
been studied in the obvious sulphur sites in amino-acids
and in proteins such as ferredoxins.

The next most similar element to selenium is arsenic.
There is already considerable evidence that arsenic levels
affect selenium levels in animals and arsenic compounds
have been used to treat animals suffering from excess of
selenium. Arsenic compounds are used in cattle fattening.

The outstanding peculiarity of selenium biochemistry
is its relationship to Vitamin E. It is postulated that
vitamin E protects membrane unsaturated lipids from peroxidation. Clearly reduced Glutathione could act in a
similar way. Once again the interrelationship of the -ols
of Group VI of the Periodic Table are quite striking;
vitamin E is an alcohol, glutathione is a thiol and probably selenium is used as a selenol. Here the general
postulate arises that selenium is a major factor in assist-
ing the clearance of electrophilic agents from cells. If
this is the case its protective effect against carcino-
genic materials could be understandable. Even the degree
to which it assists in glutathione reactions could be
readily linked to detoxification.

There is some suggestion of links between copper and
selenium (and molybdenum) balances. It must be noted that
an enzyme with two selenol anions close together would
be carrying an extremely powerful chelating agent for
heavy metal ions.

Just as is the case for all other trace elements excess
of selenium is toxic. Selenium is responsible for 'alkali

disease' and 'blind staggers' in cattle. Unfortunately
many plants accumulate the element and are unaffected by
relatively high concentrations. It is only in animals
that selenium seems to be particularly toxic.

References

Forst, D.V. and Lish, P.M. (1975). *Annual Reviews of Pharmacology* **15**,
    259-284.
Ladenstein, R., Epp, O., Romisch, A. and Wendel, A. Presented at
    TEMA III Meeting, Freissing, July 25, 1977.
Schrauzer, G.N. (1976). *Bioinorganic Chemistry* **5**, 275-281.
Scott, M.L. (1964). *In* 'Mineral Metabolism', eds. C.L. Comer and
    F. Bronner, Vol. 2B. Academic Press, New York, pp. 543-558.
Stadtman, T.C. (1974). *Science* **183**, 915-922.
Stadtman, T.C. (1977). *Nutrition Reviews* **35**, 161-166.

9 PHOSPHORUS AND BONE

F.G.E. Pautard

M.R.C. Mineral Metabolism Unit, Leeds Infirmary

Introduction

In man, the average adult skeleton contains some 500g of
phosphorus in the form of inorganic calcium phosphate. The
nature of the salt and the arrangement of bones that go to
make up the support for the soft tissues is a reflection
of a long history of vertebrae that became hardened with
mineral about 400 million years ago. Before that time,
there was probably a period of perishable backbones of
cartilage and muscle which are still to be found in modern
sharks and lampreys. The fossil and recent record present
a clear picture of gradually evolving movable parts on the
one hand and an impregnation of those parts with calcium
phosphate on the other.

Parallel with the events which have led up to modern
mammalian bone is the equally well-documented history of
the hardening of invertebrate supporting structures with
calcium carbonate, commencing about 600 million years ago
in the shells of bivalves and molluscs. However, in spite
of the fact that the abundance of carbon in the lithosphere is less than phosphorus, the accumulation of inorganic phosphate of biological origin is small while the
accumulation of carbonate is large. For every thin bed
of phosphorite there are many limestone cliffs thousands
of feet thick; it is not surprising that the mass and

diversity of calcium carbonate and the scarcity and similarity of calcium phosphate has led to the belief that
bone is a unique substance, a partnership of a form of
apatite with the fibrous protein collagen, conferring
mechanical advantages and providing a reservoir of calcium
and phosphorus for the soft tissues.

There is increasing evidence, however, that the presence of calcium phosphate in bone is neither a unique
event in evolution nor a simple arrangement of an inorganic salt with an organic macromolecule. Calcium phosphate is found widely in nature from single cells throughout all phyla to the vertebrates. Among the first animals
to leave fossil traces were the brachiopods, some of
which had shells calcified with phosphate in much the
same way as modern *Lingula*. Most animals containing carbonate also contain phosphate, in some cases varying considerably from one part of the skeleton to another. Phosphate and carbonate apparently perform the same function
in similar animals. The shrimp *Palaemon*, for instance, has
a shell hardened with calcium carbonate; a close relative,
the mantis shrimp *Squilla*, often caught in the same net,
has a shell hardened with calcium phosphate. There is no
unique association of calcium phosphate with collagen
either — the same salt with the same structure can be
found in association with such diverse macromolecules as
keratin, β-protein and α- and β-chitin. Even the mechanical advantages of calcium phosphate in the vertebrate
skeleton are doubtful, since the earliest fishes calcified with phosphate were the agnathans, with mineral in
plates outside and presumably with soft skeletons inside.
Their modern relatives such as the lamprey *Petromyzon*
still have no mineral in the cartilaginous bars of their
skeletons, yet their aquatic performance is no less effective than the bony eel *Anguilla* which may live in the same
waters.

All these exceptions and anomalies suggest that the

polar phosphate anion in bone and in a wide variety, perhaps all, animals may have a function not solely connected
with support or as an acceptable counterion for exchangeable calcium. The extensive literature on skeletal calcium
exchange and homeostasis in man has only recently been
augmented by related studies on the exchange of phosphorus
in the same sites. The view of bone as a 'calcified'
tissue is being counterbalanced by the idea of bone as a
'phosphatized' tissue and the nature of bone salt is receiving more attention in terms of the complex chemistry
of phosphorus in biological function. Studies of calcium
and its metabolism, traceable through comparatively simple
pathways, are being augmented by studies of the metabolism
of phosphorus through the cycles of energy transfer, synthesis, and the polymerization and degradation of organic
macromolecules.

Most of the evidence about bone salt, however, has been
obtained by methods predominantly in use in geology: that
is, gross chemical analysis of the mineral fraction correlated with the crystallography, which in turn has been
compared with the structural and ultrastructural biology.
The net result has been a description of bone in terms of
rock, and while this image is now changing, a geological
approach still persists, particularly as the principal
form of calcium phosphate in the lithosphere is fluorapatite, the lattice model for the hydroxyapatite witnessed by the same procedures in the skeleton. There is an
inherent misconception in these conclusions. In a pearl,
the aragonite lattice is likely to be the same inside and
outside the oyster, for the mineral, although laid down in
a precise way and in relation to an organic framework, is
outside the mantle and free of cells. In bone, the situation is different. The tissue contains a large population
of cells in close contact with the mineralized matrix
which surrounds them and there is a continual exchange of
substances throughout the tissue. Unlike the pearl, bone

is a living structure, constantly changing, to be regarded
in the same way as liver or brain. Yet the information
about the material *in vivo* is small and little biochemistry
has been carried out on bone in the same way as it has on
the liver or the brain. With one doubtful exception, no
crystallographic study of bone in a living animal has been
described so far and there is no reliable optical record
of mineral formation or function *in situ*. All the information that is available has been accumulated from measurements on dry excised bone or deduced from uncertain physiological experimentation which has been directed towards
the balance of ions in the system rather than their precise location when they are in one state or another.

The purpose of the chapter which follows is first to
summarize the present views about the nature of bone and
its mineral component and then to present an alternative
appraisal, with special reference to phosphorus in the
system. In company with a growing conviction elsewhere
that bone cells are intimately involved with calcium
phosphate, the evidence from our laboratory suggests that
we might profitably regard bone salt not as a random precipitate but as part of extremely complex subcellular
structures with a considerable biochemistry and a biological behaviour equal to the cells which create them. In
this context, the extracellular matrix of bone may be no
more than the most highly evolved expression of a cooperative arrangement of self-contained units which may
be found throughout animals (and possibly plants) and
which may have been present from the earliest stages of
the simplest organisms.

Phosphorus in Nature

Phosphorus has a special function among the substances
assimilated by living things: carbon, hydrogen, oxygen
and nitrogen are the elements of biological frameworks,
phosphorus is the instrument of their manufacture. Labile

phosphate transfers energy, synthesizes monomers and con-
structs polymers, and destroys both. In insoluble conformation with nucleosides it is part of the storage of information; the related structure, solubilized, transfers
the information or fragments it into specific instructions.
Within the boundaries of muscular contraction, substrate
and polymer synthesis, and nuclear direction are the principal 'living' properties of movement, manufacture and
reproduction. All of these functions are linked directly
and indirectly through phosphate to a common metabolism
which continues, after each multiplication of the organism,
only by parcelling the existing phosphorus between the new
progeny or by the acquisition of further supplies from
dead rocks or living prey.

From the earliest time, phosphorus, unlike easily
soluble gaseous ammonia and carbon dioxide, could only
have been gathered from its solid origins. The primary
igneous rocks would have weathered in various ways, but
only under favourable aqueous conditions would the phosphate have been acceptable to organisms. Once the phosphate was incorporated into a functioning biochemistry,
the biological cycles thus started would have contributed
increasingly to further selection and sedimentation. Since
the principal insoluble phosphate in igneous, sedimentary
and metamorphic rocks is the calcium salt, it is not surprising that its counterpart in living things is the same
substance. However, the transposition of archaean calcium
phosphate (as fluorapatite) through soluble phosphate to
the calcium phosphate of bone (as hydroxyapatite) is not
a simple matter of solubility products, whatever the
temptation to regard a femur as another pebble from an
accessory dyke. The overall geology of the apatites seems
to be the interaction of two processes — the stable end
products of volcanic activity, hydrothermal change, sedimentation and metamorphosis continuing from the earliest
periods of the rocks, and the stable end products of

biological activity exposed to the same forces.

Over long periods of time there may be no great difference between calcium phosphate of biotic or abiotic
origin; the central question is whether, during that brief
period when the salt was part of the living, it continued
to resemble the dead.

Geology

Phosphorus is not a rare element. From geochemical
data (for example, Rösler and Lange, 1972) the Clarke
Value of the element in the lithosphere is listed eleventh
at 930 ppm above carbon at 230 ppm. The concentration of
phosphorus remains surprisingly constant at about 1000
ppm in meteorites, igneous rocks and most soils, but there
are substantial changes in many sedimentary rocks of biological origin (Table I). These differences generally

TABLE I

The abundance of phosphorus

Source	P (ppm)	Ca (%)	Source	P (ppm)	Ca (%)
Lithosphere	930	2.96	Slates	700	2.2
Soil	800	1.37	Sandstones	170	3.9
Meteorites	1,050	1.3	Weathering solutions	708	2.17
Ultrabasic rocks	200	2.5	Seawater	< 1.0	0.04
Basic "	1,400	7.6	Marine lime mud	650	0.21
Intermediate "	1,600	1.5	[1]Limestone	<40	>38.8
Acid "	600	2.5	[2]Senegal phos-		
Granites	700	1.6	phorite	164,500	35.8

[1]McGregor, 1963
[2]Anon., 1971
Other values from Rösler and Lange, 1972.

appear to be the result of past activity in marine conditions, low values appearing where there is a preponderance
of calcium carbonate and high values where carbonate is
low and there is probably a local accumulation of vertebrate or other skeletons of calcium phosphate. The loss
and gain of accessible phosphate is evident today in sea

water. Marine concentrations of phosphorus are low
(uusually less than 1 ppm) in contrast to the concentration in weathering solutions entering the oceans (about
980 ppm). In a study of the elements in three hot salt
springs, for example, Uzumasa (1965) recorded amounts of
phosphorus in two of them close to the Clarke Value for
weathering solutions. In the third spring, the phosphorus
present was little more than that in sea water (Table 2),
perhaps the result of abstraction by thermal bacteria.

TABLE II

*Phosphorus, calcium and carbonate concentrations
in 3 hot springs compared with seawater*

Location	P (ppm)	Ca (%)	CO_2 (%)
1	770	4.94	1.87
2	730	2.33	0.24
3	6.3	5.35	0.50
Seawater	<1.0	0.04	0.0028

From Uzumasa, 1965.

The evidence suggests that the distribution of phosphorus in the primary rocks and after alteration is relatively uniform, but there are substantial differences in
the composition of sedimented rocks of biological origin
as a result of abstraction and recycling.

Igneous phosphate The principal phosphate in igneous rocks
is the calcium salt in the form of apatite. Numerous
variations and isomorphs have been reported, but fluorapatite is by far the commonest molecule and is the basis
for detailed measurement of the unit cell and the model
for replacement by alternative cations and anions. The
types of igneous apatite are increased by subsequent
thermal metamorphosis or by hydrothermal remodelling and
substitution. The structure of the fluorapatite lattice
is illustrated in Fig. 1 together with variations in
cations and anions in various conditions. Two features are

of importance to the discussion of biological apatites
and related structures that follows below. The first is
the anomalous position of carbonate in relation to the
phosphate. In igneous and metamorphosed apatite, 5% of
carbonate may be present: in bone and other similar
tissues the percentage may be higher. The second is the
replacement of phosphate by other anions (sulphate, silicate etc.) with compensation of the valencies, a feature
which may occur in bone salt.

An interpretation of the isomorphism of the apatites
has been given by McConnell (1938) and more recently by
Posner (1961) and Elliot and Young (1966).

Sedimentary and metamorphosed phosphate During the course of
weathering, igneous phosphates become fragmented and
the products may be changed or exchanged. After solution,
both ambient and thermal, sedimentation is likely to take
place and new apatite can form under favourable conditions of pH, to become metamorphosed or weathered again
and the cycle repeated.

About 400 million years ago, a proportion of the phosphate in the igneous cycle was converted to biological
apatite in increasing amounts and from that time the
nature of the sedimented apatites and their alteration
have become complex and confusing. From Devonian times,
with well-mineralized vertebrates in evidence, appreciable
beds of phosphorite (see Table I) were formed among the
larger deposits of limestone. In time, new systems of
apatite arose but probably of biological origin. The quartz-
apatite veins in Montana, for instance, may not be pluto-
nic, but may have developed from igneous intrusion into
Permian phosphate beds (Lowell, 1955). In this way, many
'apatitic' rocks may have started from phosphorite of
animal origin (Bezrukhov, 1939).

The transformation of biological apatites by various
routes is shown in Fig. 1 in relation to the apatites of
geological origin. Unfortunately, a persistent demand for

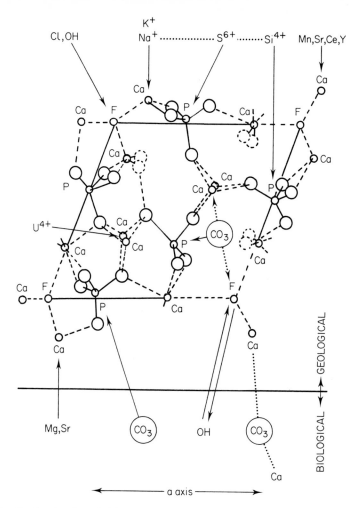

Fig. 1. Geological and biological apatites. The lower half of the unit cell of fluorapatite (after Bragg, 1937) is shown and the number and likely lattice positions of the substituents have been confined to those elements most commonly found in nature.

gross chemical analyses of heterogeneous metamorphosed phosphates gives us little clear information about the nature of the apatite species which are present. Rock analyses almost always appear in the literature without separation or fractionation and it is usually impossible to determine the quantitative changes which might have

taken place. While the lattice constants measured in a
single crystal of igneous fluorapatite can be compared
with the chemistry of the same crystal with some confidence, the tabulated list of elements in the altered
apatites allows us not conclusions as to the structure
of the individual crystals. The difficulties of useful
calculations of the chemistry of the altered calcium
phosphates of biological origin are illustrated in Fig. 2.

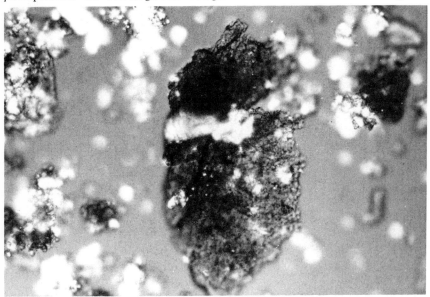

Fig. 2. Sample of powdered phosphorite from Khouribga, Morocco,
viewed in partially polarized light. The 'isotropic' component
appears as dark granular aggregates. A general analysis of this
material is set out below (from Anon., 1971).

General Analysis

P_2O_5	CaO	Fe_2O_3	Al_2O_3	MgO	Na_2O	K_2O	SiO_2	CO_2	SO_3	F_2
37.21	54.24	0.13	0.39	0.10	0.27	0.06	0.97	2.64	0.62	4.17

The optical micrograph of phosphorite from Khouribga in
Morocco shows, in partially polarized light, that part of
the rock is crystalline and part is isotropic. While it is
now generally agreed that this material may have commenced
as vertebrate phosphate, it is not possible to determine
from the chemical analysis shown in the accompanying table

how the various anions and cations are related to the
numerous crystalline species which are present. In particular, the micrograph illustrates the well-known feature
of cryptocrystallinity in the isotropic regions ('collo-
phane') which will be considered later in relation to bone.
From the work of Ames (1959) it might be inferred that the
fragments shown in Fig. 2 may be largely, if not entirely,
the result of the replacement of calcite in alkaline phosphate solutions, but the presence of fish scales and bones
throughout this phosphorite bed suggests that the apatite
was probably of animal origin in the first place and may
have undergone a variety of changes (see Fig. 1) or may
have retained some of its original structure.

Biology

Phosphorus is essential to all forms of life. With a
few rare exceptions (the phosphonic acids in the invertebrates and creatine phosphate) the element is linked
throughout organic biological structures via covalent
bonds to oxygen or as inorganic salts via polar bonds to
cations. The coupling and uncoupling of pentavalent phosphorus as orthophosphate is the prime mechanism whereby
substrates and macromolecules are formed and dismantled.
At the same time, a number of pyrophosphates and poly-
phosphates are found in relation to inorganic phases in
bone, where they may be of functional significance.

In animals with skeletons calcified with phosphate, all
but a small proportion of calcium is found in the hard
parts. The proportion of phosphorus outside the skeleton,
on the other hand, is usually larger and in man there is
a substantial pool of the element in the soft tissues
(Table III).

Organic The biochemistry of phosphate has been extensively reviewed throughout animals and plants (see, for
example, Elroy and Glass, 1961; Irving, 1964). While the
complex interactions of phosphorylated molecules are

TABLE III

*Distribution of calcium and phosphorus
in the average adult human*

	Percentage of total	
	Ca	P
Skeleton	99	85
Teeth	0.6	0.4
Soft tissues	0.6	14
Plasma	0.03	
Blood		0.3
Extravascular fluid	0.06	0.3
Total weight	1300 g	700 g

From Nordin, 1976.

difficult to trace to calcium phosphate and there is lack
of information about the actual formation of the salt in
bone, a general diagram for the metabolic pathways of
phosphate and their likely connection with bone mineral
is illustrated in Fig. 3.

Inorganic Inorganic phosphate is found in animals predominantly as the calcium salt. There is a proportion of
carbonate associated with the phosphate and other elements
which may replace cations and anions. Of the cations,
magnesium is the most significant inclusion, possibly
replacing calcium. The proportion of magnesium associated
with bone salt seldom exceeds 1%, in spite of the fact
that the element is 60% of the abundance of calcium in the
lithosphere.

Calcium phosphate is widely distributed in all animals
in varying amounts: in most invertebrates it is contained
within the larger bulk of carbonate whereas in the vertebrates the reverse is the case. The location, quantity
and arrangement of the salt ranges from microscopic, sub-
cellular non-crystalline deposits to hard tissues almost
entirely composed of well-defined species of mineral in
assemblies of relatively large characteristic crystals.

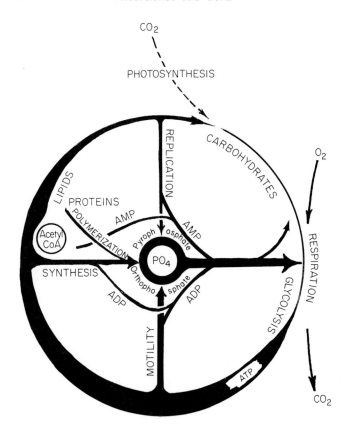

Fig. 3. Metabolic pathways of phosphate.

Where a diffuse, non-specific precipitate becomes a cytoskeleton and collections of cells become associated with
sufficiently large amounts of salt to constitute a skeleton is a debatable point which has been considered elsewhere (Pautard, 1970). The overall evidence does suggest,
however, that the vertebrate skeleton is no more than a
more highly organized expression of a common theme of
calcium phosphate manipulation which can be traced in
animals past and present from the Protozoa onwards. While
the simpler and earlier forms are not 'bone' in the anatomical sense, the resemblances at the molecular level

are close enough to suggest that the chemistry and biochemistry is likely to be similar throughout all forms of
life, even though the multicellular patterns that arise
may be widely different. This same likelihood applies to
calcium carbonate and in the interposition of the two
skeletal salts there is no clear dividing line; phosphate
may appear in one animal where all related species are
hardened with carbonate; or the same type of organic
matrix may be mineralized with carbonate or phosphate in
different animals (Pautard, 1966). Even in the mammals,
where bones and teeth are the principal site of calcium
phosphate, the same salt is found hardening hairs, hoofs,
claws and feathers, all of which have a fibrous matrix of
keratin (Pautard, 1963). Calcium carbonate is also found
in some parts of some vertebrates in greater proportion
than in normal bone: in medullary bone, otoliths and in
some keratins (Pautard, 1966). In the invertebrates, the
situation may be reversed; in the Crustacea, for example,
the otoliths and tendons may be mineralized with phosphate
(Pautard, 1966, 1976) when the bulk of the inorganic
material in the carapace is the carbonate.

Calcium phosphate is therefore to be found in many
places in many animals and such crystallographic and ultrastructural studies that have been reported suggest that
the types of deposit fall generally into three categories.

1. Larger crystals of the 'enamel' type usually found
in exterior tissues of epidermal origin.

2. Smaller crystallites of the 'bone' type with elongated c axes but with a width usually of the order of 50 Å.
These entities are usually found in endoskeletons and
often within cells.

3. Non-crystalline deposits of varying chemistry,
usually within cells and often in subcellular organelles.

The general relationships of these three types of
phosphate are summarized in the diagram in Fig. 4.

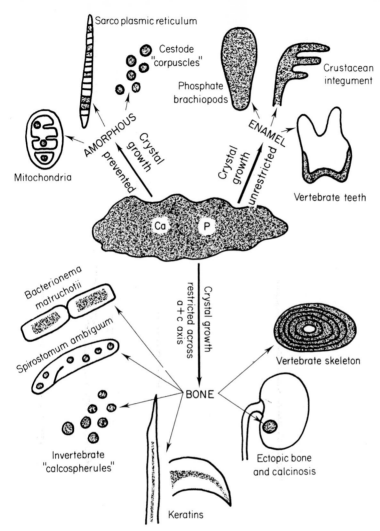

Fig. 4. General distribution of calcium phosphate in animals.

Phosphate in Bone

The nature, genesis and function of calcium phosphate in bone has been the subject of major controversy for three decades. Since the improvement in X-ray diffraction and infra-red absorption, and the parallel development of the electron microscope, the literature has been dominated by a large number of opinions about bone, a smaller

discourse about tooth enamel and little about calcium
phosphate elsewhere.

While there is no great disagreement about the nature
and formation of tooth enamel, bone has remained an un-
settled subject, partly because of metabolic changes of
medical importance but mostly because of the diverse disciplines employed in its study. The roots of the debate
lie in opposed interpretations; the chemistry and biochemistry carried over and extended from an earlier period,
the crystallography of the inorganic phase and the substances and forms, inside and outside the cell witnessed
in the electron microscope. What has been concluded from
sparse lattice information of uncertain stoichiometry has
been correlated with a general chemistry. What has been
isolated by biochemistry has been correlated with the
ultrastructure. What has been seen in the electron microscope has been correlated with the crystallography. All
these comparisons have been equated, one way and another,
with numerous physiological experiments calculated to
provide chemical evidence and biochemical reason for the
transport of the calcium and phosphorus in bone salt.

The problem of reliable information about bone is not
made easy by the difficulties in preparing the tissue.
Bone is a living structure; the apparently 'dead' organ
after removal conceals a sensitive biology which reacts
rapidly to insult. In its mature state, bone is exceptionally compact and impervious. Long after the cell
population has died, even after fixation, the biochemistry
may continue to function. Bones do not decay rapidly and
it is understandable that they are often regarded as inert
rocks. Experiments with living bone are restricted by its
position *in situ* and by its behaviour *in vitro*. As a
result the information about bone mineral and its formation has been deduced from dead and static states. The
mineralization of bone *in situ* has not yet been described
adequately with the optical microscope although there are

many descriptions of the event after fixation, sectioning and staining. Similarly, there have been numerous measurements and calculations of the structure of bone salt from the crystallography of oriented samples or dry powders, but so far, with one doubtful exception, X-ray diffraction patterns of living bone have not been reported.

These doubts have been reflected in recent literature, and an increasing awareness has resulted in greater caution about the nature of bone salt, a diminished confidence in gross chemistry, and a reappraisal of the methods of electron microscopy. A summary of the problems of interpretation and the conflict of views as a result of methodology is set out in Table IV.

Composition and Structure of Bone

The dry skeleton of a 70 kg adult contains about 60% of mineral and 40% of organic matter, of which the bulk (30% or thereabouts) is the fibrous protein collagen. The remaining organic substances consist of lipids, polysaccharides and proteins other than collagen, but since the bone cells pervade the mineral mass with many fine filaments, it is difficult to determine which of the non-collagenous components are associated with the bone salt and which are associated with the cells.

The proportion of inorganic and organic substances varies from bone to bone and from subject to subject, but the gross analysis of each of the components falls within certain limits. That is, a given bone in say, man, may have differences in the amount of calcium phosphate and the amount of collagen, a relationship which might be greatly altered during disease or deficiency, but the analyses of salt and protein are similar, and the attendant lipids, polysaccharides and proteins other than collagen vary only in conditions of extreme pathology. A similar distribution of elements and molecules is found in dentine, a mesodermal tooth tissue resembling bone

Interpretation and limitations in the

Method	Observed Composition	Observed size of crystal
Chemistry	Basic calcium phosphate with 5% carbonate and Mg, Na, K, Fe & F	
X-ray diffraction	Ca-deficient CO_3/HA OCP/HA/CO_3 mixtures TCP/HA/CO_3 mixtures	< 100 Å wide elongated and oriented c-axis
Infra-red absorption	CO_3 probably replaces PO_4 in lattice. OCP + HA	< 100 Å wide elongated and oriented c-axis
Electron microscopy		Diffuse haze, dense structures up to 200 Å × 350 Å. General shape 50 Å wide, various lengths
Electron diffraction	HA in selected regions	> 100 Å in selected areas

ACP amorphous calcium phosphate   HA hydroxypatite

except in its anatomy. On the other hand, enamel, the hard, thin epidermal cap of most vertebrate teeth not only differs greatly from bone in its anatomy, but also in the composition of the sparse matrix, although the mineral resembles that of bone but of greater crystallinity. A comparison of bone compositions with dentine and enamel is shown in Table V. Some minor elements are not listed but total less than 1% of the tissues.

Chemistry of the inorganic phase

The gross chemistry of bone mineral has been extensively reported but there is little reliable evidence as to the elements which are present in the actual inorganic

IV

study of bone salt by different methods

Physical state	Growth	Limitations
	Synthetic models of ACP-HA transformation favour TCP as the first salt	Does not distinguish size and species
Lyophilized bone is about 50% crystalline 50% amorphous	Earliest deposits amorphous	Can evaluate only crystalline species
Lyophilized bone is about 50% crystalline 50% amorphous		Partial information only on structure
	From sites on collagen or as clusters of dense filaments	Damage and distortions by electron beam. Loss of water.
Crystalline in some regions, amorphous in others	Crystal sites on collagen. Early 'nodules' are amorphous	Damage and distortions by electron beam. Loss of water.

OCP octacalcium phosphate   TCP tricalcium phosphate

moiety, instead, that is, of being associated with the outer shell of the mineral or with surrounding organic substances and cells. The contact between collagen fibres and other organic molecules and the calcium phosphate is so intimate that it is difficult to prepare mineral free from contamination. Methods that have been developed which depend on the complete removal of organic substance by chemical means to leave bone ('anorganic bone') with low values of nitrogen usually result in severe physical changes which might be reflected in the chemical composition. Similarly, methods which have attempted to remove the organic matter by selective enzyme treatment have made little progress because of the impermeability of

TABLE V

*Inorganic composition of bone, dentine and enamel
from adult humans*

	Percentage of total dry fat-free weight		
	Bone	Dentine	Enamel
Ash	57.1	70.00	95.7
Ca	22.5	25.9	35.9
P	10.3	12.6	17.00
CO_2	3.5	3.17	2.35
Mg	0.26	0.82	0.42
Na	0.52	0.25	0.55

From Zipkin, 1966.

the bone matrix.

Two current methods for the analysis of the mineral composition are ashing at favourable temperatures (which rearranges the crystallinity and the relative ion affinities) and demineralization with acid or chelating agents (which preferentially remove the inorganic ions while leaving covalently-bound elements attached to organic molecules). Neither method is entirely satisfactory. Ashing does not discriminate between any of the elements in the inorganic phase and in the case of phosphate it does not allow any conclusion as to the state of the ion. Similarly, demineralization procedures depend largely on the nature of the agent used and the way in which the tissue is treated. Bowden (1960) showed, for example, that extraction of bone with trichloracetic acid or hydrochloric acid produced different values in the analysis for citric acid, which must reflect, by cation binding, differences in the release of elements and hence the gross composition of the 'inorganic' phase.

Various analyses for the same bone, and for different bones, using different methods of preparation, are shown in Table VI. There are variations in these estimations, but it is not possible to establish whether they may be explained by errors in methodology or by biological

TABLE VI

Inorganic analyses of bones by different authors and on different subjects. The relative percentages are compared on a wt/wt basis

Sample bone	Percentage dry, fat free					
	Ash	Ca	P	CO_2	Mg	Na
Human rib						
Follis (1952)						
Age 2.5 years with marrow (BM)	–	22.67	10.67	2.8		
without " (B)	–	25.56	11.86	3.05		
17 " (BM)	–	22.95	10.75	2.72		
(B)	–	25.12	11.49	3.09		
37 " (BM)	–	22.87	10.51	3.04		
(B)	–	25.75	11.71	3.43		
66 " (BM)	–	23.11	10.71	3.18		
(B)	–	25.62	11.60	3.73		
Agna *et al.* (1975)	–	25.0	10.45	3.26		
Zipkin *et al.* (1960)						
Fluorine 1 ppm	56.04	21.64	9.80	3.03	0.28	0.44
" 4 ppm	57.38	22.01	10.18	2.98	0.34	0.39
Various bones (from Zipkin, 1970)						
Human sternum	52.1	19.3	9.0	2.3	0.22	0.33
iliac crest	50.4	21.9	9.8	4.0	0.28	0.51
vertebra	52.6	19.4	9.3	2.6	0.25	0.43
Ox femur	75.0	25.8	11.9	5.8	0.50	–
Rat femur	67.3	24.4	12.3	2.76	0.60	–
Turkey femur	70.0	26.5	12.6	4.5	–	–

differences in the sample, or by a combination of both.

Crystallography of the inorganic phase

Laue diffraction patterns of bone consist of a few diffuse reflexions with the arc of the 'apatite' 002 spacing transverse to the fibre axis (Fig. 5a). The pattern (in this case of a fish rib bone) suggests crystallites of small width (about 50 Å) and of indeterminate length preferentially oriented parallel to the fibre axis

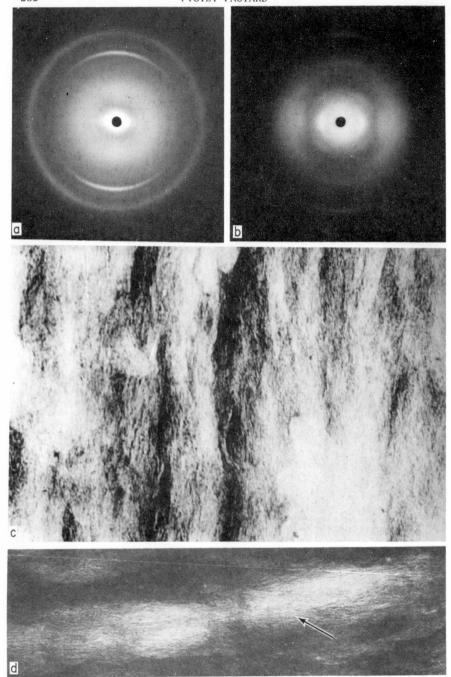

of collagen (Fig. 5b). Such lattice dimensions that can
be calculated from the spacings visible in native bone, or
enhanced after heating, are consistent with a unit cell
similar to that of hydroxyapatite (Fig. 1). Since de Jong
first observed these general relationships in 1926, there
has been an increasing volume of literature seeking to
refine the crystallography in relation to the chemistry.
Unfortunately, the idealized apatite lattice, based on
single crystal analysis (cf. Fig. 1) is a realistic model
for biological apatite only on the assumption that bone
is a homogeneous distribution of ions and the chemical
values for each element can be apportioned equally to
each crystallite in the structure. In the case of enamel,
electron diffraction patterns allow direct comparison of
individual crystals with various apatite species. In the
case of bone, the small size of the crystallites allows
no such comparison. Indeed, the entire conclusion as to
the nature of bone salt has depended on the tacit assumption that because bone mineral is apparently a random precipitate, albeit perhaps with cellular control, the crystal beds in the tissue may be treated statistically in
the same way as an inorganic preparation in the laboratory.
     The immediate consequence of the analyses of bone min-
eral such as those in Table VI has been a concerted
attempt to fit the proportions of cations and anions into

Fig. 5 (opposite). Features of calcium phosphate in rib bone from the
bream *Abramis brama*. (a) X-ray diffraction pattern, rib axis vertical,
CuKα radiation, nickel filtered, 4 cm specimen-to-film distance. The
broad arc of the 002 reflection is across the axis, suggesting that
the c axis of small crystallites is parallel with the rib axis.
(b) Sample as in (a) after decalcification with EDTA for 24 hours.
The pattern of calcium phosphate has been replaced by the features
of collagen with the fibre axis vertical. (c) Electron micrograph,
printed as a negative to show the elongated mineral particles lying
generally parallel to the rib axis, which is vertical in the illustration, X 64,000. (d) Electron micrograph of a region from the same
sample as (c), printed as a positive to show an area where the collagen fibres have been exposed to reveal the characteristic periodicity
of about 640 Å (arrowed). The rib axis (and the collagen fibre axis)
is horizontal in the illustration, X 56,000.

TABLE

Models of the mineral

Model	Formula
Hydroxyapatite	$Ca_{10}(PO_4)_6(OH)_2$
Tricalcium phosphate hydrate and calcium carbonate	$Ca_3(PO_4)_2 \cdot 2H_2O + CaCO_3$
Carbonate apatite	$Ca_{10-x-u}(PO_4)_{6-x}(CO_3)_x(OH)_{2-x-2u}$
Calcium-deficient apatite, hydrogen bonding between orthophosphate oxygen	$Ca_{10-x}H_{2x}(PO_4)_6(OH)_2$
Octacalcium phosphate lamellae with hydroxyapatite	$Ca_8H_2(PO_4)_6 \cdot 5H_2O + Ca_{10}(PO_4)_6(OH)_2$
Calcium-deficient apatite, hydrogen bonding reduced, hydroxyls missing	$Ca_{10-x}H_x(PO_4)_6(OH)_{2-x}$
Amorphous calcium phosphate and apatite	$Ca_3(PO_4)_2 \cdot 2H_2O$? $+ Ca_{10}(PO_4)_6(OH)_2$

a model of hydroxyapatite, particularly with respect to
the carbonate which is present. The Ca/P molar ratio of
an ideal hydroxyapatite should be 1.67 and the observed
value does, in fact, fall between 1.57 and 1.74, but since
a proportion of the lattice must accommodate carbonate
if the structure is a carbonate apatite or calcium carbonate is a separate phase, a subtraction of the balancing
Ca^{2+} leaves the Ca/P ratio less than the correct value.
There have been several explanations for the discrepancy.
There may be more phosphate, less calcium or the lattice
may be mixed in some way. The evidence for and against
each of the hypotheses that have been proposed is set out
in Table 7. Of the various possibilities among the main

VII
phase of bone

Evidence	Contra-evidence	General reference
X-ray diffraction	Carbonate present. Chemical composition not equivalent	de Jong (1926)
Chemically acceptable	X-ray diffractions not identical	Dallemagne (1945)
X-ray diffraction	Carbonate on surface (Elliot, 1965)	McConnell (1952)
Infra-red spectra	Pyrophosphate at 600° too low. (Berry and Leach, 1967)	Posner and Perloff (1957)
Correct composition X-ray diffraction equivalent pyrophosphate acceptable	Infra-red and thermal analysis not consistent (Berry and Leach, 1966)	Brown *et al.* (1962)
Infra-red spectra Pyrophosphate values		Winand (1965)
X-ray diffraction intensities and broadening		Termine and Posner (1967)

theories, the calcium-deficient lattice of Winand (1965) is in best agreement with the experimental data presented so far. The merits of the arguments have been examined by Eanes and Posner (1970) and will be referred to later in relation to the substructure of bone and the position of phosphate and carbonate in it.

Amorphous calcium phosphate

Although it has been suspected for a long time from crystallographic (e.g. Pautard, 1964) and ultrastructural studies (Robinson and Watson, 1955) that not all the mineral in bone might be crystalline, an important step in the understanding of bone salt stems from the studies of

Harper and Posner in 1966 and Termine and Posner in 1967.
These authors first established that the X-ray diffraction
pattern of bone hydroxyapatite was insufficient to account
for all the mineral present. Many subsequent studies have
established that a proportion of bone salt does not give
a detectable X-ray diffraction pattern. Whatever the definition of 'amorphous' and 'random', 'non-crystalline'
and 'sub-crystalline', one aspect of the evidence is dis-
turbing. This is that the percentage of crystallinity in
a sample depends on the way in which it is prepared.
From the data presented in Table VIII it is clear at once
that any crystallography of bone in relation to the chemistry must be regarded with caution. With more recent
statements by Termine (1972) and the doubtful crystallinity of some bone *in situ* (Pautard, 1972) there is now no
clear evidence as to the true structure of calcium phosphate in living bone. The chemistry and physics of the
dry, partly-crystalline material that has been examined
does suggest that the most likely species of calcium
phosphate with an observable lattice is an apatite and the
salt which becomes crystalline after drying may have been
sub-crystalline before treatment. The facts of change,
however, make uncertain any deduction about the extent
to which this is relevant to the structure of the salt.
If the post-mortem changes are those of crystal growth
alone, the lattice calculations are reasonably valid.
If, however, the post-mortem changes involve biochemical
and chemical change (as now seems to be the case in some
instances) then the lattice calculations are not meaning-
ful.

Tissue structure

Bone is made, maintained and continually repaired and
remodelled by large numbers of mesodermal cells. In the
'simplest' types of bone such as the acellular tissue of
fish scales, cells manufacture the structure but are not

TABLE VIII

Changes in the amorphous/crystalline ratio of bone salt in mature rabbit femur after different methods of preparation

Method of preparation	% Crystalline apatite	% Amorphous calcium phosphate
Air dry 25° C	76.2	23.8
Air dry, 110° C	70.1	29.9
Air dry, 250° C	71.4	28.6
Alcohol-ether; air dry, 50° C	63.1	36.9
Alcohol-ether; vacuum dry, 50° C	57.0	43.0
Technical grade ethylenediamine; H_2O wash; air dry, 110° C	77.5	22.5
Technical grade ethylenediamine; vacuum dry, 100° C	69.6	30.4
Analytical grade ethylenediamine; H_2O wash, air dry, 110° C	65.3	34.7
Analytical grade ethylenediamine; vacuum dry, 100° C	56.9	43.1
Lyophilization	56.0	44.0
Anhydrous liquid grinding	54.9	45.1

(From Termine and Posner, 1967.)

contained within it. In mammalian bone, the cells remain in contiguity in various patterns and at different distances within the dense mineralized regions throughout the life of the tissue. The extent to which bone cells participate in the translocation of the mineral to those areas where the collagenous matrix is calcifying is a matter of as much scattered opinion as the nature of the bone salt itself. The main cause of disagreement is the extremely complex activity of cells during bone growth, the variation between bone cells once growth has ceased and the wide panorama of histological changes which take place in bone cells and their environment as each part of each bone ages and responds to stress, injury and disease.

At any one time, therefore, the fabric of bone is an equilibrium between new bone forming, old and damaged bone being removed and intermediate conditions of repair and reorganization. The precise arrangement of cells in the mineralized matrix depends on the position and func-

tion of the tissue. Cells and mineral may be arranged in
sheets (laminar bone), in sponge-like assemblages of bony
bars (trabecular bone) or in compact, abutting systems of
concentric layers (haversian bone). A bone may be composed of one or more of these types in any combination or
proportion, but the basic cell/mineral arrangement,
irrespective of the anatomical pattern dictated by the
organ, is the same in every case — layers of cells interposed by layers of mineralized matrix. When the bone regions are mature, these cells (the osteocytes) are en-
tombed completely within their environment of calcium
phosphate and collagen and continue to function for the
life of their part of the tissue, which may be days, weeks
or years. For their support, the osteocytes are serviced,
through their fine cytoplasmic filaments in tunnels (the
canaliculi) in the matrix, by an extensive vascular
system which supplies oxygen, metabolites and other substances and which removes waste products. A photomicrograph of bone showing the various types of cellular
pattern is shown in Fig. 6 and each feature is enlarged
into a diagram to illustrate the relationships.

In any gross chemical analysis of bone mineral, the
variable nature of bone must be considered; current views
about cytological comparisons have been comprehensively
reviewed by Aaron (1976).

Ultrastructure

The present uncertainty about the nature of bone salt
and its association with the organic matrix has been enhanced in part by the evidence gathered from electron
microscopy over the past two decades. Bone is a most
difficult tissue to prepare for the electron microscope.
It must be fixed rapidly, sectioned with as little disturbance as possible and stained to make specific features
visible. In the undecalcified state, none of these steps
can be carried out without the risk of errors — fixatives

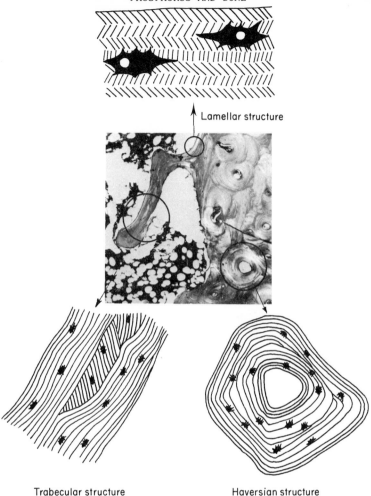

Lamellar structure

Trabecular structure   Haversian structure

Fig. 6. The general structure of bone.

and stains do not penetrate the dense mineralized regions
and the hardness of the material makes sectioning, even
with diamond knives, liable to artifacts of displacement.
The techniques of optical microscopy have changed little
and are referrable to the living state by direct observation. The techniques of electron microscopy have been
developed largely for soft tissues and the results are not
referrable to the living state except through the optical
microscopic equivalent for which, unfortunately, there is

no reliable comparison in bone. There have arisen, in the
study of bone, separate methodologies for each instrument
and its related discipline − confirmation of physics and
chemistry at the ultrastructural level on the one hand
and comparison of living and dead situations on the other.

Within the limitations of the method, however, there
is some agreement at the ultrastructural level as to the
dimensions of the bone mineral. The general appearance of
bone in thin section (shown in Fig. 5c) verifies that the
dense mineral particles are generally of the same width
and oriented in the direction of the collagen fibres
(Fig. 5d). There have been numerous descriptions of bone
'crystallites' (see Ascensi and Bonucci, 1972, for recent
discussion and references); among authors there is a wide
acceptance that the mineral particle is a cylinder with
a width of about 50 Å but with some indication of lateral
aggregation to give platelets up to 200 Å wide across one
axis. All these observations confirm the sizes predicted
by crystallography.

The location of the amorphous component of bone is
less certain. There have been scattered reports of 'hazy'
or 'diffuse' deposits from the earliest description by
Wolpers in 1949, but the identification of such features
with non-crystalline structure is uncertain. Even after
more reliable methods of tissue preparation (for example,
freeze-sectioning − Gay, 1977) the images which appear in
the electron microscope still leave us with no clear idea
as to how we might relate the physical evidence with what
is seen at the ultrastructural level. Some general conclusions from electron microscopy of bone mineral are set
out in Table IX.

Metabolism

Although it has been known since the declaration by
Tomes and de Morgan in 1853 that bone undergoes continual
changes of resorption and remodelling, it was not until

TABLE IX

Some general conclusions from electron microscopy of bone

Cell	Calcification front	Mature bone	Mineral detail	Conclusions
Featureless	Patches of mineral appear in association with collagen. Crystals develop on specific sites.	Mineral in the form of elongated crystals generally oriented with collagen.	Crystals about 50 Å wide, or as plates.	Mineral nucleates on collagen outside cells, for example Hohling *et al.* (1974)
Featureless	Amorphous deposits of bone appear in clusters up to 1μ in relation to vesicles.	Mineral is mostly random and continuous; some alignment with collagen.	Dense features about 50 Å wide surrounded by a diffuse haze.	Mineral commences as unoriented patches of bone salt outside cells, for example Gay (1977)
Osteocyte 'loads' with mineral in juxtanuclear region.	Spherical structures 0.1–1μ appear and proliferate in the seam space.	Spherical structures distorted into domains with boundaries. Some alignment of mineral.	Width (about 50 Å) of mineralized filaments does not vary with density. Substructure is present.	Mineral commences inside cells, migrates and matures in matrix, for example Aaron (1978)

1958 that the studies of Mclean and others made it clear
that bone not only exchanges calcium and phosphorus with
the soft tissues, but also monitors the amount of the
elements stored and released through a chain of intermediate agents. Bone turnover depends on many factors:
growth and decline; insult and recovery; nutrition and
excretion; stress and weightlessness, infection and prophylaxis. In any of these events, the skeleton responds
to the demands placed upon it by changes in the concentration of calcium and phosphorus, from soluble ions in
the osseous vascular system to insoluble calcium phosphate
from the densest regions of compact bone. In time, the
factors of physical stress (or lack of it), metabolic
imbalance and faulty translocation may produce permanent
damage to the skeleton with consequent deterioration of
its exchange function and mechanical failure of weakened
areas.

There have been numerous studies of bone turnover and
most of them have concentrated on the exchange of calcium.
This is not because phosphorus is mobilized less from the
skeleton, but because calcium is easily measured, has
specific locations and pathways and is less intimately
connected with the complex biochemistry which demands
much larger quantities of phosphorus. Although an interest
in phosphorus in relation to bone metabolism is increasing, there is an inclination to regard calcium as the
clinically significant ion ('calcium homeostasis') in
metabolism. However, although calcium malfunction does
have profound medical effects, it is clear from the extensive literature that phosphate is essential to the transport of calcium throughout biological structures and it is
surprising that much of the methodology familiar to soft
tissue physiology and biochemistry is not in common use in
bone studies.

One reason for the neglect of phosphorus in metabolic
experimentation may be that until recently, the bio-

chemistry of the bone cell has been separated from the
mineralized substance of the matrix. It has been generally
assumed that although the hardened fabric in a bone is
inhabited by colonies of cells, the 'working life' of the
tissue is characterized by a cellular biochemistry (enzyme activity, respiration, synthesis and lysis) only
during formation and resorption, but in the period between
(which may be several years) calcium and phosphorus exchange is a passive process. There is, however, increasing evidence that the intercellular biochemistry may out-
weigh the cellular biochemistry for much of the life of
bone, with the result that any appraisal of the role of
phosphate (which may be directly concerned in the biochemistry) will depend on the contact between cells and
matrix during formation and resorption, and the activity
within the matrix independent of the cells, or in association with them, during exchange.

Formation

   Bone is laid down either by migrating cells which mineralize a cartilaginous model, or by sheets of cells
which, in concert, fabricate a web of collagen and polysaccharide (the osteoid) in well-defined, optically-
translucent borders (the seam space) which become progressively filled with calcium phosphate in a characteristic way (the calcification front). The actual way in
which the mineral develops in the matrix is central to
any understanding of the nature of bone in general and
the role of phosphate in particular. It is therefore helpful to illustrate the overall event and to place the well-
documented anatomical events in the context of current
theories as to how, at the molecular level, the calcium
phosphate is laid down and what happens to it afterwards.
   A characteristic seam space, stained with toluidine
blue to show the calcification front, is shown in the
photomicrograph in Fig. 7. Diagrams of participating cells

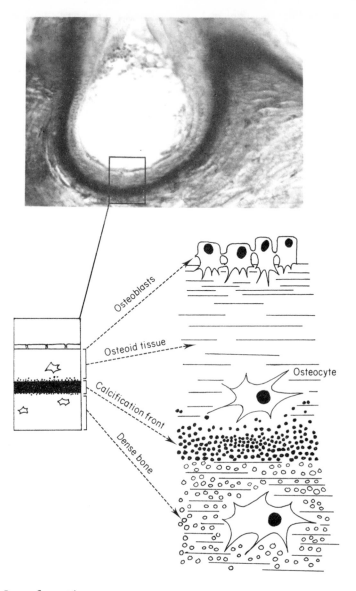

Fig. 7. Bone formation.

and their progress are drawn in relation to the picture
to illustrate the steps in mineralization. The development of dense bone from the mesothelial cell progenitors
follows much the same path in every type of bone. Only

the details and the final pattern vary according to the
tissue. First the osteoblasts are differentiated to form
sheets which produce quantities of collagen and polysaccharide filling the space between the new cuboidal
cells and the old mineralized front. Next, the osteoblasts gradually change into osteocytes with filaments
filling the canaliculi which interlace the new osteoid.
Mineral begins to appear in discrete clusters at a distance from the edge of the cells and the intervening gap
is usually free of mineral. Then, as the osteocyte matures,
the filamentous extensions increase, with the canaliculi,
in number and complexity and the intercellular matrix
fills with mineral until it is adjacent to the cell and
reaches its final density.

To account for the sudden appearance of calcium phosphate outside the cell, the changing history of techniques which have been used to study the phenomenon have
produced corresponding changes in the ideas about bone
formation, Table X. Since all of these ideas are relevant,
in their way, to each aspect of calcium phosphate and
are significant, in the discussion to follow, to the role
of phosphate in the system, a brief description of the
main thread of the evidence is of importance.

Alkaline phosphatase

Robison discovered the enzyme alkaline phosphatase in
1923 in calcifying cartilage and later observed that
rachitic cartilage calcified *in vitro* in the presence of
phosphoric esters. The theory inevitably developed that
calcification results from the simple precipitation of
calcium phosphate from phosphate ions released by local
enzyme action in the presence of calcium. For a long time,
the idea of a cell-mediated calcification was accepted
and even when serious objections were raised, the theory
was vigorously defended. The principal obstacles to
Robison's theory were that the pH value of the maximum

TABLE X

Evidence, contra-evidence and defence
for theories of mineralization

Evidence and support	Contra-evidence	Defence
Alkaline phosphatase		
Found in calcifying systems, absent in rickets, bone forms *in vitro* with substrate	pH too high for serum fluids	Not applicable if intracellular
	Concentration of esters in serum fluids too low	Not applicable if intracellular
	Found in soft tissues	May be connected with a fundamental calcium phosphate cycle
Epitaxy		
Bone recalcifies in metastable solution	Bone salt is amorphous, therefore nucleation is not possible	The salt may really be sub-crystalline and still 'nucleated'
Bone collagen, and reconstituted collagen, calcifies	Bone salt found in keratins with no collagen present	Some group or side-chain common to all calcifying proteins and polysaccharides might be present
	No orientation *in vitro*	Perhaps a secondary medical effect
Globules and vesicles		
Seen in calcifying cartilage, contain bone salt, bind calcium, contain alkaline phosphatase	Not all calcify	A matter of certain types only
	No crystals present at early stage	Calcified outside the cell or dissolved in preparation
	Not seen in bone	Short time before mineral accumulation
Mitochondria		
Bone salt found ultrastructurally, present during pathological calcification	Not seen outside the cell	Might be modified or degenerate
	No optical evidence for bulk of intracellular mineral	Might be transferred or restricted
Intracellular synthesis of mineral		
Optical evidence with calcium and phosphorous stains and tetracycline	Mineral extracellular	Optical evidence positive. Map clearly shows intracellular distribution

TABLE X cont.

Evidence and support	Contra-evidence	Defence
Map shows 'switch on' system Soft tissues included Electron micrographs of clusters within cell similar to structures outside osteocytes and inside other cells in other tissues	No electron microscope evidence except in the osteoclast 'Vesicles' that are seen are devoid of structure Epitaxy not possible if mineral already formed	Deposits soluble. Errors due to use of uranyl acetate. Silver image is positive Contents lost during chemical preparations Secondary phenomenon after leaving the cell, or artefact due to rapid crystal formation after death

activity of alkaline phosphatase was too high for normal physiological function and the amount of phosphoric ester in serum fluid was too low to permit any appreciable precipitation of calcium phosphate at a given site. Again, alkaline phosphatase is present in many tissues which do not calcify. Interest finally diminished when Pritchard showed in 1952 that alkaline phosphatase is present in the mesenchyme even before the osteoblasts differentiate. The enzyme appears first in the nucleus of the cell, then in the cytoplasm as the osteoblast develops and later in the extracellular matrix. It has been assumed since that this sequence of events cannot be closely connected with mineralization, even though the enzyme is always present during bone formation and is absent in the extracellular matrix in rickets. As we shall see later, the distribution of alkaline phosphatase may indeed reflect the sites of calcium phosphate accumulation which may be confined neither to bone nor to bone cells.

Epitaxy

The two properties of bone mineral, namely small crystallites of regular size oriented with respect to collagen, already anticipated by the optical studies of Schmidt in 1923 and verified by X-ray crystallography and electron

microscopy twenty years later, persuaded many authors to
propose new theories as to how the mineral and the fibres
were put together. To explain the organized association
between calcium phosphate and protein, the idea of
epitaxy was developed.

Epitactic models of calcification in bone all depend on
two important features. The first is that calcium and
phosphate in serum fluids are always metastable with
respect to bone salt and hence simple precipitation from
ions is not thermodynamically feasible. Under these conditions, crystals of apatite (or other phosphates) can
form only where there is a site of correct geometry to
initiate a lattice of given dimensions. The second fact
is that most experimental evidence suggests that decal-
cified bone, or collagen alone or in combination with
polysaccharide, will initiate crystallization at Ca/P
products below those required for spontaneous precipitation.

Epitactic mechanisms such as those proposed by Strates
and Neuman (1958), Glimcher (1959), Sobel (1964), Bachra
(1964) and others all involve stereotactic sites in some
part of the organic matrix. These proposals are particularly attractive since they explain the sudden appearance
of mineral in the extracellular seam space and they are
consistent with the equilibrium conditions between serum
fluids and bone salt. Moreover, the appearance of regular
arrays of crystals on collagen fibres has been reported in
bone (Fitton Jackson, 1957) and in mineralizing turkey
leg tendon (Nylen *et al.*, 1960). Whatever explanations are
put forward for the genesis of bone salt, they must take
into account the observed orientation of the crystallites,
their sites on collagen fibres and the specific nature of
the association of the mineral.

Globules and vesicles

A recurrent theme in more recent theories of mineralization, related and unrelated to epitactic mechanisms,
is the part played by spherical, membrane-bound bodies of
cellular origin which are found in numbers in the extracellular space during the deposition of bone salt. These
objects were originally observed in 1967 in calcifying
cartilage by Bonucci, who called them 'globules' and
established that they originated from a parent cell and
contained structure and biochemical properties. Later, in
1969, Anderson observed similar objects in developing
endochondral bone and suggested that the accumulation of
crystals within these 'vesicles' might be connected with
calcium phosphate deposition, thus relating the intracellular precursors to the epitaxy previously ascribed to
collagen. The property of calcifiability was further
emphasized by Slavkin and others in 1972, when similar
vesicles from dentine were found to contain, besides
calcium phosphate, various enzymes, including alkaline
phosphatase. The persistent association of alkaline phosphatase, Robison's calcification enzyme, with vesicles
has been detailed by de Bernard (1975) who has isolated
an aggregable glycoprotein which binds calcium and be-
haves as a phosphatase.

While caution is required in any assessment about
vesicles because they are a common feature of most cells
and their environment, their presence in a mineralizing
region may be significant, although most authors who hold
this view consider that they are merely the agents for
concentrating calcium phosphate outside the cell in relation to the epitactic sites on the collagen fibre.

Mitochondria

Another feature associated with the deposition of
calcium phosphate within cells is the special role of the
mitochondrion in the process. It is well known that under

certain pathological conditions mitochondria in animal
cells (Lehninger, 1965) and plant cells (Hodges and
Hanson, 1965) will accumulate dense deposits of calcium
phosphate. While this phenomenon has not been observed
so far in forming bone, it is certainly a feature of
calciphylaxis and in chondrocytes undergoing hypertrophy
(Martin and Mathews, 1969).

However, in considering the possible role of the mitochondrion and the vesicle as cell-directed sources of
calcium phosphate, there is some inconsistency between
the various observations. The mitochondrion is essentially
an organelle contained within the plasma membrane of the
cell; the vesicle is a cell-made structure frequently
found outside the cell. The association of calcium phosphate with the mitochondrion and the vesicle is obverse
to their location, at least as far as the ultrastructural
observation is concerned. That is, the calcified deposits
of mitochondrial origin are all found inside the cell and
are not transported to the calcification front. Such
accumulations of salt may be of significance in the intracellular concentration of calcium and phosphorus (Mathews
et al., 1971) but they may have little relationship with
the extracellular matrix. On the other hand, the vesicles
are usually empty until they reach the calcification
front, when they will presumably concentrate calcium and
phosphorus from the surrounding matrix. It would seem to
make better sense if the mitochondria within the cells
were transporting (invisibly) the calcium and phosphorus
within the cell and the vesicles were transporting
(visibly) the calcium and phosphorus out of it. An explanation for this anomalous state of affairs may be that
the means of preparing bone tissue may precipitate calcium
phosphate in the mitochondria and remove it from the
vesicles, a likelihood which will be discussed later in
relation to cell behaviour.

Calcification by bone cells

The current interest in vesicles and mitochondria is
one aspect of a growing body of evidence that the bone
cell is involved in the mineralization of the matrix. The
earlier studies of Robison (1923) implicated cells in the
formation of calcium phosphate but the precise relationship was obscure. Most important to the origins of bone
mineral is the serious discrepancy in the literature between the observations which have been reported by optical
microscopy and those which have been reported by electron
microscopy. The studies of Bohatirchuk in 1965, Kashiwa
in 1966 and Rolle in 1969 all suggest that during bone
development, the optical microscope with its attendant
staining techniques clearly indicates that calcium and
phosphate, combined in some discrete form, are present
within bone cells. At the same time, contemporary studies
by electron microscopy of similar regions of bone failed
to confirm the presence of calcium and phosphate at the
ultrastructural level.

An explanation of these anomalies is essential. If the
calcium phosphate is deposited *outside* the cells, then on
the assumption that the extracellular space is biologically 'inert', a calculation as to the equilibrium between
bone salt and extracellular fluids is likely to be valid.
If the calcium phosphate is deposited *inside* the cells,
then since there is, at present, no way of knowing what
the equilibrium conditions are when the salt is formed,
although it might still be likely that once the formed
salt is extruded to the calcification front it maintains
equilibrium with the extracellular fluids. Recent optical and electron microscope studies by Aaron (reviewed in
1976) have offered an explanation of the differences
between the optical and electron microscope and further
evidence as to the site where calcium phosphate may be
assembled within the cell. Observations of the developing

mouse calvarium suggest that during bone formation the
cell primarily involved in mineralization is the osteocyte, which 'loads' at some stage of its development,
first with calcium, then with phosphate in a complex
series of events until, within the saccules of the golgi
apparatus, characteristic structures are formed which
later appear in the extracellular space. If the tissue is
prepared by methods recommended for electron microscopy
(that is, by fixation with osmic acid followed by staining, particularly with uranyl acetate) the optical picture is then as devoid of evidence for intracellular
calcium phosphate as the electron picture. On the other
hand, if methods used for preparing the tissue for optical
microscopy are adapted for the electron microscope, then
the pattern of intracellular mineral is confirmed at the
ultrastructural level.

There seems little doubt that certain cells, at certain times during bone formation, are themselves densely
mineralized, and direct proof is accumulating (for example tetracycline staining and labelling, Aaron and
Pautard, 1973; 1975) that the mineral inside the cells is
related to mineral in the matrix and the structures made
by the golgi apparatus are not retained solely for metabolic or cytoskeletal purposes. Whatever the mechanism of
mineral formation, however, the mature bone salt is associated in some way with the organic matrix and part of
it, at least, is exchanged with the vascular fluids. With
the most recent evidence, to be discussed below, there is
the added possibility that the biological control of the
calcium phosphate is not lost in the extracellular environment and the mineralized structures, once formed,
may be capable of an independent biochemistry and biophysics.

There is no single item of evidence from any source in
the literature that entirely rules out the possibility
that other theories may be valid and it is advisable to

regard each of the proposed mechanisms as an explanation
for one aspect of what is clearly a heterogeneous system.
A general summary of the evidence, contra-evidence and
defence for each theory is set out in Table X.

Resorption

As the formation of new bone takes place in several
ways — endochondral and osteoid mineralization, callus
formation and fragment repair, so resorption takes place
also by a variety of processes. However, unlike new bone,
mature tissue is demineralized not only by the destruction
of the hard matrix by special multinucleate cells (osteoclasts) but also by the partial demineralization of
otherwise intact regions which may mineralize at a later
date. Three of these processes, which do not seem to require the agency of osteoclasts, have been extensively
debated and are an important factor in bone metabolism
(see Aaron, 1976 for review) and in any interpretation of
the structural or ultrastructural state of calcium phosphate.

The first process, described by Frost (1960) and
termed 'feathering', is a reduction in the density of
mature regions of bone matrix with attendant histological changes. The second process, described by Belanger in
1969 and termed 'osteocytic osteolysis' is the enlargement of the lacunae, the cavities in the bone matrix in
which the cells live, suggesting the removal of mineral
from the lacunar walls. The third process, recently described by Aaron in 1977, and termed 'autoclasis', is the
fragmentation and dispersion of specific regions of bone
as a result of activity inherent in the osteocytes and
possibly in the bone matrix itself.

The resorption systems are summarized in Fig. 8. Here,
the central optical micrograph shows the characteristic
appearance of bone undergoing progressive removal by
osteoclasts from the edge of the tissue. The alternative

F.G.E. PAUTARD

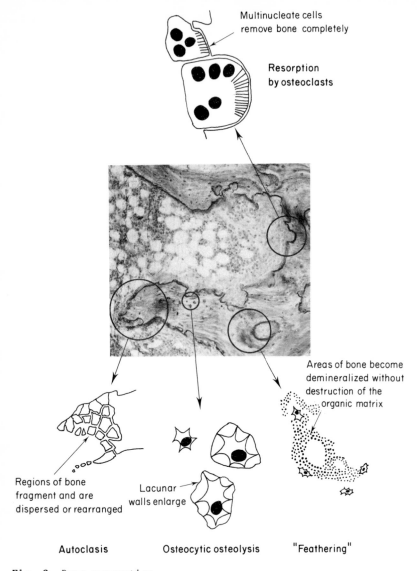

Multinucleate cells remove bone completely

Resorption by osteoclasts

Areas of bone become demineralized without destruction of the organic matrix

Regions of bone fragment and are dispersed or rearranged

Lacunar walls enlarge

Autoclasis   Osteocytic osteolysis   "Feathering"

Fig. 8. Bone resorption.

processes are drawn in relation to those areas likely to have such changes with time.

Calcium and phosphate exchange

Although it is likely that calcium and phosphorus might

enter an animal through the epidermis in solution or as
a solid (and in lower orders this may form a significant
amount) the principal route of entry of both ions in
higher animals is through the walls of the internal
epithelium as solubilized or dispersed food. Similarly,
the loss of ions may take place by the sloughing of the
epidermis (a factor in the Crustacea and some vertebrates)
or by glandular exudation, the principal route of exit is
through the urine or the faeces. In man, for all practical
purposes, the gain and loss of calcium and phosphorus
is measured by the total food intake and the total excretion. In equilibrium conditions (zero balance) the ion
intake is equal to the ion output but this condition is
seldom realized as the balance fluctuates positively and
negatively by small amounts, even in health. In pathological conditions the balance of each ion can alter considerably with consequent effects on many tissues, which
respond to the deficiency or surplus in characteristic
ways.

Since both calcium and phosphorus are closely involved
with many physiological functions (nerve conduction, muscular contraction, energy transfer, synthesis and degradation) but in disproportionate amounts (usually with phosphate the greater and calcium the lesser) and in varying
concentrations, each tissue, and probably each cell, must
monitor and exchange both ions to maintain electrical
balance. The reactions are so complex and are often so
rapid that individual measurement and summation are
clearly impossible without isolation and the pathways of
calcium and phosphorus are therefore measured in terms of
absorption and excretion in the intestine and kidneys;
that is, the intestinal epithelium absorbs and excretes
the ions in the gut and the kidney epithelium excretes
and absorbs the ions in the tubules. Inside the epithelium, however, the vascular system continually redistri-
butes the calcium and phosphorus, and since the bulk of

both elements resides in bone, the tissue is the reservoir which the ions enter and leave in response to vascular demand. The overall metabolism of calcium and phosphorus under conditions of zero balance is shown in the
diagram in Fig. 9, which illustrates the average flux of

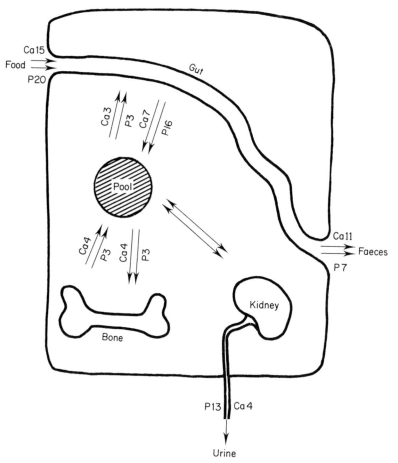

Fig. 9. Exchange of calcium and phosphorus in man. The values (from Wilkinson, 1976) are expressed as mg/kg/day.

ions from bone in relationto the soft tissues, which are
not detailed. The accuracy of the values (recently reviewed
by Wilkinson, 1976) depends on reliable measurements in
food, digestive juices and excretory products. The differences between the net exchange of calcium and phosphorus

are interesting. While the value of the exchange of phosphorus to and from bone is 3 mg/kg/day, calculated in
relation to the calcium value of 4 mg/kg/day, the conclusions are supported by some evidence based on isotope
exchange (Triffit et al., 1968). The molar Ca/P ratio of
1:1 would suggest either that the bone salt exchangeable
with the vascular fluid is more nearly tricalcium phosphate than hydroxyapatite or that the bone salt is apatite
and more phosphorus is turned over than calcium.

The exchange of calcium and phosphorus in bone is complicated by two factors which alter the site of translocation. The first is that the removal and replacement of
bone is a continual process, so that at any instant, calcium phosphate is being absorbed and reformed. The second
is the loss of ions without the destruction of the tissue.
Neither factor can be measured directly and so the relative movement of calcium and phosphorus is calculated by
isotope labelling and recovery over periods of time. The
separation of the information for the two factors of bone
formation/resorption (turnover) and exchange has led to
considerable disagreement as to the steps from vascular
fluid to bone mineral. The speed of translocation of calcium and phosphate has made simple equilibrium studies
unacceptable. The proposal by Carlson in 1951 that there
were two distinct phases of ion activity, one in which
calcium and phosphorus were incorporated into new bone
and one in which both were in continual exchange with
some other site in bone, stimulated ideas of fixed pools
and exchangeable pools. Bauer et al. (1955) devised a
simple formula on the assumption that when, say, ^{45}Ca is
introduced, the total pool of ions would have the same
specific activity as the plasma in a finite period of
time. The studies of Nordin et al. (1963), however,
showed that the exchangeable pool actually increases with
time and numerous models (some of which are illustrated
in Fig. 10) have been proposed to explain, by the best

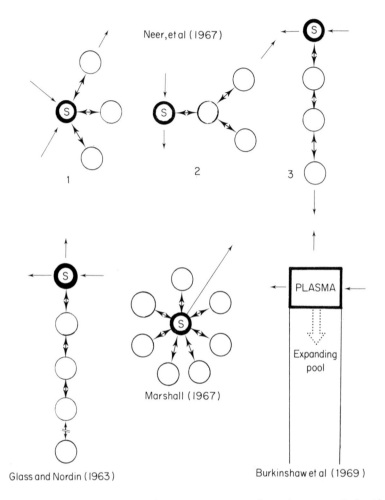

Fig. 10. Examples of models of compartments and pools to explain the kinetics of exchange in bone (after Marshall, 1976).

approximation, the observed history of each ion with time. The evidence for calcium activity, based on isotope exchange experiments by numerous authors (see Marshall, 1976, for recent review), suggests that a proportion of the element is incorporated into new bone and a proportion is held in sites which gradually release it later to be incorporated into further new bone. The course of

events is summarized in the graph in Fig. 11, which shows
that about half of the isotope is excreted in 80 days and
little escapes thereafter from the bone environment.

While the mechanism of phosphate exchange in response
to stimuli may closely follow that of calcium (for example Day and McCollum, 1939; Baylink *et al.*, 1971;
Matthieu *et al.*, 1972) there are less data to place the
events on a reasonable mathematical footing. The result
is that the homeostasis of phosphorus has been well established without any clear indication as to how much is
removed from bone and what proportion in bone is removed
from the inorganic calcium phosphate. With such a large
pool of phosphorus outside the mineral, the amount of

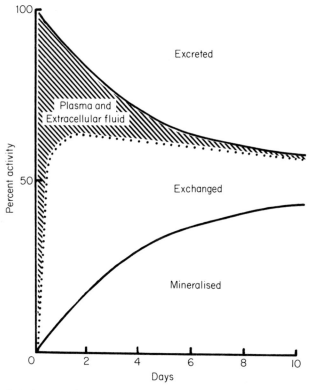

Fig. 11. Calcium activity in man after intravenous injection. The
shaded area is that part of the total exchange pool occupied by the
spaces (plasma etc.) outside cells (from Marshall, 1976).

bone salt involved in direct exchange may be small and the
major factor in determining solution is the concentration
of the calcium ion.

Anomalies and alternatives

The information above represents a general consensus
of views as to the nature of bone salt, particularly in
relation to geological models. With increasing evidence
that calcium phosphate is intimately connected with cell
activity, there is considerable conflict between chemistry
and physics on the one hand, and biology and biochemistry
on the other. The simple chemical and physical models of
bone formation, resorption and exchange need to be reviewed as it becomes more likely that the mineral in bone
is a complex substance in dynamic, rather than static,
equilibrium with its environment.

The current ideas about bone salt often depend on
isolated experimentation, and there are anomalies which
need to be explained to reconcile the opposing views.
A summary of the principal discrepancies is as follows:

1. The chemistry of bone salt is usually the result of
gross analyses which vary because of differences in subject and preparation. Such decisions as to the formulation of appropriate molecules in relation to the variance
depend on the assumption that every inorganic ion is
equally distributed throughout the mineral phase. The
evidence, however, both from the biology (formation and
resorption, Figs. 7 and 8, for example) and from the state
as deduced by various methods (Table IV) does not support
a conclusion for a homogeneous salt and the balancing of
cations and anions into a single formula is therefore not
warranted.

2. The extensive crystallographic arguments (Table
VII) that have been put forward for the structure of bone
salt are based almost entirely on X-ray diffraction evidence for the few reflexions that are visible in dry bone.

Since the same methodology shows that the proportion of
crystalline material depends on the preparation of the
sample (Table VIII) the stoichiemetric assignment of
elements to the unit cell from chemical assay is not reliable.

3. The 'amorphous' phase of calcium phosphate in bone
is either a matter of state or of composition. If the
'amorphous' component is an early, or stable mature, form
of subcrystalline hydroxyapatite, then calculations from
the observable crystalline phase (Fig. 5a) may be applied
throughout the bone salt, although the relatively few
lattice planes may have small effects on the composition.
If, however, the 'amorphous' phase is not an apatite
(recent evidence, for example that of Meyer and Eanes,
1978, tends to suggest that tricalcium phosphate is the
most likely amorphous salt) then calculations as to crystalline species and their variations (Table VII) are not
profitable. In particular, the position of carbonate in
the system cannot be determined without isolation of the
components.

4. There are opposed observations (Table X) as to the
actual site where the mineral is formed. Electron microscopy suggests that bone salt appears outside the cell
and in some relation to collagen; the event may be controlled by the agency of cell-made vesicles that generate,
or assist, the local precipitation of salt. In direct contrast, optical microscopy suggests that calcium phosphate
appears first inside the cell and later outside the cell.

5. Calculations of calcium and phosphorus exchange are
made on the basis of specific ion activities in the plasma
in equilibrium with bone salt (Figs. 9 and 10). However,
the only route from the vascular system to the mineral is
through palisades of cells (Fig. 6) and blood itself contains cells and solids which are themselves exchanging
calcium and phosphorus with the plasma.

These anomalies represent the focus of current views

that lie between two extremes. At one extreme, the chemical model predicts the precipitation and crystallization of a deficient, poorly crystalline apatite outside the cell under certain well-defined conditions. At the other extreme, the biological model predicts that the mineral is largely non-crystalline, is assembled within the cell and extruded to selected regions of the matrix. The consequences of these conclusions are important. The chemical model does not assume any dynamic relationship between the species of salt, since the metabolic pathways (Fig. 3) are confined to the cells alone, and only at the periods of formation and resorption. The biological model, on the other hand, may depart to any degree from a simple static collection of ions. It may be no more than a membrane-limited precipitation of salt; or it may be an extremely complex arrangement of mineral and macromolecules with an extensive biochemistry and a capacity to translate calcium and phosphorus in response to external stimuli.

Of all these possibilities, the most recent evidence leads me to suggest that the mineral in bone may not only be cellmade, but is also part of the architecture of large numbers of subcellular organelles which fill the extracellular space.

Subcellular calcium phosphate

Calcium and phosphorus within cells are not always either freely ionized or organically bound. There are many examples of discrete deposits of calcium phosphate throughout the animals. Indeed, before the turn of the century, there were numerous studies of mineral inclusions within organisms. In 1894, for example, Schewiakoff carefully recorded, in *paramecium caudatum*, the appearance and disappearance of calcium phosphate in vacuoles in specific parts of the cell. The size of the mineralized structures must have been at the limit of resolution of the micro-

scopes of his day, yet his drawings and anlyses patiently
anticipate recent evidence (Aaron and Pautard, 1973) which
shows the same phenomenon in most soft tissues stained
with tetracycline and viewed with modern equipment.

There is, however, a tendency to dismiss any accumulation of insoluble salt within cells as 'excretory',
even though it may be part of a complex structure. This
philosophy has been nurtured by the optical evidence,
in both plants and animals, of large crystals of calcium
salts which seem free in the cytoplasm and appear to be
functionless. Yet ultrastructural examination usually
reveals cellular fabrics closely associated with such
objects. In plants, for example, the long bundles of
single crystals (raphides) of calcium oxalate monohydrate
are a common feature and are usually considered as waste
products. Yet the structural intricacy of these intracellular deposits may be so great that the ramifications
can be followed only with the greatest difficulty (Arnott
and Pautard, 1970, Figs. 31–34). It is usual, therefore,
that when mineral appears within a cell, it is rarely an
inorganic salt alone but is usually modified or enclosed
by organic boundaries or septa of various kinds.

Distribution Calcium phosphate has been reported within
cells of many kinds. In Schewiakoff's (1894) description
of the inclusion of the salt within vacuoles, he cites
F. Stein in 1859 as the first observer, although it is
probable that there were even earlier notes on the subject. A recent review by Fauré–Fremiet in 1957 of calcified objects within cells includes ultrastructural information.

Of the present investigations of intracellular calcium
phosphate elsewhere than in mammals, the observations of
Ennever and others (reviewed by Ennever and Creamer, 1967)
on the calcium phosphate in the bacterium *Bacterionema
matruchotii* is of considerable interest. In this organism,
which is a common inhabitant of dental debris and plaque,

the calcium phosphate within the cell is crystallographi-
cally and ultrastructurally similar to bone salts. Although the mineral might accumulate in the bacterium by
various routes (e.g. Pautard, 1970, p. 125) its disposition, environment and metabolism may constitute a model
for its characteristic form in a number of subjects,
including bone.

Filamentous clusters One form of subcellular calcium
phosphate seems to occur repeatedly in a wide variety
of cells. The salt is not permanently associated with
specific cell features but is apparently part of a sub-
cellular organelle of characteristic size and shape which
is usually found free in the cytoplasm. These objects are
generally spherical in shape and have been given various
names (spherules, spherulites, calcospherules, modules,
granules, corpuscles) by different authors. They vary in
size from 0.1–2 µ and they give chemical tests for calcium
phosphate and often, but not always, show X-ray diffraction patterns of an apatite of low crystallinity. The
ultrastructure of these objects is characteristic — a
network or radially-arranged assembly of dense ribbons or
cylinders of indeterminate length but with a width aver-
aging 50 Å. The density of the filaments varies and there
is often a less dense region in the centre of the spherical, or near-spherical systems. In some locations in cells,
the shape may be distorted to become elliptical or even
elongated.

While it is customary to regard these mineralized
particles as precipitates of salt, there is much evidence
that they contain organic substance and are enclosed
within a membrane or envelope. Because of the uncertainty
as to the detail and relationship of these subcellular
fabrications, they have been generally described as 'filamentous clusters' and the term will be retained here to
avoid confusion with other undefined forms of calcium
phosphate which may or may not be related. Filamentous

clusters have been described with and without an outer
envelope, as largely inorganic with a small amount of
organic material, and vice versa. Similar structures have
been reported to contain calcium carbonate.

Of the various versions of filamentous clusters, three
of them, in the protozoan *Spirostomum ambiguum*, in mammalian keratin and in bone itself, seem to represent the
accumulation of calcium phosphate in a characteristic way.

Spirostomum ambiguum

Spirostomum ambiguum is a large protozoan commonly
found in most ponds and probably in most soils. It is an
elongated cell up to 1 mm long with an extended mega-
nucleus and a stomum about the central region. The animal
swims by means of cilia; it also possesses a sinistral
band of contractile myonemes which are probably used to
force passages through silt and wet soil, which may be its
principal habitat.

A striking feature of this animal, details of which
appear elsewhere (Pautard, 1959; also Pautard, 1970, 1976)
is that it gives a pronounced diffraction pattern of an
apatite (Fig. 12a) closely resembling that from bone but
without the corresponding collagen pattern when the salt
is removed (Fig. 12b; cf. Figs. 5a and b). The source of
the pattern can be traced in the animal (shown in Fig.
12c) to numerous small spherical particles (Fig. 12d)
varying in diameter from 0.2–2 μ distributed throughout
the cytoplasm. The isolated particles stain variably with
tetracycline, suggesting different amounts of calcium
phosphate, give electron probe analyses for calcium and
phosphorus and histochemical analyses for protein and
polysaccharide. In the electron microscope, the particles
are seen to be formed in vacuoles (Fig. 12d), appearing
as arrangements of filaments radiating from the periphery,
with less dense central regions and indications of zonation. A characteristic feature of the filaments is that

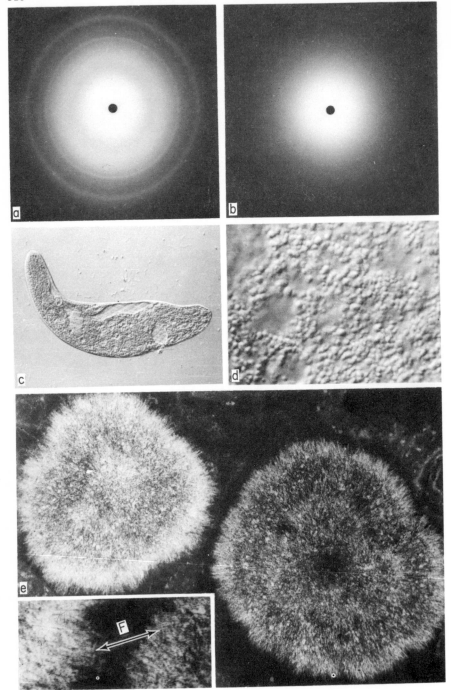

they appear to maintain a relatively constant width of about 50 Å irrespective of density (Fig. 12d, inset; see also Pautard, 1970 for details) and the central region appears to be less filamentous in nature. Those structures may be extruded from the animal from time to time, but they always appear to be present, although, during division, the mineral is absent, to reappear later in the new cells after a few days.

At the time when the intracellular disposition of filamentous clusters was first observed (Pautard, 1959) in *Spirostomum,* the only comparable objects in bone were the 'islands' described by Scott and Pease (1956) and by Robinson and Cameron (1956). In recent literature, however, filamentous clusters have been observed with increasing frequency as 'nodules' (Bernard and Pease, 1969, Gay, 1977) although the appearance has always been described as 'extracellular'.

Fig. 12 (opposite). Calcium phosphate in *Spirostomum ambiguum.* (a) X-ray diffraction pattern of whole, washed cells mounted on collodion film. Ni-filtered CuKα-radiation, specimen-to-film distance 4 Cm. The reflexions are closely comparable to those from bone (cf. Fig. 5a). (b) X-ray diffraction pattern of (a) above after demineralization in 10% EDTA. The diffuse reflexions suggest non-specific protein, not comparable with similar preparations of demineralized bone, where the collagen spacings are apparent (cf. Fig. 5b). (c) Optical micrograph of a live *Spirostomum ambiguum* slightly flattened under a coverslip to reduce motility. The cell is normally thinner and more elongated. The spiral band of myonemes can be seen faintly crossing the large vacuoles, X 140. (d) Central region from (c) above photographed at higher power, using high speed film and interference contrast optics. Groups of particles of various sizes can be seen in the cytoplasm, X 1,140. (e) Electron micrograph of two particles from a similar subject to (d) above. The illustration, of formation within a vacuole is printed as a negative to show the varying degree of mineralization and the characteristic filamentous edge of these subjects. There is a suggestion of zonation and the two examples demonstrate the similarity of appearance irrespective of density, X 48,000. **Inset:** Higher magnification of the area marked in (e) above. The filaments (arrowed F) extend into the vacuolar space, which in positive printing reveals further details of membranes and attachments, X 146,000.

Baleen

Baleen is the horny plate (popularly known as 'whale-
bone') fringed with stiff hairs that the rorqual whales
use to gather food. In company with many other epidermal
proteins (Earland *et al*.., 1963; Pautard, 1963) the baleen
plate and fibres contain deposits of calcium phosphate
oriented parallel to the plate and the fibre axis (Pautard,
1962, 1965). The X-ray diffraction pattern of the fringe
fibres (Fig. 13a) closely resembles that from bone, but
after demineralization the pattern which remains (Fig.
13b) is characteristic of α-keratin and not collagen (cf.
Fig. 5b). The fringe fibres can be disrupted into their
component cells and fractionated (Fincham, 1966) to yield
preparations with varying amounts of calcium phosphate.
The denser fractions contain as much as 40% of mineral,
which is present inside the flattened cells in the form of
particles similar in size to those in *Spirostomum ambiguum*. In the electron microscope, the particles appear as
filamentous clusters but less regular than those in the
protozoan and often grouped together and elongated (Fig.
13c; cf. Fig. 5e).

Baleen therefore represents the 'alternative' mammalian tissue to bone, where a similar form of calcium phosphate is associated with an organic matrix not containing
collagen but composed of a typical epidermal keratin
(Fincham *et al*., 1965). The anomalous position of baleen
in the tissues of higher animals calcified with phosphate
has, until recently, been regarded as an octopic form of
intracellular deposit not related to bone. The clearly
functional distribution of mineral within the fringe
fibres suggest a close biological control over the disposition and amount of salt within each cell and with increasing observation of similar intracellular structures
in bone, increasing doubt has developed as to the true
nature of the calcium phosphate in the extracellular
matrix.

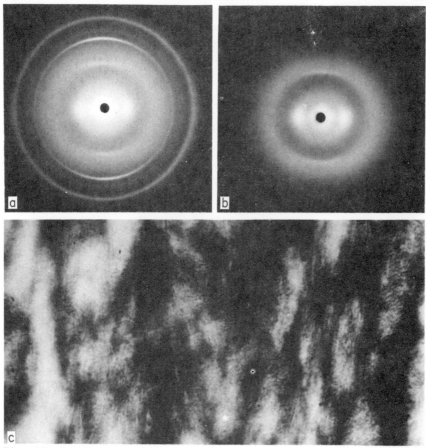

Fig. 13. Calcium phosphate in baleen. (a) X-ray diffraction pattern
of fringe fibres of the baleen plate of the Sei Whale *Balaenoptera
borealis*. The fibre axis is vertical. Ni-filtered, CuKα-radiation,
specimen-to-film distance 4 cm. The reflexions closely resemble those
from bone (Fig. 5a) and from *Spirostomum ambiguum* (Fig. 12a). (b) X-
ray diffraction pattern from (a) above after removal of mineral with
10% EDTA. Although the pattern is oriented parallel to the crystallites of apatite seen in (a) above, the reflexions are typical of
α-keratin and collagen spacings (cf. Fig. 5b) are not present.
(c) Electron micrograph of fringe fibres of Sei Whale baleen, fibre
axis vertical. The mineral is disposed in patches which are not as
spherical as those in *Spirostomum* (Fig. 5e) but are usually elongated
and often joined (cf. Fig. 5c), X 16,800.

Bone

   Apart from occasional indications that mature, dense
bone may contain discretely mineralized regions, it is
only recently that more convincing evidence has been

forthcoming that the extracellular salt may be entirely
contained within structures fabricated by the cell, as
in the case of the spherical filamentous clusters in
S. *ambiguum* and the intracellular clusters of less regular
shape in baleen cells. The two crucial questions which
need to be answered in any assessment of the nature of
calcium phosphate in bone are 'Where is the calcium and
phosphorus put together?' and 'What happens to it after
it is lodged in the extracellular environment?'. If the
mineral arrives as a result of a concentration gradient
or stereotactic nucleation outside the cell, the entire
mineral complement is likely to be continuous and homogeneous. If the mineral commences within the cell, as in
Spirostomum and baleen and is then extruded and allowed
to become confluent or continuous, then the salt is also
likely to be homogeneous. If, however, the calcium phosphate is made inside the cell and after extrusion remains
in the same structural state, then it is less likely that
the mineral will become homogeneous and most likely that
it will remain as a heterogeneous collection of separate
accumulations of ions which may have widely differing
properties.

Origin of calcium phosphate

While there is a long history of optical observations
of mineral particles in bone since Queekett recorded in
1846 that 'ossific matter consists of small granules',
the appearance of discrete mineral deposits within cells
is still a matter of conflict between the optical evidence and the electron evidence, summarized above. It is
clear that dense deposits of mineral occur inside cells
in the mineralizing zone, but it is difficult to show
that the intracellular deposits are the same as, or precursors of, the mineral which appears outside the cell,
and are not a separate unrelated cytoskeleton. The experiments of Aaron (1973) on developing mouse calvaria, how-

ever, suggested that the mineral was laid down in a systematic way, the osteocytes becoming first 'loaded' with calcium, followed by the appearance of calcium phosphate, then by the disappearance of calcium phosphate from the cells and its appearance in particulate form outside the cells. Subsequent studies of tetracycline staining and labelling (Aaron and Pautard, 1973, 1975) suggested that the 'loading' and 'unloading' cycle may be as rapid as four minutes. Moreover, the optical evidence can be reproduced at the electron microscope level by suitable modifications of optical staining (Aaron and Pautard, 1972). The site of mineral accumulation appears to be within the saccules of the golgi apparatus (Aaron and Pautard, 1973, 1975; Pautard, 1975; Park and Kashiwa, 1975).

The evidence from our laboratory is summarized in Fig. 14. The optical micrograph in Fig. 14a illustrates the 'loading' of cells in the mouse calvarium, with 'unloading' in a time sequence which is probably less than 10 min. The important feature of this illustration is that the staining procedures were carried out in whole mount, that is with the calvarial bone undisturbed. Since the intercellular regions are unstained yet are already densely mineralized (as judged by microradiography and by staining after sectioning) the stain can therefore only penetrate those regions which are accessible. Moreover, if the same tissue is treated with electron microscope reagents, the optical evidence for mineral disappears. The most likely explanation is that the first-formed mineral inside the cell, and soon after it leaves the cell, is in direct contact with the vascular system, thought which the stain will enter the tissue. At a time after the mineral leaves the cell, it becomes impervious to reagents. Hence the dense bone is not accessible to optical stain but becomes so after sectioning, suggesting some 'packaging' step. At the same time, the removal of the accessible mineral by electron microscope reagents

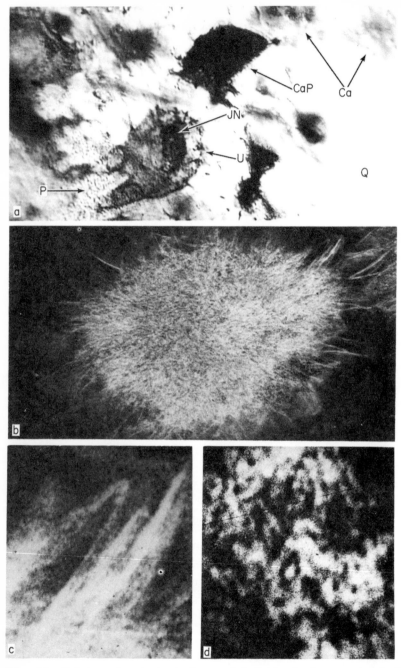

Fig. 14. Calcium phosphate in bone. (a) Optical micrograph of the central plate of a 6-day-old mouse calvarium stained with GBHA for calcium, and with silver for phosphate. While the black and white

which do not affect the sectioned material suggests that
the first formed calcium phosphate is highly soluble in
such reagents as uranyl acetate, osmic acid and even distilled water. These possibilities explain the fortunate
optical illustration of the forming mineral and the
absence of electron microscope evidence for it.

Figure 14b illustrates a filamentous cluster immediately outside a cell in the epiphyseal plate of developing
rabbit bone. These are commonly observed objects in the
calcification front, but they rapidly become confluent
and the boundaries are difficult to discern, although
photographic overexposure can improve the detection
(Arnott and Pautard, 1967, Fig. 1). The cluster in Fig.
14a is printed as a negative to improve the viewing of
the mineral. The enlargement of detail in Fig. 14e is the
typical 'crystallite' found at the edge of these structures, but there is some doubt as to the reliability of
such objects witnessed in ultrathin section after staining or washing and drying on an electron microscope grid.
The more usual appearance of the mineral phase may be in
the nature of a labyrinth of interconnected filaments and
foci, as shown in a preparation from the same specimen as

reproduction does not show the differential staining of the black
(phosphate) and red (calcium) components, it does illustrate clearly
the course of events during the mineralization of this tissue. Outside the zone, the clear area (Q) is quiescent. Nearby, there are
two cells with pale calcium 'loading' (Ca). The next cell in sequence
is so densely loaded with calcium phosphate (CaP) that it is difficult to discern cell detail. The cell arrowed U shows the typical
'unloading' pattern with dense salt still remaining in the juxta-
nuclear area (JN) and numerous particles appearing at the edge of,
and outside, the cell membrane (P), X 1,080. (b) Electron micrograph
of a resin-embedded section of a filamentous cluster from developing
rabbit femoral bone. Printed as a negative to show the filamentous
edge, this is comparable to similar structures in *Spirostomum ambiguum* (cf. Fig. 12e) although it tends to be flattened in shape, as
in baleen (Fig. 13c), X 84,000. (c) Detail of filament substructure
from area outlined in (b) above, X 168,000. (d) Electron micrograph
of detail from the same subject as (b) above, but prepared by scrap-
ing samples from broken bone immediately after removal and freeze-
drying directly onto grids, X 365,000.

Fig. 14b, but rapidly freeze-dried from the fresh scraped
bone and placed directly on the grid without sectioning.

Calcium phosphate in the matrix

   With few exceptions, the mineralized bone matrix has
usually been regarded as a homogeneous congregation of
small particles of calcium phosphate. The fractionation
of powdered bone samples by flotation in liquids of different density has been carried out by a number of
methods (e.g. Fincham, 1968) and the results tend to show
that there is a spread of density with age. However, the
general conclusion is that this is a consequence of different proportions of mineral to collagen rather than
differences in the nature of the mineral. An exception to
this view stems from the experiments of Quinaux and
Richelle (1967) who showed that the fractionation of bone
powders resulted not only in differences of crystallinity
with increasing density, but also in the nature of the
salt itself, as judged by the X-ray diffraction patterns
after controlled pyrolysis. At the lowest specific gravity,
tricalcium phosphate was the most likely abundant salt;
as the specific gravity increased to the maximum value,
hydroxyapatite appeared progressively. Carbonate was present in each fraction.

   While these experiments suggest a variation in composition with density, the presence of collagen in the fragments possibly obscures more marked differences in the
mineral. The isolation of bone mineral free from collagen
is difficult in mature bone, where the bulk of the salt
is extracellular. However, mineralized fractions can be
isolated from bone cells by homogenization and fractionation. Hirschmann and Nichols (1972) reported differences
in enzymes and other substances in calcium- and phosphate-
rich subcellular fractions. Isolation of mineralized
particles from developing mouse calvarium (Fig. 14a) can
be carried out by freeze-drying the tissue and treating

with collagenase for 24 hours (Pautard, 1976). The resulting suspension can be fractionated in solvents to give
preparations of isolated particles (Fig. 15a) which contain over 90% of the calcium phosphate in the tissue.
Demineralization with EDTA leaves 'ghosts' of protein and
polysaccharide, although the yield is too small to permit
more detailed examination.

While particles can be separated from young bone,
mature cortical bone does not respond to the same treatment. However, if aqueous suspensions of dense, mature
cortical bone are milled for brief periods, large numbers
of intact spherical particles are displaced from the fragments (Fig. 15c). Treatment with collagenase then removes
the collagen and the suspension of particles can be fractionated in solvents. However, in the case of mature
bone, the fractions contain not only spherical particles
from $0.1-1$ μ in diameter (Fig. 15d) but also mineralized
assemblies composed of dense cores joined by bridges
(Figs. 15e, f). These remarkable objects often show intricate looping and convolution (Fig. 15g).

This surprising result in mature bone suggests that the
mineral within the matrix is not homogeneous but is part
of another microskeleton of great complexity. Although
it is now possible to isolate large numbers of microskeletal structures, analysis of their composition and
function must await the fractionation and classification
of the various types. Preliminary studies (Aaron and
Pautard, in preparation) suggest that the spherical structures are made within the cell and mature in the extracellular space, usually in domains of several components
united together. The bridged assemblies appear to be related to dense bodies found inside the cell; outside the
cell they have been observed to be connected by short
bridges. The absence of free bridged assemblies after
milling, but their appearance in numbers after collagenase treatment, makes it likely that they are attached to

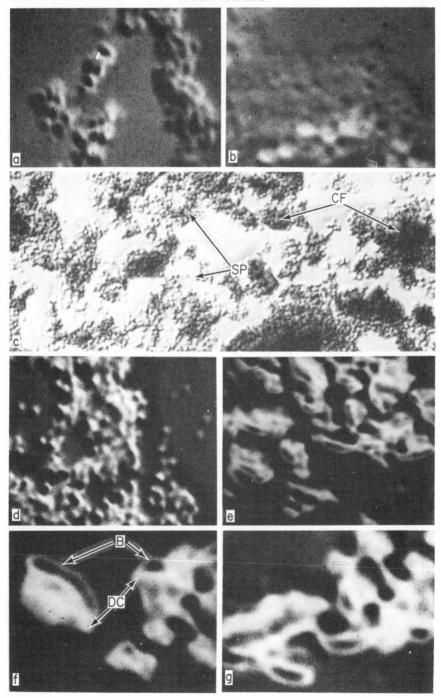

collagen fibres. We suppose that the looped complexes
are the result of fusion and co-operation by numbers of
dense bodies (or their proliferation) in the proximity
of collagen, forming collars for one or more fibre bundles, and with sockets or recesses in which adjacent
spherical or elongated structures, either unattached or
in domains, may be located. The survival of the bridges
after considerable attrition suggests that they have high
tensile strength and may be crystalline. We would expect
the inner surfaces of the loops to carry a negative
'print' of the collagen periodicity; such adhesion would
serve to explain the enhancement of collagen surfaces in
unstained sections seen in the electron microscope (Fig.
5d). Moreover, it is likely that the appearance of mineral
deposits on collagen fibres (Glimcher, 1960) and on tendons (Nylen *et al.*, 1960) is a reflexion of 'looping'
activity at that point.

It may be imagined that the apparently dense, continuous mass of bone mineral is, in reality, an articulated
microskeleton as complicated as the gross skeleton itself.
Indeed, it is even likely that within each structure there
are further arrangements of mineral, also heterogeneous,
even articulated: and all programmed when fabricated
by the cell to perform a specific function in a specific

Fig. 15. Optical micrographs of isolated mineral components of bone.
(a) Six-day-old mouse calvarial bone, freeze-dried, treated with
collagenase and density fractionated. The particles are mostly
spherical, X 2,460. (b) Preparation as in (a) above but demineralized
in 5% EDTA and stained to show the 'ghosts' of the structures, X
2,460. (c) Ox cortical bone, defatted, dried, powdered and wet milled
for 30 min. The spherical particles (SP) have been displaced from the
denser mass of the collagen fragments (CF), X 1,200. (d) Preparation
(c) above after treatment for 24 hours with collagenase, and density
fractionated to separate the mineral components. The region photographed is largely composed of spherical particles, X 2,250. (e) A
different area from (d) above showing a preponderance of bridged
assemblies, X 2,250. (f) Detail from area (e) above showing the
bridge association (B) with dense cores (DC), X 4,100. (g) Detail of
area (e) showing the degree of complexity of the bridges and their
connections with many dense cores, X 4,100.

part of the tissue. Tendons have to be locked into place
in strategic parts of the bone surface; yet they may need
to be rotated in some modes and bent in others. The looped
structure in Fig. 15g may be as mechanically complex as a
vertebra. It may be one of millions to each cubic centi-
metre of tissue. Each structure may be different, or there
may be recognisable variations. It is likely that they
may change with growth or stress, moving their position,
rearranging themselves or migrating to new sites. Present
indications are that there is a biochemistry available to
cope with such metabolic demands.

With a panorama of such magnitude to consider, it is
impossible to relate the biology to the chemistry of
calcium phosphate until the isolated microskeletal components can be analysed, preferably one at a time. With
these new developments in mind, however, it is nevertheless possible to reconsider briefly some aspects of formation, resorption and exchange summarized above.

Alternative interpretation

With the possibility that the calcium phosphate in bone
laid down as part of an organized structure in the golgi
apparatus of the osteocyte is extruded to specific parts
of the matrix and is programmed to carry out a variety of
extracellular functions, there can only be remote connections between the calcium and phosphorus in vascular fluids
and the same ions that are finally lodged in the microskeletal structures. With the added likelihood that both
the location and transposition of any ion at any time
within each structure may depend on the local demands and
the available biochemistry, much further information is
required about the isolated structures before any reliable
conclusion is possible. It is useful, however, to consider
current ideas in the light of these new events.

Formation Each general theory of mineralization (for
example, Table X above) is based on experiments which are

significant in themselves but which need explanation to
avoid stalemate or conflict. A tentative review of the
main hypotheses is set out below with reference to the
diagram in Fig. 16.

In **Hypothesis 1** (e.g. Glimcher, 1960; Hohling *et al.*,
1974), the cell does not participate in the mineralization,
but the stereotaxy of the collagen fibre induces crystal
formation. The 'nodules' not apparently connected with
collagen but filling the interfibre space are presumably
induced by the crystals that have already been seeded.
From numerous experiments on nucleation, there is little
doubt that some mineral is closely connected with some
part of the collagen fibre, but the circumstances need
further investigation. The predominant evidence is for
intracellular mineral formation (Fig. 14a) with no convincing histochemistry that calcium or phosphate are present in the extrcellular space in any form other than as
discrete deposits of salt. The data varies as to the way
in which mineral is associated with the collagen fibres.
In fowl bone (Fitton-Jackson, 1957) the deposits first
appear as small particles on the collagen interband. On
the other hand, the deposits in turkey leg tendon (Nylen
et al., 1960) appear larger than the usual size (their
Fig. 4) with single crystal diffraction patterns, and
while this tissue and the nature of the phosphate associated with it may not be typical of bone ultrastructure,
the morphology may be important in some local function.
In studies of tendon calcified *in vitro*, Luben *et al.* also
observed small particles (their Fig. 1a) but their study
also showed that the developing site (their Figs. 3, 4,
5) is concentrated in a discrete region only while the
adjacent fibrils remain deposit-free. Since the preparation used in these experiments consisted of freeze-dried,
shredded beef tendons, the ultrastructure, and probably
the biochemistry, were probably well preserved. This would
perhaps explain the local character of the larger deposits,

since those regions of the collagen which became calcified
in the solutions used may have been covered by subcellular
entities. Indeed, the topography of the deposits illustrated might well be predicted from the profile of a looped
assembly (Fig. 15g above) which would retain most of its
properties after the mild treatment used.

An interesting comment on the conditions in which demineralized bone will remineralize is provided by the
experiments of Bachra (1972), who compared the behaviour
of native bone matrix with reconstituted collagen in
calcifying solutions. Bachra found that when collagen was
prepared from rat tail tendon by extraction with acetic
acid and precipitation with sodium chloride, the reconstituted fibrils did not nucleate calcium phosphate from
the calcifying solution, although when the buffer was
renewed, salt was deposited among the fibrils. It was
concluded that the loss of CO_2 with a corresponding eleva-
tion of pH was responsible. On the other hand, when sheep
bone collagen was prepared by powdering the bone and demineralizing it exhaustively with ice-cold EDTA, considerable deposition of apatite took place on the collagen
fibres. However, electron microscopy showed that the
mineral was laid down in distinct 'islands' about 1 μ in
diameter, which later enlarged or became confluent.

At a maximum mineral content of 30% by weight (about
half that of mature bone) a considerable area of collagen
was devoid of salt, i.e. that although the dense areas
were fully mineralized the inorganic component occupied
half, or less, of the collagen volume that it would have
done in its original state. Bachra offered the explanation that nucleation occurred only in certain regions of
the collagen (in agreement with the results of Luben *et
al.*, 1973) and these 'active sites' formed clusters after
further nucleation by the initial crystals until the com-
pression of the collagen restricted further growth. Bachra
also proposed that the difference between the non-

mineralizing, acid-soluble tendon collagen, and the mineralizing, demineralized bone matrix might be the result of
differences in collagen structure or some substance
associated with it. An alternative explanation for these
results might be that in demineralized bone matrix, particularly when prepared under good conditions, the microskeletal structures (and probably their biochemistry)
are retained intact and they will remineralize to their
original density and morphology, whereas tendon collagen,
which will have no, or fewer, looped assemblies associated
with them, will mineralize less, or not at all if the
collagen is reconstituted and free from the attaching sub-
cellular structures. The failure of demineralized bone to
remineralize completely may be the result of loss of
mobile or soluble structures (cf. Fig. 15e) not firmly
attached to the collagen. A further experiment in this
series seems to be significant. Bachra heated the demineralized bone matrix for 6 hours in 10% NaCl and in-
cubated the gelatinized, partly renatured, collagen in
calcifying solutions. He found that although the material
mineralized almost to the same degree as the original
unheated matrix, the 'islands' were missing and the enhancement of the collagen periodicity was absent. Such a
result precludes the possibility that limitation of
mineralization is the result of 'island' formation;
Bachra (1975) comments 'It was tentatively concluded
that if a non-collagenous material was involved in the
nucleation catalysis, at least part of it should be bound,
possibly covalently, to the collagen'. There seems little
difference of view between a protein polysaccharide sheath
on the collagen which nucleates and contains the calcium
phosphate, and the same substance, in a geometric
arrangement, which performs the same function, except
that it is commenced in the cell instead of on the
collagen surface. Any conclusion as to whether the cohesion of the mineral with collagen is a function of the

collagen or the microskeletal structure, or both, remains to be seen. It would be instructive to find out if
the collagen *before* it was biologically calcified will
calcify *in vitro*, if the mineralized 'islands' will remineralize a *second* time to the same topography, and if
the remineralized 'islands' can be isolated by removal
of the collagen (as in Fig. 15d, e).

In **Hypothesis 2**, Fig. 16 (e.g. Taves, 1965; Talmadge,
1969; see also Robertson, 1976) the cell is presumed to
supply ions by a 'pump' mechanism to the extracellular
space, but the same mechanisms of calcium and phosphorus
translocation might equally well apply to the accumula-
tionof ions within the cell. However, the concentration of
ions leaving the cell is limited by their solubility products and the conditions for deposition will be the same
as for Hypothesis 1 above. On the other hand, accumulation within the cell is mediated by the biochemistry
binding and releasing calcium and phosphorus from specific
sites and by barriers of protein, lipid and polysaccharide
which compartmentalize the reactions. In such situations
the overall concentration of calcium and phosphorus can
have no meaning at a molecular level.

In **Hypothesis 3** (for example, Anderson and Reynolds,
1973), the biochemistry is transferred from the cell in
vesicles to the extracellular space, where the intercellular ion conditions are the same as in Hypotheses
1 and 2 above, see Fig. 16. While this view satisfies the
observation that some of the mineral is not associated
with collagen but is in the form of clusters or nodules,
and it accounts for the extensive biochemistry found in
the matrix, it conflicts directly with the evidence that
a large proportion (if not all) the mineral is assembled
within the bone cells. Even in calcifying cartilage, the
tissues most studied for its vesicle content, the accumulation of intracellular granules, aggregates and 'globules' of calcium phosphate is evident from optical micro-

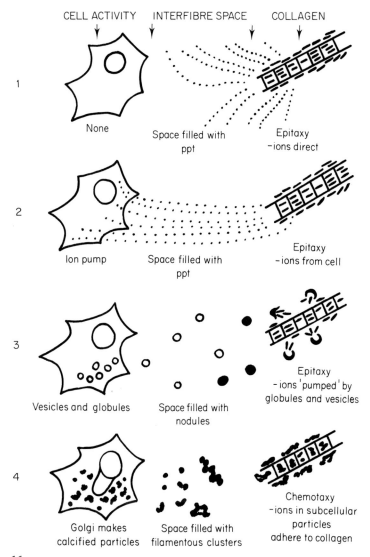

Fig. 16

scopy (Silbermann and Fromner, 1974), in direct contrast to the electron microscopy (e.g. Anderson, 1969). These differences can be resolved by assuming that the extracellular, uncalcified vesicles are a different cell product from the intracellular calcified objects, or that the empty vesicles were originally mineralized but have become demineralized as a result of preparative methods. In this connection, it is of interest that the procedure of Ali *et al.* (1970) for isolating matrix vesicles will also yield the bulk of the mineralized structures in the denser fractions (Pautard 1976). It is unwise to relate extracellular vesicles entirely to the mineralizing mechanism, since membrane-bound cell fabrications are commonly found throughout all tissues (Slavkin, 1975).

In **Hypothesis 4**, Fig. 16, the speed and complexity of mineralization within the cell seems to be the psychological objection to a completely cell-oriented mechanism. There can be no doubt that bone cells do fabricate mineral; the idea that each cell may make up to half a million structures and pack them exactly around it and in specific ways with collagen appears to ask too much of the biology. Yet elsewhere, such manipulations are commonplace and not regarded as exceptional by most biologists. In particular, the formation of coccoliths within the golgi apparatus of the dinoflagellates is a good example of the speed and organization required to fabricate complex mineralized objects by single cells. Coccoliths are structures of intricate morphology containing calcium carbonate; they are made inside the cells and extruded to form sheaths and coverings on the outer surface. In *Coccolithus huxleyi*, for example, each coccolith is composed of a number of units of calcite which, although of complicated shape, is a single crystal of the salt. The units develop in the cytoplasm and pass through the plasma membrane, fitting together in a pattern over the entire surface (Watabe, 1967). Each coccolith may contain some 30 units and each flagellate may contain some 200 coccoliths. On a continuous basis, the formation rate is equivalent to 200 units an hour, or one complete coccolith every 8 minutes, a mass of mineral as great and as complex as that 'loaded' in the bone cell. It is not surprising that such organisms can produce a million tons

of calcium carbonate as coccoliths in one year in one
Norwegian fiord alone and the residual skeleton is the
main constituent of chalk cliffs a thousand feet and more
in height. Since these organisms have been studied in
some detail, the fabrication of the mineralized coccoliths is better understood than the calcium phosphate
clusters in bone cells. In *Hymenomonas carterae*, there is
clear evidence that the golgi apparatus is responsible for
both coccolith (Outka and Williams 1971; Pienaar, 1971)
and body scale (Pienaar, 1969) formation. Although there
have been indications in the past that the golgi apparatus in bone cells may have similar activity (Taves, 1965;
Mathews *et al.*, 1968, their Fig. 2) it is only recently
that it has become clear that the major bulk of mineral in
bone is made in this way.

Resorption Osteoclastic remodelling, and possibly osteocytic osteolysis (Fig. 8 above) appears to remove the
mineralized matrix completely, presumably reducing the
components to fragments, monomers and ions. In contrast,
the removal of mineral from the intercellular space without disturbance to the organic fabric ('feathering') or
by the fragmentation and translocation of the mineralized
matrix by autoclasis without destruction of the mineral
or the organic matter associated with it, raises many
questions as to how the mineral is removed, replaced or
redirected in relation to the geometry of the tissue.
Since cells do not appear to be involved in either the
demineralizing or disrupting steps, the most likely agents
of change would seem to be the mineral structures themselves, either alone or in some combination with collagen.
Such a process of calcium and phosphorus transport,
coupled with severance of connections between mineral
boundaries and collagen, would require an organized biochemistry. The attachment and activation of enzymes for
solution and dissolution can be anticipated from the
evidence in the matrix (Kuettner, 1968), from calcium

rich fractions in cells (Hirschmann and Nichols, 1972)
and from extracellular vesicles (Slavkin, 1975). Current
histochemical evidence on bone matrix and on isolated
microskeletal structures (Aaron and Pautard, in preparation) suggests that acid phosphatase, RNA and DNA may be
present.

Exchange Until further information is available from
isotope experiments and other studies as to the involvement of the microskeletal structures, particularly with
respect to the nature of their calcium phosphate, it is
not possible to relate any of the present data about the
movement of calcium and phosphorus to and from the vascular system. Since the mineral is contained in countless
isolated 'pools' which may have different properties, it
is only possible to consider, in a general way, the pathways of the ions in relation to the formation and resorption cycles. These are set out in Fig. 17.

Phosphorus in bone - a reappraisal

In an extracellular population of separate structures
of such diversity, the nature and history of the calcium
phosphate within them may be so varied that we shall be
faced with the impossible task of examining each structure
individually. While it is clear that the average number
of calcium and phosphorus atoms remains within certain
limits (Table VI above) the relationship of each element
cannot be decided by simple formulation from the overall
analysis. Moreover, thermodynamic calculations as to
species of phosphate (e.g. Meyer and Eanes, 1978) may
not apply equally to each structure. Indeed, the changes
with preparation (Table VIII above) make it likely that
much of the physical state of calcium phosphate in bone
is transient and is both energy dependent and unstable.
Since cellular metabolism is closely bound up with mineral
formation in the beginning and with its removal at the
end, we need to enquire what metabolic activity is involved

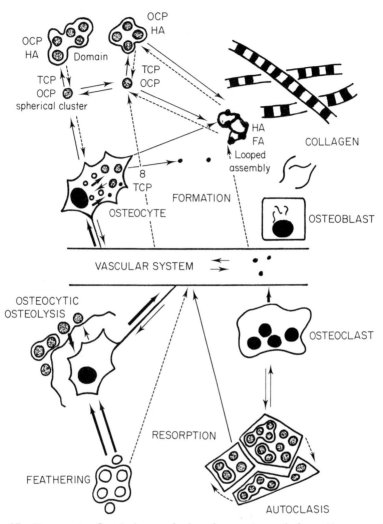

Fig. 17. Movement of calcium and phosphorus to and from the vascular system in bone formation and resorption. The relative amounts of the element exchanged are arrowed densely (principal routes), lightly (intermittent routes), broken (less likely routes).

The most likely calcium phosphate salts are indicated from the probable Ca/P ratio.

B	Brushite	$Ca_2 H PO_4$
TCP	Tricalcium phosphate	$Ca_3(PO_4)_2$
OCP	Octacalcium phosphate	$Ca_8H_2(PO_4)_6$
HA	Hydroxyapatite	$Ca_{10}(PO_4)_6OH_2$
FA	Fluorapatite	$Ca_{10}(PO_4)_6F_2$

during the life of bone salt and what proportion of such
activity actually resides within the framework of the
mineralized particles.

The position of phosphorus in this arrangement does
not seem to be solely a simple arrangement of a structur-
ally advantageous element, a convenient means of trans-
locating calcium, or a waste product hardening an advanced
articulated skeleton. The nature of the calcium phosphate
that is made and its maintenance during the life of a
region of bone, together with the characteristic form of
the structures it occupies may reflect properties of shape
and conservation of function which may have been present
from an early evolutionary time, such that a modern vertebrate skeleton is a symbiosis of 'simpler' arrangements.

Perhaps the common feature of the phosphate microskeleton, and thence to the cytoskeleton and bone, is
that unlike calcium carbonate and silica, two other
inorganic substances used to harden animals and plants,
it is a microcosm of all the vital functions of
phosphorus—energy transfer, motility and replication. In
this way, we might regard the location and exchange, function and evolution of calcium phosphate as a dynamic
inorganic model for the complex organic phosphates that
have developed into substrates for living processes.

Location and exchange

Since the initial site of mineral deposition is within
the cell, the principal step in the assembly of the organic elements is their transport within the saccules of
the golgi apparatus. While it is possible that phosphate
accumulates passively by diffusion from the blood, the
concentration of calcium and phosphorus required to avoid
precipitation will be too low to account for the speed
with which the fully-formed mineral appears in the cytoplasm. The route from the nearest capillaries to the juxta-
nuclear vacuole (the golgi apparatus in bone cells) may

therefore be a matter of 'facilitated transport'. In the case of phosphate, the choice is wide and complex. From the phosphate cycle illustrated in Fig. 3, all the processes likely to deliver phosphate to the golgi complex are available.

However, the rapid transport of phosphate to the mineral site raises many problems. The large amounts of phosphorus which need to be marshalled to concentrate the element can only be accomplished by the accelerated translocation of the free ion from pools of bound phosphate. To avoid local precipitation of calcium phosphate the free ions must be accelerated to the final site, or the free ions must be concentrated and isolated from cations, or the bound phosphate must be conveyed as such to the golgi vacuole, to be dephosphorylated at the site. All of these routes require selection and organization if the final result is to be a structure rather than a random heap of crystals.

Since most of the evidence in the literature refers to the metabolic behaviour of bone as a whole, there is little indication as to which factor (ion concentration, free and bound phosphate, inhibition of phosphate transfer, etc.) is related to the precise mechanism of phosphate accumulation within the filamentous clusters inside the vacuoles. Of the possibilities, dephosphorylation at the mineral site is the most likely from the scant evidence that is available. It is difficult to see, nevertheless, how the arrangement of mineral and organic fabric can be united without the precipitation of calcium phosphate in unwelcome places. The mitochondrion is often cited (e.g. Brighton and Hunt, 1976) as a source of calcium phosphate, but the osteocyte, where the greatest mineral activity has been observed (Fig. 14a above), contains few mitochondria, and not apparently in any relation to the golgi apparatus. The accumulation of calcium phosphate by mitochondria, however, may reflect

a similar general mechanism within the structure of the
filameatous clusters (Fig. 12e above). The dephosphorylation of substrate will also be attended by the problems
of removal to the residue. In the case of the most likely
substrate, ATP, the residue of ADP (or AMP if the pyrophosphate is subsequently split by pyrophosphatase) will
need to be removed from the site and the energy released
will have to be dissipated or transferred.

From the present evidence, the most reasonable assumption is that the system is first 'loaded' with calcium
(Aaron and Pautard, 1973; 1975, see Pautard, 1976) and
then calcium phosphate (Fig. 14a) through the nuclear
region. However, the translation of this observation into
terms of fabrication and mineralization of large numbers
of filamentous clusters in a few minutes must await much
further information of the sequence of events in the
golgi vacuole.

From the persistent observation of the presence of a
calcium-binding protein and alkaline phosphatase in the
bone matrix and in matrix 'vesicles' (for example, de
Bernard, 1975) it might be assumed that the organic
framework of each filamentous cluster contains alkaline
phosphatase which first binds calcium in particular
regions of the molecule. With rise of pH and influx of
substrate, phosphate is split and combines with calcium
in competition with the protein. However, such a mechanism
would be self arresting, since substrate residues and
aggregating ions would block the entry of fresh organic
phosphate from the vacuolar sap and prevent access to the
reactive enzyme site. This difficulty would seem to require a flow of material to and from each site to ensure
that the entire structure was systematically mineralized,
although the calcium/protein complexes would themselves
form further phosphate foci if they were unobstructed.
The constant width of the mineralized region, together with
the sponge-like character of the organic labyrinth,

suggests that the architecture of the arrangement is at once a template, a framework for enzymes and a mechanical pump for the redisposition of the ions and substrate residues. Certainly, the zonal nature of the mineralization (as in *S. ambiguum*, Fig. 12e, for example), the motility of the spherical particles at the calcification front and the phosphatase activity all suggest that the mineral may be part of a phosphate engine, gaining energy from phosphate substrates and transferring it in various ways throughout each microskeletal element.

It is possible that some alternative pathways of mineralization take place outside the cell, particularly in the looped assemblies. These structures have not been observed within cells; their large size makes it unlikely that they would be missed. The final shape of these objects may be the result of extracellular activity. In this respect, the presence of dense bodies within bone cells, and outside the cells both singly and associated in chains with bridges, might be significant. Similar dense bodies, also detected within golgi bodies, have been reported in developing enamel (Deporter and Ten Cate, 1976; Deporter, 1977). These calcium-rich objects in the ameloblast do not have the fine structure observed in filamentous clusters but they may be the source of the mineral in enamel, which is markedly more crystalline than bone. The dense bodies are apparently extruded through the secretary pole of the ameloblast, to discharge their contents into the enamel matrix. It thus seems that bone itself may use at least two systems, filamentous cluster formation in the golgi apparatus and extruded from the cell as such, and dense bodies, also made in the golgi apparatus, which are extruded from the cell and undergo extracellular changes, during which they become conjoined by bridges which probably interpenetrate and enwrap the collagen fibres. Such a dual system would explain the distinct difference in the mode of formation both inside

and outside the cell, as well as the large crystals observed in association with collagen in bone and tendon
(Nylen *et al.*, 1960 for example).

   With these variations of type it is not possible to
assign any simple compound for calcium phosphate. The
best that can be judged from the scanty evidence presented
above is that the first formed structures, and possibly
most of the filamentous clusters, will have lower Ca/P
ratios than apatite, while the looped assemblies may be
largely hydroxyapatite or fluorapatite, as conjectured in
the scheme for phosphate exchange set out in Fig. 17.

Function

   There have been numerous theories proposed for the
origin of vertebrate calcification (reviewed, for example,
by Halstead, 1969) but the function of phosphate as the
preferred anion has been considered as less important
than calcium. On the assumption that bone mineral is a
homogeneous precipitate of calcium phosphate, the present
consensus of opinion favours the idea that the salt has
mechanical advantages in association with collagen and
provides a store of calcium and phosphorus. Since there
is an abundance of phosphate in man, the most studied
animal, it has become widely accepted that the phosphate
is a safe counterion to toxic calcium, which can be controlled and mediated by the biochemical pathways of phosphorus. The evolutionary view that phosphorus might have
been retained because of the lack of the element in competing animals increasing in dynamism (Pautard, 1962) may
be relevant in an early marine environment, but it does
not seem to be adequate to explain the abundance of phosphate in terrestrial animals, where carbonate might be an
equally satisfactory anion, particularly as it is found
in amounts greater than the average 5% (Table VI above)
in medullary and other bones (Biltz and Pellegrino,
1969).

Some recent evidence from our studies of the developing mouse calvarium suggests that phosphatase activity is not confined to the initial formation of the mineral deposits but takes place within the matrix *after* the dense bone has been laid down. In particular, we have been surprised to find acid phosphatase distributed in patterns of activity that not only appear optically to be in the matrix, but at the ultrastructural level coincident with the topography of the microskeletal elements. Since acid phosphatase is usually associated with degradation (Kuettner *et al.*, 1968) it seems likely that the 'phosphate engine' of the microskeletal components continues to function in the mature state.

With the increasing possibility that many of the 'matrix' properties actually reside in some part of the microskeletal structures and that these patterns of activity may be related to the gross performance of a bone, it is difficult to escape the conclusion that the main function of phosphorus in bone is to translate information about stress (or lack of it) into an appropriate response. With large members of microskeletal structures articulated and correlated with each other, the gross skeleton can not only grow safely, but also react rapidly in ways not feasible with carbonate.

In most invertebrates, calcium carbonate is the principal hardening salt, and the method of growth is either by continual apposition, or by the ecdysis of the old shell and its replacement by a larger, soft model which becomes mineralized. Where an exoskeleton is composed of phosphate instead of carbonate, the alternative salt may be a matter of abrasion resistance. On the other hand, the advantages of phosphate in other invertebrates may reflect the metabolic flexibility of the anion. In *Lingula*, for example, the calcium phosphate is oriented within the shells, which are not hinged, but are free to move relative to each other. Each valve is connected by a complicated web of muscles attached to the animal, which can manipulate and close the shells for feeding and safety. Since the animal is not hinged, the shells grow concentrically and the muscles have to be continually repositioned; the orientation of the mineral particles is in relationto the points of attachment of the muscles to the valves (Kelly *et al.*, 1965). A reflection of this carbonate/phosphate/musculature relationship is to be found in the Crustacea, where the carapace salt is usually calcium carbonate, while the tendons, which transmit the muscular forces to the shell, contain calcium phosphate.

In the vertebrates, the skeleton usually has to grow to a finite size. To do this within the animal, it cannot develop by apposition alone or by ecdysis. It has to increase internally with the minimal disturbance to mechanical function, and, like *Lingula*, it has to reposition the muscles and tendons continually outside each bone, with a corresponding change of alignment within the mineralized tissue. The explanation for the extensive local phosphatase activity may be that changes in stress are signals for scission of the phosphate links which anchor the mineral to the collagen and bond the boundaries between components, thus allowing slip, relieving strain and allowing the new position to be re-anchored in relation to the forces generated by the muscles.

Once growth has ceased, however, the fully hardened bone has to face a lifetime of mechanical insult. The long-term demands on the performance of a bone may be met by the familiar features of osteoclastic resorption and osteoblastic remodelling. Such slow responses, however, cannot accommodate the second-to-second situations which arise from isotonic and isometric effort and it is here that the arrangement of microskeletal components can provide mechanical precision with electronic speed. The energy released from nucleotides in the contracting muscle will be transmitted to the mineral environment in the bone

via the collagen of the muscle fascia, tendons and matrix.
As the demand rises, so the boundaries of all the microskeletal components will need to adjust accordingly. The
mineral/matrix and mineral/mineral interfaces will need to
distribute the load, and the direction of the load, according to the movement; too strong a bond and crystalline
fracture would result; too weak a bond and hysteresis would
result. Since the bonding forces of all kinds may be
needed in a single event, the microskeletal structures
must 'know' how to provide them. Moreover, the system
must permit failure by fracture and slip under extreme
conditions yet recover mechanical performance by repair
and rearrangement. The present interest in microfissures
in bone (Frost, 1960; reviewed by Aaron, 1976) is an ac-
knowledgement that there is a constant shift in the internal structure of bone at microscopic level. The well-
documented facts of mineral loss with bed rest, limb
immobility and weightlessness in space may be the rearrangement of the microskeletal population in the absence
of mechanical stress while the appearance of march fractures in young soldiers and wrist fractures in older women
may be a failure to repair the same microfissuring after
unaccustomed demand on a specific bone.

The success of the vertebrate skeleton may therefore
be the result of the mobility of phosphorus in both inorganic and organic phases, each of which may monitor and
engineer changes in a population of independent components. The capacity of the microskeleton to react may depend on the way in which a congregation of structures
can translocate phosphorus. To test such a possibility,
we need to examine the biochemical behaviour of bone and
its components under stress, about which there is little
information at present.

On land, the largest invertebrates, with rigid carbonate skeletons, move slowly and risk fracture; the
most agile vertebrates suffer little damage. A clumsy

land crab of 5 kg is the goliath of the carbonate carapace; a 5 kg lizard can run rapidly over rocks and look
back to an ancestor weighing 50,000 kg.

Evolution

   Bone is a relatively modern tissue and probably arose
some time after backbones developed from the invertebrates. The selection of calcium phosphate as the preferred salt may have taken place at the same time to
enable the articulated skeleton to function more effectively in ways outlined above. Although the phosphatic
brachiopods left fossils in the rocks some 200 million
years before the earliest fishes, the advanced anatomy
and muscular performance of the animals relate them close
to the prechordates and thus support the idea that calcium phosphate was 'found' at the same time as the advantages of a movable skeleton.

   However, the resemblances between the microskeletal
components in bone, the particles in *Spirostomum ambiguum*
and the inclusions in *Bacterionema natruchotii* are too
close to ignore the possibility that organized deposits
of calcium phosphate may have existed long before the
vertebrates and possibly in archaean times before the
invertebrates and even before the arrival of cells. The
investment of relatively stable extracellular structures
in bone with biochemical behaviour may be the modern
counterpart of such assemblies which may have been present
from the earliest time. It seems appropriate to conclude
a discussion about the role of phosphorus in bone by
summarizing the theme that the element, in some inorganic
state with calcium, may have played a central part in the
success of simple proto-organisms, which may have been
able to improve their capabilities because of the protective architecture of the mineral in the first place
and available phosphorus in a form that allowed translation of energy, construction of organic macromolecules

and distribution of information.

The steps which seem to be the most likely from all the information presented above might first have involved the formation of large numbers of calcium phosphate, or calcium phosphate/carbonate, structures which became progressively more successful as they accumulated, ordered, and were ordered by, the organic molecules trapped within them.

In the first step, the inorganic changes associated with the weathering of igneous phosphate would have produced spherulites of calcium phosphate. Such features of crystallization and metamorphosis are common to many minerals. Silica, for example, may aggregate into 'lepispheres' (Wise and Kelts, 1972) from 0.1–10 μ in diameter with rough or smooth surfaces according to the environment (Oehler, 1975). Explanations for spherulitic crystallization have been put forward by Keith and Padden (1963), who considered that the radiating fibrous habit was the result of instability in the first-formed crystal in the beginning and stabilization of the fibrous form during enlargement of the sphere. Similar explanations have been offered for the origin of phosphate spherules in bone, particularly with respect to the transition between amorphous and crystalline phases in synthetic systems (Nylen *et al.*, 1972; also Eanes, 1975; Füredi-Milhofer, 1975 and discussion thereon). Recent ultrastructural observations of similar structures in bone (Gay, 1977) seek to correlate the experimentation *in vitro* with the observations *in vivo* on the assumption again that the mineral is ultimately dispersed randomly throughout the matrix.

The second step may have been the stabilization of the geological microspheres by the inclusion of organic molecules, either from the aqueous environment, or by synthesis from gases trapped within the growing sphere. There would have been an increasing tendency for the evolving association of mineral and organic molecules to depart increasingly from a simple radial arrangement of crystals towards a labyrinth of mineral with a large surface area.

The third step may have been the advantages of a labyrinth of correct topography and dimension promoting the formation of energy-rich organic phosphate esters. Since, in favoured microspheres, neither the substrate nor the phosphate would escape from the labyrinth, the architecture of the labyrinth walls and the cavities between them would provide spatial templates for the specificity that we now associate with organic macromolecules. The ability to split phosphate and recycle it within the labyrinth would provide the machinery of photo- and thermo-phosphorylation, synthesis and polymerization familiar to the cell.

The fourth step would seem to involve a more complex interaction between groups of microspheres, perhaps in a symbiosis recapitulated by the microskeletal assemblies in bone, allowing exchange of products, greater flexibility and probably locomotion by the polarization of the system.

The final step would be the gradual removal of the mineral labyrinth as the macromolecules formed within took over the function of the inorganic precursor. It would be of interest to consider if cytoplasm itself bears any archetypal fingerprints in this respect.

In this scheme, however possible, the biology is missing. In the train of events which may have led to complex macromolecules, enzymes and ribosomes, there is absent the one step which is the essential hallmark of living structures, namely the replication of forms and their disposition in space. It is not difficult to foresee how, with so much phosphate available in the labyrinth, nucleoside phosphates or their inorganic precursors could have arisen. Moreover, the replication of such molecules by synthetases or by some geometry of the mineral is equally likely, even unavoidable. The problem is that process

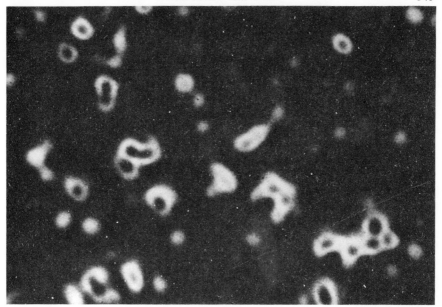

Fig. 18. Senegal phosphorite, dispersed as in Fig. 15c but without treatment with collagenase. The spherical particles and looped assemblies and bridges are clearly illustrated in this negative print, × 3,150.

whereby the information for a given event could be translated into the migration of specific substances in specific directions to form a copy. When we arrive at the final step it is difficult to imagine the fabrication of the simplest structure, even though the raw material is close at hand. Part of the solution, at least, must lie within the golgi saccule of the cell.

The attraction of an evolutionary role for calcium phosphate is that the chemistry and physics of the salt allows the production of large numbers of similar objects while providing phosphorus in safe havens for a potential biochemistry. We are thus spared the mathematically impossible task of contemplating periods of archaean time too long for chance organic collisions in a planet with a finite history. We may well find that the inorganic/ organic dialogues are still in progress in some rocks, while countless early attempts still remain unfulfilled in others.

Addendum

While this chapter was being completed, some current observations in our laboratory are an interesting comment on the discourse that has been presented. In his original paper on bone salt, de Jong classified it in the same group as the geological phosphorites. From that time, the debate has waxed as to the structure of bone mineral on the one hand and phosphorite on the other. From the earlier view that marine phosphorites are of ceprolitic origin, recent information on glauconitic phosphorites in forming beds (Oarker, 1975) tend to support the experiments of Ames (1959) that calcium phosphate may arise by the replacement of calcium carbonate.

In the sample of Senegal phosphorite shown in Fig. 18, however, we have observed, both in preparations treated as in bone in Fig. 15c and without treatment, that the bulk of the material is in the form of looped assemblies closely resembling those shown in Fig. 15f and we conclude that at least in this sedimentary rock, the mineral is most likely to be the intractable remains of the bones and scales of marine animals, thus completing the biological and crystallographic circle of deJong's conclusions.

Acknowledgements

Colleagues in the M.R.C. Mineral Metabolism Unit supplied much information. Prof. B.E.C. Nordin has been most helpful, and I am indebted to Dr. J. Aaron for many drawings.

References

Aaron, J.E. (1973). *Calcif. Tiss. Res.* **12**, 259.
Aaron, J.E. (1976). *In* 'Calcium, Phosphate and Magnesium Metabolism' (B.E.C. Nordin, ed.), p. 298. Churchill Livingstone, Edinburgh, London, New York.
Aaron, J.E. and Pautard, F.G.E. (1973). *In* 'The Cell Cycle in De-

velopment and Differentiation' (M. Balls and F.S. Billett, eds.),
    p. 325. Cambridge University Press, Cambridge.
Aaron, J.E. and Pautard, F.G.E. (1975). *In* 'Calcium Metabolism, Bone
    and Metabolic Bone Diseases' (F. Kuhlencordt and H.P. Kruse, eds.),
    p. 211. Springer-Verlag, Berlin.
Agna, J.W., Knowles, H.C. and Alverson, B. (1958). *J. Clin. Invest.*
    37, 1357.
Ali, S.Y., Sadjera, S.W. and Anderson, H.C. (1970). *Proc. Nat. Acad.
    Sci.* **67**, 1513.
Ames, L.L. (1959). *Econ. Geol.* **54**, 829.
Anderson, H.C. (1969). *J. Cell. Biol.* **41**, 59.
Anderson, H.C. and Reynolds, J.J. (1973). *Develop. Biol.* **34**, 211.
Anon. (1971). *In* 'Phosphorus and Potassium', No. 55, p. 21. The
    British Sulphur Corp., London.
Arnott, H.J. and Pautard, F.G.E. (1967). *Israel J. Med. Sci.* **3**, 657.
Arnott, H.J. and Pautard, F.G.E. (1970). *In* 'Biological Calcifica-
    tion' (H. Schraer, ed.), p. 375. Appleton-Century-Downs, New York.
Bachra, B.N. (1967). *Clin. Orthop.* **51**, 199.
Bachra, B.N. (1972). *Calcif. Tiss. Res.* **8**, 287.
Bachra, B.N. (1975). *In* 'Physico-chimie et Cristallographie des
    Apatites D'intérêt Biologique', p. 247, Editions du Centre National
    de la Recherche Scientifique, Paris.
Bauer, G.C.H., Carlsson, A. and Lindquist, B. (1955). *Kgl. Fysiograph.
    Sällskap. i. Lund. Förh.* **25**, 1.
Baylink, D., Wergedal, J., Stauffer, M. (1971). *J. Clin. Invest.* **50**,
    2519.
Belanger, L.F. (1960). *Calcif. Tiss. Res.* **4**, 1.
Bernard, G.W. and Pease, D.C. (1969). *Am. J. Anat.* **125**, 271.
Berry, E.E. and Leach, S.A. (1967). *J. Inorg. Nucl. Chem.* **29**, 317.
Bezrukhov, P.L. (1939). *Doklady Acad. Sci. URSS.* **24**, 282.
Biltz, R.M. and Pellegrino, E.D. (1969). *J. Bone Jt. Surg.* **51**, 456.
Bohatirchuk, F. (1965). *Am. J. Anat.* **117**, 287.
Bonucci, E. (1967). *J. Ultrastruct. Res.* **20**, 33.
Bowden, J. (1960). *Arch. Oral. Biol.* **2**, 319.
Bragg, W.L. (1937). *Atomic Structure of Minerals*, Cornell University
    Press.
Brighton, C.T. and Hunt, R.M. (1976). *Clin. Orthop.* **100**, 406.
Burkinshaw, L., Marshall, D.H., Oxby, C.B., Spiers, F.W., Nordin,
    B.E.C. and Young, M.M. (1969). *Nature (Lond.)* **222**, 146.
Carlsson, A. (1951). *Acta Pharmacol. (Kbh)* **7** (Suppl.), 1.
Dallemagne, M.J. (1945). *Actualities Biochem.* **2**, 5.
Day, H.G. and McCollum, E.V. (1939). *J. Biol. Chem.* **130**, 269.
Deporter, D.A. and Ten Cate, A.R. (1976). *Arch. Oral Biol.* **21**, 7.
Deporter, D.A. (1977). *Calcif. Tiss. Res.* **24**, 271.
Eanes, E.D. (1975). *In* 'Physico-chimie et Cristallographie des
    Apatites D'intérêt biologique', pp. 295-301, Editions du Centre
    National de la Recherche Scientifique, Paris.
Eanes, E.D. and Posner, A.S. (1970). *In* 'Biological Calcification'
    (H. Schraer, ed.), p. 1. Appleton-Century-Downs, New York.
Earland, C., Blakey, D.R. and Stell, J.G.P. (1962). *Nature (Lond.)*
    196, 1287.
Elliot, J.C. (1965). *In* 'Tooth Enamel' (M.U. Stack and R.W. Fearn-
    head, eds.), p. 20. John Wright & Sons, Bristol.
Ennever, J. and Creamer, A. (1967). *Calcif. Tiss. Res.* **1**, 87.

F.G.E. PAUTARD

Faure-Fremmiet, E. (1957). *J. Protozool.* **4**, 96.

Fincham, A.G., Graham, G.N. and Pautard, F.G.E. (1965). *In* 'Tooth Enamel' (M.U. Stack and R.W. Fearnhead, eds.), p. 117.

Fincham, A.G. (1969). *Calcif. Tiss. Res.* **3**, 327.

Fitton-Jackson, S. (1957). *Proc. Roy. Soc. (London)* **146**, 270.

Follis, R.J. (1952). *J. Biol. Chem.* **194**, 223.

Frost, H.M. (1960). *J. Bone Jt. Surg.* **42A**, 447.

Frost, H.M. (1960). *Henry Ford Hosp. Bull.* **8**, 25.

Füredi-Milhofer, H., Bilinski, H., Brečević, Lj., Despotović, R., Filipović-Vincekovic, N., Oljica, E., and Purgarić, B. (1975). *In* 'Physico-chimie et Cristallographie des Apatites D'intérêt biologique', pp. 303-310. Editions du centre Nationale de la Recherche Scientifique, Paris.

Gay, C.V. (1977). *Calcif. Tiss. Res.* **23**, 215.

Glass, H.I. and Nordin, B.E.C. (1963). *Phys. in Med. Biol.* **8**, 387.

Glimcher, M.J. (1959). *Revs. Modern Phys.* **31**, 359.

Glimcher, M.J. (1960). *In* 'Calcification in Biological Systems' (R.F. Sognnaes, ed.) p. 421. American Association for the Advancement of Science, Washington.

Halstead, L.B. (1969). *Proc. Malacol. Soc., Lond.* **38**, 552.

Harper, R.A. and Posner, A.S. (1966). *Proc. Soc. Exptl. Biol. Med.* **122**, 137.

Hirschmann, P.N. and Nichols, G. (1972). *Calcif. Tiss. Res.* **6**, 67.

Hodges, T.K. and Hanson, J.B. (1965). *Plant Physiol.* **40**, 101.

Hohling, H.J., Ashton, B.A. and Koster, H.D. (1974). *Cell. Tiss. Res.* **148**, 11.

Irving, J.T. (1964). *In* 'Mineral Metabolism. An advanced Treatise' (C.L. Comar and F. Bronner, eds.), Vol. II, p. 249. Academic Press, New York and London.

de Jong, W.F. (1926). *Rec. Trav. Chem.* **45**, 445.

Kashiwa, H.K. (1970). *Clin. Orthop.* **70**, 200.

Kashiwa, H.K. and Komorous, J. (1971). *Anat. Rec.* **170**, 119.

Keith, H.D. and Padden, F.J. (1963). *J. Appl. Phys.* **34**, 2409.

Kelly, P.G., Oliver, P.T.P. and Pautard, F.G.E. (1965). *In* 'Calcified Tissues' (L.J. Richelle and M.J. Dallemagne, eds.), Universitié de Liege, Liege.

Kuettner, K.E., Guenther, H.L., Ray, R.D., and Schumacher, G.F.B. (1968). *Calcif. Tiss. Res.* **1**, 298.

Lehninger, A.L. (1965). 'The Mitochondrion'. W.A. Benjamin, New York, p. 164.

Lowell, W.R. (1955). *Econ. Geol.* **50**, 715.

Luben, R.A., Sherman, J.K. and Wadkins, C.L. (1973). *Calcif. Tiss. Res.* **11**, 39.

Mathieu, H., Cuisinier-Gleizes, P., Debove, F., Benest (1972). *D. Arch. franc. Ped.* **29**, 603.

McConnell, D. (1938). *Amer. Min.* **23**, 1.

McConnell, D. (1952). *J. Dent. Res.* **31**, 53.

McElroy, W.D. and Glass, R. (1951). 'Phosphorus Metabolism', Vols. I and II, Johns Hopkins Press, Baltimore.

McGregor, D.J. (1963). 'High-Calcium Limestone and Dolomite in Indiana', p. 1, Bulletin 27, Indiana Geological Survey.

McLean, F.C. (1958). *Science* **127**, 451.

Marshall, J.H. (1967). *In* 'Compartments, Pools and Spaces in Medical

Physiology', A.E.C. Symp. Ser. 11. U.S.A.E.C. Division of Techni-
cal Information.
Marshall, D.H. (1976). *In* 'Calcium, Phosphate and Magnesium Metabol-
ism' (B.E.C. Nordin, ed.), p. 257. Churchill Livingstone, Edin-
burgh, London and New York.
Martin, J.H. and Matthews, J.L. (1970). *Clin. Orthop.* **68**, 273.
Matthews, J.L., Martin, J.H., Arsenis, C., Eisenstein, R., Kuettner,
K. (1971). *In* 'Cellular mechanisms for Calcium Transfer and
Homeostasis' (G. Nichols and R.H. Wasserman, eds.), p. 239.
Academic Press, New York.
Meyer, J.L. and Eanes, E.D. (1978). *Calcif. Tiss. Res.* **25**, 59.
Neer, R., Berman, M., Fisher, L. and Rosenberg, L.E. (1967). *J. Clin.
Invest.* **46**, 1364.
Nordin, B.E.C., MacGregor, J. and Bluhm, M.M. (1963). *Clin. Sci.*
24, 301.
Nordin, B.E.C. (1976). *In* 'Calcium, Phosphate and Magnesium Metabol-
ism' (B.E.C. Nordin, ed.), p. 1. Churchill Livingstone, Edinburgh
London and New York.
Nylen, M.U., Eanes, E.D., Termine, J.D. (1972). *Calcif. Tiss. Res.* **9**,
95.
Nylen, M.U., Scott, D.B. and Mosley, V.M. (1960). *In* 'Calcification
in Biological Systems' (R.F. Sognnaes, ed.), p. 129. American
Association for the Advancement of Science, Washington.
Oehler, J.H. (1975). *J. Sediment. Petrol.* **45**, 252.
Outka, D.E. and Williams, D.C. (1971). *J. Protozool.* **18**, 285.
Park, H.Z. and Kashiwa, H.K. (1975). *Calcif. Tiss. Res.* **19**, 189.
Parker, R.J. (1975). *J. Sediment. Petrol.* **45**, 230.
Pautard, F.G.E. (1959). *Biochim. Biophys. Acta* **28**, 514.
Pautard, F.G.E. (1962). *Clin. Orthop.* **24**, 230.
Pautard, F.G.E. (1963). *Nature (Lond.)* **199**, 531.
Pautard, F.G.E. (1964). *In* 'Modern Trends in Orthopaedics' (J.M.P.
Clark, ed.), p. 13. Butterworths, London.
Pautard, F.G.E. (1965). *In* 'Calcified Tissues' (L.J. Richelle and
M.J. Dallemagne, eds.), p. 347. University of Liege, Liege.
Pautard, F.G.E. (1966). *In* 'Calcified Tissues' (H. Fleisch, H.J.J.
Blackwood and M. Owen, eds.), p. 108. Springer, New York.
Pautard, F.G.E. (1972). *In* 'The Comparative Biology of Extracellular
Matrices' (H.C. Slavkin, ed.), p. 440. Academic Press, New York
and London.
Pautard, F.G.E. (1975). *In* 'Physics-chimie et Cristallographie des
apatites D'interet Biologique', p. 93. Editions du Centre National
de la Recherche Scientifique, Paris.
Pautard, F.G.E. (1976). *In* 'The Mechanisms of Mineralization in
Invertebrates and Plants' (N. Watabe and K.M. Wilbur, eds.).
University of South Carolina Press, Columbia, South Carolina.
Pienaar, R.N. (1969). *J. Phycol.* **5**, 321.
Pienaar, R.N. (1971). *Protoplasma* **73**, 217.
Posner, A.S. and Perloff, A. (1957). *J. Res. Nat. Bur. Standards* **58**,
279.
Posner, A.S. (1961). *In* 'Phosphorus and its compounds' (J.R. van
Wazer, ed.), Vol. 2, p. 1429.
Pritchard, J.J. (1952). *J. Anat.* **56**, 259.
Quekett, J. (1846). *Trans. Microsc. Soc. Lond.* **2**, 46.
Quinaux, N. and Richelle, L.J. (1967). *Israel J. Med. Sci.* **3**, 677.

Robertson, W.G. (1976). *In* 'Calcium, Phosphate and Magnesium Metabolism' (B.E.C. Nordin, ed.), p. 230. Churchill Livingstone, Edinburgh, London and New York.
Robinson, R.A. and Cameron, D.A. (1956). *J. Biophys. Biochem. Cytol.* 2 (suppl.), 253.
Robinson, R.A. and Watson, M.L. (1955). *Ann. N.Y. Acad. Sci.* 60, 596.
Robison, R. (1923). *Biochem. J.* 17, 286.
Rolle, G.K. (1969). *Calcif. Tiss. Res.* 3, 142.
Rösler, P.J. and Lange, H. (1972). 'Geochemical Tables'. Elsevier, Amsterdam, London, New York.
Schewiakoff, W. (1894). *Zeit. Wiss. Zool.* 58, 32.
Schmidt, W.J. (1923). *S.B. niederrhein. Ges. Nat. -u. Heilk.* Vol. 1.
Scott, B.L. and Pease, D.C. (1956). *Anat. Res.* 126, 465.
Silberman, M.S. and Frommer, J. (1974). *Clin. Orthop.* 98, 288.
Slavkin, H.C. (1975). *In* 'Physico-chimie et Cristallographie des apatites D'intérêt Biologique', p. 162. Editions du Centre National de la Recherche Scientifique, Paris.
Sobel, A.E. and Burger, M. (1964). *Proc. Soc. Exp. Biol. Med.* 87, 7.
Strates, B. and Neuman, W.F. (1958). *Proc. Soc. Exptl. Med.* 97, 688.
Talmadge, R.V. (1969). *Clin. Orthop.* 67, 210.
Taves, D.R. (1965). *Clin. Orthop.* 42, 207.
Termine, J.D. and Posner, A.S. (1967). *Calcif. Tiss. Res.* 1, 8.
Termine, J. (1972). *In* 'The Comparative Biology of Extracellular Matrices' (H.C. Slavkin, ed.), p. 443. Academic Press, New York and London.
Tomes, J. and deMorgan, C. (1853). *Phil. Trans. B.* 143, 109.
Triffit, J.T., Terepka, A.R. and Neumann, W.F. (1968). *Calcif. Tiss. Res.* 2, 165.
Watabe, N. (1967). *Calcif. Tiss. Res.* 1, 114.
Wilkinson, R. (1976). *In* 'Calcium, Phosphate and Magnesium Metabolism' (B.E.C. Nordin, ed.), p. 36. Churchill Livingstone, Edinburgh, London and New York.
Winand, L. (1965). *In* 'Tooth Enamel' (M.V. Stack and R.W. Fearnhead, eds.), p. 15. John Wright & Sons, Bristol.
Wise, S.W. and Kelts, J. (1972). *Trans. Calif. Coast Geol. Soc.* 22, 177.
Wolpers, C. (1949). *Grenzgeb. d. Med.* 2, 527.
Young, R.A. and Elliot, J.C. (1966). *Arch. Oral Biol.* 2, 190.
Zipkin, I., McClure, F.J. and Lee, W.A. (1960). *Arch. Oral Biol.* 2, 190.
Zipkin, I. (1966). *In* 'The Science of Nutrition and its Application in Chemical Dentistry' (A.E. Nizel, ed.). W.B. Saunders Co., Philadelphia.
Zipkin, I. (1970). *In* 'Biological Calcification' (H. Schraer, ed.), p. 69. Appleton-Century-Crofts, New York.

10 HEAVY METALS IN MEDICINE

R.D. Gillard

Department of Chemistry, University College,
P.O. Box 78, Cardiff, CF1 1XL, Wales

Introduction

Scope of the Article

So far as we know at present, heavy elements are of rare natural occurrence in biology. For elements with atomic mass numbers greater than 100 this is true for all terrestrial natural occurrence (geological or biological). Apart from molybdenum and iodine, there seems to be no demonstrated natural (biological) function for any element with an atomic number greater than 40.

In this sense, most heavy elements are 'unnatural' in human beings. However, this is not to say that they do not occur or are not potentially useful as additives. Indeed, the subject and scope of this article will be concerned with these 'foreign' elements of no currently known biological function. It will therefore exclude molybdenum and iodine.

The Analytical Composition of Human Beings

Not surprisingly, there is a great deal of information on the levels of most elements in human beings. Before attempting to summarise some of this information, it would be as well to make the caveat that many analyses are of doubtful validity. As Bowen (1966) pointed out, analytical

errors may be enormous. For example, the formalin used for
preservation of samples is often heavily contaminated with
metal compounds, and these vitiate analyses of the preserved samples. From the published work on the analytical
composition of mammalian soft tissues much refers to human
tissue, and Table 1 gives some typical results for some of
the heavier elements, with a few 'normal' elements for
comparison.

Among the elements in Table 1, there are apparently
a number of special sites of accumulation in the body,

TABLE I

Analyses^a of mammalian soft tissue

Element	Brain	Lung	Liver	Kidney
Silver	0.04	0.005	0.03	0.005
Barium	0.012	0.67	0.007	0.06
Gold	0.5[b]	0.3[b]	0.0001[c]	0.5[b]
Mercury	–	0.03[e]	0.022[e]	0.25[d]
Lead	0.24	2.3	4.8	4.5
Calcium	320	480	140	390
Iron	200	1300	520	290

[a]Results are in ppm of *dried* tissue: they are taken except as
noted from I.H. Tipton and M.J. Cook, *Health Phys*. 1963, **9**, 103.
[b]Results from International commission on Radiological Protection,
1964. Publication 6 on permissible doses for internal radiation.
[c]R.M. Parr and D.M. Taylor (1963) *Physics Med. Biol*. **8**, 43.
[d]A. Stock (1940) *Biochem. Zeit*. **304**, 73.
[e]R.M. Forbes, A.R. Cooper and H.H. Mitchell (1954) *J. Biol. Chem*.
209, 857.

including the concentration of barium by bone (Sowden
and Price, 1958) and by the choroid of the eye (Garner,
1959).

This survey of the use of heavy metals in medicine will
first mention some of the classical uses, both of the
metals themselves, and of their compounds. The discussion
of the compounds will be concerned with both the beneficial classical uses, and also will mention in passing
some of the problems of toxicity.

Finally, the chapter will describe some of the newer developments in the use of compounds of heavy metals in medical contexts.

Classical Uses of Heavy Metals

The Metals

Perhaps the most common use of the metals themselves is as a source of inert strong materials. Chief among such applications have been those in dentistry: typical formulations include those for fillings, where the commonest constituent is mercury. The dangers and difficulties of this do not need further explanation here, although it is perhaps worth remembering that the Mad Hatter in 'Alice in Wonderland' by Lewis Carroll (derived from the proverbial phrase 'mad as a hatter') allegedly owed his symptoms to mercury poisoning encountered among the workers in the hat industry, who used solutions of mercuric compounds as a depilatory agent in preparing the pelts for making hats.

The cosmetic dental alloys which are used in so-called 'gold' fillings are nowadays made of a surprising variety of metals. For example, two recently patented compositions are as follows. One low intrinsic value alloy for dental use consists (How Medica Inc, 1977) of gold (0 to 45%), platinum (0 to 30%), palladium (0 to 20%), copper (30 to 55%), gallium (5 to 10%), zinc (0 to 1%), and iridium (0 to 0.01%), with the actual composition controlled by the fact that the total of gold, platinum and palladium should be at least 35%. A similar alloy, containing less gold than usual, is described in another patent (Johnson Matthey Co. Ltd., 1977) as containing palladium (45–62%), silver (5–22%), gold (25–42%), up to 2% iridium, and up to 1% tin, and a maximum of 5% gallium.

Although such techniques are now less common than formerly, there was a clear need many years ago for strong malleable materials to act as strengthening and reinforcing bodies in surgical operations. For example, in cases of fracture of bones in elderly patients, the healing time is lengthy, and therefore some extraneous support is often useful. For this purpose, the metal silver was often employed, in the form of pins. Some heroic surgery was also carried out, particularly for severely war-wounded soldiers, using silver plates as structural components.

The utilization of such metal implants often results in surprising rates of dissolution. For example, metallic copper dissolves in bovine serum albumin and other proteins. This was discovered as a result of the observation (Osler, 1971) that copper is lost (at about 60 μg per day) from intrauterine devices coated with copper. Clearly, any habitual contact with objects made of a particular metal may lead to an increased levels of compounds of that metal in the tissues (as in the use of copper bracelets for rheumatic conditions) and it would be good to have statistical analysis of the possible (negative?) correlation between wearing gold jewellery and arthritic conditions (or platinum and cancer).

In a similar way, an analysis would be valuable of the level of occurrence of affective disorders (like schizophrenia) among cigarette smokers, since the lithium content of tobacco ash is known to be high. The point here is that lithium carbonate and other simple salts have, during the past ten years, proved remarkably useful in reducing the frequency of occurrence of unipolar affective disorders. One might perhaps expect the regular ingestion of lithium salts to give smokers a decreased incidence of schizophrenia or other such mental problems as unipolar mania or depression.

Compounds

The beneficial use of compounds of the heavy metals, which do not occur naturally, has, until recently, been

relatively restricted, except in so far as will be discussed in the therapeutic connection in the next section.
However, there have been a few long-standing applications,
including the use of 'barium meals' as a source of opacity
for X-rays. The basic principle of the use of barium sulphate in this application is the extreme insolubility of
the salt under physiological conditions of pH. (The high
scattering factor for the barium ions of course provides
the X-ray opacity). However, because of the successive
equilibria of the sulphate ion with protons, in acid
media, the concentration of sulphate ions is lowered, and
in strongly acid solution, this leads to a dissolution
of the barium sulphate. This phenomenon was apparently
known (Gillard, 1976) to Sherlock Holmes!

Since barium salts are classified as 'Schedule 1
poisons' in the United Kingdom, ingestion of any soluble
salt of barium is obviously to be avoided. The origin of
the toxicity seems not to be known, although in view
of the general similarity of the chemistries of barium and
calcium, one might expect the function of barium in
toxicity to be as an inhibitor of calcium metabolism.
Indeed, there is not even universal agreement as to the
toxicity of barium. There is a suggestion (Rygh, 1949)
that both barium and strontium may be essential elements
for mammals. This suggestion seems not to have been subsequently supported, although there are statements in the
literature (Odum, 1951; Vinogradov, 1953) that the sul-
phates of strontium and barium are used in the exoskeleton of, respectively, the radiolarian *acantharia*,
and the rhizopod xenophyophora, respectively.

In interpreting these remarks, it is as well to remem-
ber that toxicity is a complex property. Thus, for example, although pure salts of strontium are toxic to
angiosperms, about two-thirds of the calcium may be
replaced (Scharer, 1955) by strontium, with little apparent ill effect. Similarly, although arsenic, in the form

of arsenate, or arsenious oxide, is a classical poison,
the toxicity of arsenate is certainly reduced considerably in the presence of an excess of phosphate. The
mechanisms of such interferences and competitions are
not clearly understood, and, in the case of arsenic, are
highly reminiscent of the extraordinary phenomenon of
the so-called 'arsenic eaters', where, in a number of
well-documented instances, humans have become accustomed
by constant dosage to an intake of arsenic which would
be lethal to the normal person. Among the better-known
examples of this weird phenomenon was Mr. Maybrick,
whose wife was imprisoned in the late 19th century for
his murder, by arsenic.

Although toxicity is not a primary concern of this
essay, there are a number of applications of compounds
of normally 'toxic' heavy metals which are utilized in
chemotherapeutic applications. The example of 'salvarsan
606', discovered by the father of chemotherapy, Ehrlich,
is too well known to merit description, but the use of
the mercurial diuretics is worth description.

These mercurial diuretics are particularly commonly
employed, and work very well, their function being to
promote the formation of urine. The general class of
structure is shown as I. The group R is

$$R-CH_2-CH-CH_2HgX$$
$$|$$
$$OH$$

(I)

typically a hydrophilic polar moiety. Although these
derivatives of propan-2-ol are all potentially resolv-
able, there seems to have been no study of the influence
of the optical hand upon their efficiency. Since it is
known (Hughes, 1972) that these mercurial diuretics
operate by means of inhibition of sulphydryl enzyme

sites, it seems that one might expect stereoselective effects in the inhibition at the enzyme active site.

The type of mercurial used is given by means of an intramuscular injection, because the absorption is too slow or too little for oral administration, and intravenous injection is ruled out by their high toxicity. The diuretics are hydrolysed or otherwise decomposed in the kidneys, and one product is the mercuric ion. It is, of course, possible that this may itself be the active agent. Stereoselective experiments would be interesting in this connection.

Since the mercuripropanols inhibit ATP-ase and other SH-containing enzymes, they have been used to counteract infection by bacteria, where presumably the same property of combination with the thio-groups is involved. The general affinity of thiol groups for such heavy (class b) metals as lead, arsenic, or mercury (hence, their old name, mercaptan) have led to their employment as chelating agents for removing the polarisable metal ions from cells. For example, British anti-lewisite is shown in (II), and functions by means of binding the arsenic atom of the chlorovinylarsine poison gas lewisite. Similarly, β-penicillamine, shown in (III) has been employed (Chisolm, 1968) in the treatment of lead poisoning.

$$CH_2OH$$
$$|$$
$$CH-SH$$
$$|$$
$$CH_2-SH$$

(II)

$$H \quad C(CH_3)_2 SH$$
$$\backslash \quad /$$
$$C$$
$$/ \quad \backslash$$
$$H_2N \quad CO_2H$$

(III)

The mode of binding of the penicillamine in such applications (and cf. the treatment of Wilson's disease) is
not always evident, since as will be clear from the
formula, it is a substituted cysteine. Linkage isomerism
in coordination complexes of bidentate cysteine is well-
known, though not often referred to as such.

Among the many other applications of heavy metals or
their compounds, the use of palladium (either as 'palladium black' or palladium chloride) as a catalyst in
bacterial culture media may be mentioned. The object
(Mylroie and Hungate, 1954) is to lower the oxidation
potential of the medium in order to recover strict
anaerobes, but in a recent study (Owens, Rolfe and
Hentges, 1976) of cost-effectiveness, the value of the
procedure has been questioned. In no case does the incor-
portation of palladium salts into the bacterial medium
seem to have given noteworthy effects on morphology or
metabolism.

Newer Developments

The section will describe in varying detail the more
recent applications of compounds of the more noble metals
in chemotherapy. As has been pointed out a number of
times, until very recently, systematic trials in therapeutic contexts of compounds containing metal ions had
been notable by their absence. Nearly all the work had
been done with organic compounds, and their metal derivatives have hardly ever been studied, until the late
1960s, in a systematic way.

There had been certain exceptions to this, notably
the study (Kirschner et al., 1966) by Kirschner of the
metal complexes of known cancerostatic compounds. The
principal finding was that metal complexes were certainly
active, although probably at no greater level than that
of their parent organic ligand.

Chrysotherapy

An area where compounds of a noble metal had been used
extensively for many years is that called 'chrysotherapy'
(Greek χρυσός = gold). Gold salts have been used in the
treatment of rheumatoid arthritis for over 40 years,
but little is known about the physiological or chemical
mechanisms whereby they alleviate rheumatoid disease
activity, although they certainly do so to considerable
effect. The gold is normally administered in the form of
a complex, usually with a sulphur-containing ligand. The
commonest treatments appear to be with sodium aurothiomalate, and with 'aurothioglucose'. The nature of these
compounds will be discussed later. Other compounds of
gold which have been used, probably less commonly in
recent times, include (Freyberg *et al.*, 1941) sodium
gold thiosulphate, and 'colloidal gold sulphide'. The
most common gold solution employed for intramuscular
injection is the so-called 'myochrysin', which is a
sodium aurothiomalate (sold in the United Kingdom by May
and Baker).

In the inflamed area giving rise to rheumatoid pain,
whatever the cause of the inflammation may be, whether
it arises from traumatic or immunological damage to the
tissues, it is commonly found that the area of inflammation accumulates large numbers of macrophages and also
neutrophile polymorphs, which seem to originate in the
bone marrow. In rheumatoid arthritis, one cardinal
feature is the inflammation of synovial tissues, and it
is known (Chayen and Bitensky, 1971) that the lysosomal
enzymes of these macrophages and polymorphs present in
rheumatoid pannus are actively concerned in the destruction of articular cartilage.

In recent studies of chrysotherapy, the phagocytic
activity of the macrophages and polymorphs was suppressed
in rheumatoid patients receiving treatments with gold
salts, and a progressive reduction in this macrophage

phagocytic activity was found (Jessop *et al.*, 1973)
during serial observations on some patients receiving
injections of gold weekly.

Another study correlated results on levels of gold
in the serum of patients with rheumatoid arthritis receiving injections of myochrysin with the response to
therapy, and the development of toxicity. In addition to
the gold serum determination, assessments were made of
the number of painful joints on full range of active
movement, duration of morning stiffness and functional
capacity. The chief criterion for clinical response to
therapy was based on a reduction by half in the number of
painful joints on full range of active movement. In this
particular study (Jessop and Johns, 1973), of 33 patients
with rheumatoid arthritis, 21 had a beneficial response
to gold. Jessop and Johns confirmed an older observation
that gold salts are more likely to be effective if given
early in the course of active rheumatoid disease. There
was no obvious correlation between the level of gold in
the serum and a response to therapy in this study (Jessop
and Johns, 1973).

A later study (Vernon-Roberts *et al.*, 1976) used a
complex of silver as a staining method to reveal the presence of gold (by means of the visible black reaction product formed by the deposition of metallic silver on the
underlying metallic gold). Gold distribution was studied
in the tissues of several rheumatoid patients who had
died after stopping chrysotherapy with sodium aurothiomalate, and in some samples of synovial tissue removed
surgically at intervals during chrysotherapy in five
other patients. Among other remarkable findings are that
gold persists in the synovial (and indeed some other)
tissues for up to 23 years after stopping chrysotherapy,
and, in particular, that gold is indeed selectively concentrated within inflamed synovial tissues during chrysotherapy. Interestingly, little or no gold is found in

joints which are not affected by the inflammatory process at any time.

It seems likely that the beneficial therapeutical
effects of gold salts are mediated through their action
on macrophages. The detailed molecular basis for any such
activity is unknown, although there have recently been
some studies of the binding of gold compounds to such components of the connective tissues as collagen.

Although the nature of the therapeutically reactive
compounds of gold presents a promising field for chemical
research, the nature of myochrysin seems not to have been
studied in any great detail, although it clearly contains
a complex (or complexes) of thiomalate with the gold(I)
ion. The chemistry of the thiosulphate complexes with
gold (occasionally used in chrysotherapy) is rather better
known.

The most common ('stable') complex in the gold-
thiosulphate system is (Clusky and Eichelberger, 1926) the
highly charged $[Au(S_2O_3)_2]^{3-}$ ion. The sodium salt is
quite soluble in water, but will give stable crystals
of its dihydrate $Na_3[Au(S_2O_3)_2].2H_2O$, which may be dehydrated around 150° allegedly without decomposition.
This sodium salt seems to have been one of the earliest
chrysotherapeutic agents, under the name 'Sanochrysin'.
It was used (H. Möllgaard, 1924) in treating tuberculosis.
It is interesting, in view of the biological activity of
gold thiosulphates, that the much better known analogous
complex compounds* of silver(I), e.g. $Na[Ag(S_2O_3)_2]$ have
an extraordinarily sweet taste.

Dwyer's Antibacterial Chelates   •

Of the 140,000 compounds screened by the United States
National Cancer Institute before 1970 for potential anti-

---

*Their formation was first noted in 1819 by J.F.W. Herschel
(the younger) in the development of his process for fixing photographic negatives by using hypo!

tumor activity, only about a dozen were (Rosenberg,
1971) inorganic, in any sense. However, there had been
a certain amount of work, chiefly in Australia, upon the
effect of transition metal complexes on the growth of
bacteria. Albert had showed, in a classical piece of
work (Albert, 1968), that the anti-bacterial activity
of 8-hydroxy-quinoline was due to the formation of metal
complexes in the culture medium. Little activity against
bacteria was found when 8-hydroxyquinoline (oxine) was
added to culture media which did not contain heavy metals,
but the addition of such heavy metals as copper or iron
(ferrous or ferric) to the culture media led to the formation of oxinate-metal complexes which were biologically
active.

In other extensive work in Australia, under the in-
spiration of F.P. Dwyer, octahedral complexes of phenanthroline and bipyridyl with certain metals were tested
against a variety of micro-organisms for bactericidal
and bacteriostatic activity. In most cases, the activities found were compared (Dwyer *et al*., 1969) with those
of the free ligands (which were added as their salts,
such as the hydrochlorides, methiodides, or ethiodides).
Commonly, the complexes of the metal ions were found to
be more active than were the free ligands, although,
occasionally, in a fixed time assay, equal activity was
displayed by the chelate compounds and their free ligands.

Metal chelates with similar bactericidal titres following a 2 day incubation with *Staphyloccus aureus*,
Streptococcus pyogenes, or *Escherichia coli* differed
greatly (Butler *et al*., 1969) in their bacterial activities in kinetic studies. In general, the complexes of
the kinetically labile metal ions (for example copper,
cadmium, zinc, or manganese) were found to be more active
than were the analogous, but kinetically inert, compounds
of iron(II), nickel or ruthenium(II). In experiments
utilising the allegedly cobaltous chelate complexes with

phenanthroline, the bactericidal action was relatively
slow, and this was thought to be due to formation of
inert cobaltic complexes in solution, since it could
hardly be a property of the rather labile cobalt(II)
configuration. This explanation was supported by an
experiment (Cade *et al*., 1970) in which the strict
anaerobe *Fusiformis nodosus* in a similar set of tests
was killed most rapidly by a methyl-substituted phenanthroline complex of cobalt(II) under conditions of growth
which clearly (strictly anaerobic) would not favour oxidation from kinetically labile cobaltous to kinetically
inert cobalt(III).

The activity of complexes containing methyl substituents in the phenanthroline or bipyridyl ligands was in
general increased, relative to the unsubstituted.
In line with general experience in a plenitude of other
anti-bacterial studies, this presumably would arise from
the increased lipophilic character of such complexes,
which would allow them more effectively to penetrate to
sites of action within the cell. However, the highest
activity was shown by the water soluble tris-complex
compounds of 5-nitro-phenanthroline, with the metal ions
copper(II), iron (II), and nickel against *Mycobacterium
tuberculosis*. Here, bacteriostasis was achieved at the
extraordinarily low molarity, 3×10^{-8} molar, in a 14 day
assay. The high activity of these particular complexes
may be indicative of a special mode of action, as outlined
subsequently.

Clinical trials were relatively disappointing, in that
selected chelates were not effective (Dwyer, 1969) against
Mycobacterium tuberculosis, *S. aureus*, and *S. pneumoniae*
in mice infected with these organisms. Such compounds
appear to have more potential for practical use as topi-
cal anti-bacterial agents. The tris complex of the nickel
ion with tetramethyl-1,10-phenanthroline is (Butler,
1970) effective in controlling staphylococcal infection

in the newly born, and in the control of a number of skin diseases, such as acne vulgaris.

The analogous complex of the manganous ion has also been found (Cade *et al.*, 1970) to inhibit the spread of dermatological infections due to dermatophytes or candida. This last piece of research described a controlled clinical trial on the treatment of dermatological infections with the manganese(II) tris-chelate complex of phenanthroline.

The mechanism of action of these complex compounds is not as yet clear. It may well be that their biological activity stems from their dissociation into the free ligand (which is the actual active species) and the metal ion. Such an explanation would be supported by the fact that kinetically labile complexes are in general more rapidly active than their inert analogues. The anti-bacterial activities of the free ligands (phenanthroline or bipyridyl) are in some cases sufficient to account for the observed activities, and it may well be that one of the functions of the metal complexes is simply to serve as an efficient transporter of the M-heterocyclic ligand to the site of action, rather along the lines which Albert discovered for 8-hydroxyquinoline.

However, there are some observations which do suggest that this cannot be the whole story. It may well be that, in some cases, the metal complex is involved in the biological activity. In all cases, clearly, a combination of both may well be possible. Among the observations suggesting that the metal complexes may have their own integral activity is that tris-1,10-phenanthrolineruthenium(II) ion displays anti-choline esterase activity, but is known (Koch *et al.*, 1957) to remain unaltered in animal tissues. This study concerned the metabolic fate of the complex, labelled with isotopic ruthenium (^{106}Ru). The compound certainly possesses anticholine esterase and curare-like activity.

Similarly, the complex tris-5-nitro-1,10-phenanthro-
linenickel(II) is active against M. tuberculosis at extremely low concentrations $(10^{-7}M)$. One feature of these
chelated systems which was not apparent at the time that
these antibacterial experiments were done should now be
mentioned, in case it may prove necessary to incorporate
it into a full description of the mode of action.

The two complex compounds which manifest activity and
which must arise from the intact metal complexes are
$[Ru(phen)_3]^{2+}$ and $[Ni(5-NO_2-phen)_2]^{2+}$. In recent work on
such complex compounds of N-heterocycles with transition
metal ions, the following equilibria have been demonstrated. K_S represents the equilibrium called 'covalent
hydration' and the subsequent acid dissociation (described by K_A) of the covalent hydrate (V) forms the
pseudo-base (VI). Among the available new results are:

i) $Ni(phen)_3^{2+}$ exists (Gillard and Williams, 1977) in
solution in equilibrium with its covalent hydrate;

ii) $[Ru(bipy)_2(py)_2]^{2+}$ treated with hydroxide yields
(Gillard and Hughes, 1977) the pseudo-base (VII) shown;

iii) orange $[Ru(bipy)_3]^{2+}$ upon oxidation *even in
acidic media* yields green-blue $[Ru(bipy)_3]^{3+}$ which
changes (Sagues, 1977) to the purple covalent hydrate
$[Ru(bipy)_2(Hbipy.OH)]^{3+}$ (an observation of interest in
view of the great amount of current work upon the photo-
catalytic decomposition of water by such tris-complexes);

iv) of all the chelating N-heterocyclic 1,2-di-imine
ligands so far studied, the equilibria are most evident
(Gillard *et al.*, 1976) for 5-nitro-1,10-phenanthroline,
the very ligand which yields such a highly active nickel
complex in tests upon *M. tuberculosis*.

In view of the recent demonstration (Henry and Hoffmann, 1977) that the free ligand 2,2'-bipyridyl itself
exists in aqueous media in equilibrium with its covalent
hydrate, we must clearly be prepared to envisage an
important role for such species in the bacteriostasis,
when it comes to be studied at the molecular level.

Platinum Compounds as Anti-Cancer Agents

This area is the most obviously exciting of all the
new bio-inorganic fields which have opened up in the last
ten or fifteen years. The origin of the discovery is
interesting, since it reveals how careful observation
can lead to serendipitous discoveries of high importance.
In his foreword to the proceedings of a recent conference
on coordination complexes of platinum, Haddow (1974) described the early history of work with compounds of heavy

metals in chemotherapy and carcinogenesis. He said that
attention was frequently drawn to the 'importance of many
metals - e.g. lead, iron, metalloid arsenic and so on.
Interest had started in platinum many years ago, following the possibility that various complexes between the
metal and mercaptopurine might possess significant chemotherapeutic properties. Various attempts to confirm such
findings ended, however, in complete failure. Interest in
platinum was revived by the fresh observations of Rosenberg and his colleagues, and here the outcome was entirely
different'.

These complexes of platinum studied by Rosenberg are
striking in the sense that the free ligands in these
cases have little or no inherent biological activity.
For instance, hexachloroplatinate(IV) inhibits the growth
of E. coli at low concentrations. It was the formation
of this complex from metallic platinum electrodes during
studies on the effect of electric fields on E. coli that
led to the enormous current interest in platinum complexes as anti-tumour agents. In the presence of ammonium
ion (which was a component of the bacterial culture medium) and light, the hexachloroplatinate(IV) ion is first
formed, and then photochemically substituted to form
cis-$[PtCl_4(NH_3)_2]$. The deliberately added hexachloro-
platinate salts were then shown to inhibit bacterial
growth, whereas the cis complex mentioned had no marked
effect on the increase in bacterial mass, but did inhibit
the division of bacterial cells, causing them to grow in
the form of filaments. The neutral complex of platinum(II),
cis-$[Pt(NH_3)_2Cl_2]$ was shown also (Howle and Gale, 1970)
to inhibit the cell division of E. coli and this was
among the first platinum compounds shown (Rosenberg et
al., 1969) to have activity against tumours. It was early
found that the corresponding trans-isomers were inactive
either in causing the filamentous growth in bacteria, or
in inhibiting tumour growth in animal studies.

While the activity of simple cis-dichloro-compounds of platinum(II) has generated an enormous amount of activity, which has been repeatedly reviewed, and which will therefore not be surveyed again in detail here, certain of the facts are worth re-emphasising. Thus, the original *cis*-dichlorodiammineplatinum(II) has been shown (Reslova, 1972) to induce prophage, to inhibit selectively (Harder and Rosenberg, 1970; Howle and Gales, 1970; Beck and Brubecker, 1973) the synthesis of DNA, to display (Beck and Brubecker, 1975) mutagenic activity, to decrease (Beck and Brubecker, 1973; Drobnik *et al.*, 1973) the survival of mutants blocked in various aspects of the repair of DNA, and to form (Roberts and Pascoe, 1972; Shooter *et al.*, 1972; Munchausen, 1974; Pascoe and Roberts, 1974; Harder, 1975) intra and inter-strand cross-links *in vivo* and *in vitro*.

Possibly the most striking area of biochemical interest at present is the exact nature of this interaction with DNA. This presumably underlies the mechanism whereby this class of platinum complex exerts its specific activity, and is understandably, therefore, the subject of intense current research. The formation of platinum bridges between strands of DNA appears to be a relatively uncommon event. In HeLa cells for instance (Pascoe and Roberts, 1974), one in 400 reactions led to a cross-link in the pesences of the cis-dichloro complex, and one in 4000 reactions led to a cross-link in the presence of the trans-isomer. Similarly, results with transforming DNA (in experiments (Munchausen, 1974) where *Haemo-philus influenzae* was the recipient) and with (Shooter *et al.*, 1972) DNA from phage Tr also showed that the formation of inter-strand cross-links by platinum is uncommon, and further, in those experiments, the inter-strand cross-links did not appear to be important in the inactivation of the DNA molecule.

Clearly, the mode of action of the initial complexes

of platinum is of high interest, and there have been
several suggestions. Most of them involve the lability
toward solvent, or similar ligand molecules of the chlorine atoms attached to platinum, suggesting the formation
of some kind of complex between the platinum and moieties
of the DNA. For example, Goodgame and his co-workers
proposed recently (Goodgame *et al.*, 1975) a model for the
binding of platinum(II) DNA based upon the X-ray crystallographic analysis of the structure of the complex with
5-IMP. Platinum is attached to the nitrogen atoms of
position 7 of both purine groups, in a square planar
configuration, as usual for platinum(II). Binding by the
amino groups of purines to platinum, in the manner often
suggested, was thought by Goodgame and his associates to
be unlikely, because the lone electron pair of these
amino groups is involved heavily in π-bonding to purine
ring. Further, inter-strand platinum bridges were thought
very unlikely in double-stranded DNA, because of steric
considerations. The preferred mode of action was suggested as platinum forming intra-strand links by binding to
double-stranded DNA in a unidentate manner, followed by
hydrogen bonding between co-ordinated water (present by
virtue of substitution at the platinum atom of the original complex) and the oxygen atom in position 6 of the
guanine moiety. With adenine, such involvement of the
amino-group at position 6 would be less likely. The separation of the DNA strands at this stage should then lead
to additional binding between the platinum atom and a
neighbouring base, i.e. two nitrogen atoms at position 7
of either adenine or guanine, or possibly to the nitrogen
of position 3 in cytosine.

All the available evidence points to DNA as the prime
target for interaction with these biologically active
platinum compounds, and perhaps the most striking feature
of the dihalo-activities is that only the cis-isomers are
strongly active. This difference between the biological

efficacies of the cis and trans isomers is almost certainly due to their different stereochemical configurations, since the primary chemistries are relatively similar. In the *cis*-isomer the chloride ligands are about three Angstroms apart, contrasted with a distance of four Angstroms for the corresponding inter-nuclear separation in the trans isomer.

Although most theories of the molecular mode of action of these dihalo-complexes have pointed out the lability of the chloro-ion, and therefore its ready replacement by any available ligating atoms (usually nitrogens) of the nucleic acid, there is a little-known property which distinguishes the two geometric isomers of dichlorodiammine-platinum(II), which may well be relevant. The *cis*-isomer forms an adduct with sulphuric acid. The trans-isomer does not.

When cis-$[PtCl_2(NH_3)_2]$ is treated with sulphuric acid (about 1:1 with water), a dark blue (almost black) slimy powder is usually obtained (Tchugaev, 1915; Drew, 1934), with the composition $[Pt(NH_3)_2Cl_2] \cdot 2H_2SO_4$. On warming with water, this new compound redissolves to give its components, and its infra red spectrum indicates strong hydrogen bonding. On one occasion only (Gillard and Wilkinson, 1964), a few acicular crystals were obtained, which showed strong dichroism, absorbing light whose electric vector was parallel to the needle axis. This axis coincides with an unmistakable Pt—Pt vector (the intermolecular distance being 3.06 Å). There is no such adduct formation with the *trans-isomer*.

It seems possible that such an interaction through hydrogen bonding between the intact *cis*-dichloro-complex and the DNA may give an ordered structure of platinum atoms and a mechanism for intra-strand links. This was suggested (Bielli *et al.*, 1974) some years ago, but has not found favour relative to chemical theories, requiring substitution at platinum.

As has been repeatedly emphasised, the platinum compounds which give anti-tumour activity invariably give
filamentation of bacterial cells, but the converse is not
necessarily true. Remarks applicable to bacterial division are not therefore necessarily relevant in anti-
tumour work. However, it is certainly supportive of a
hydrogen-bonding phenomenon *in platinum interaction with
bacteria* that the active compounds not only contain the
moiety (VIII) but also the moiety shown as (IX)

(VIII)

(capable of hydrogen bonding in the same way as *cis*-
$Pt(NH_3)_2Cl_2$ above). Indeed it might be said that the
structural element (IX) gives a more complete description

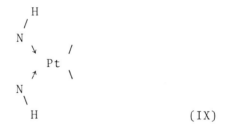

(IX)

of bacterially active platinum compounds than does structure (VIII). For example, $[Pt(NH_3)_2(C_2O_4)]$, $[Pt(en)(ox)]$,
and $[Pt(en)(mal)]$ (all highly active) are covered by (IX)
but not (VIII). Similarly, $[Pt(py)_2Cl_2]$ and $[Pt(bipy)Cl_2]$
included in (VIII) but not (IX), appear to be inactive.

Although in the earlier stages of the discovery and
extension of the range of active compounds in platinum
chemotherapy, it was common practice to study effects
upon bacteria as well as in anti-tumour applications,
this practice has become much less common. This is perhaps

unfortunate, since in view of the early work of Dwyer, we might well expect to find a number of anti-bacterial applications of coordination complexes, particularly of the kinetically inert type, which could be valuable. In this context, a recent study (Fibardi *et al.*, 1978) of the effect of *cis*-dichloro-diammineplatinum(II) on blood stream trypomastigotes is of interest. Although the anti tumour drug did not cure mice infected with *Trypanosoma cruzi* tripomastigotes, it was found that incubation of stored blood with the drug, prior to its inoculation into test mice, protected them against *T. cruzt*. The suggestion is made that this activity may be useful, particularly in view of the fact that Chagas' disease (a problem in Brazil) often occurs because of blood transfusion using accidentally infected blood.

Naturally, one of the major concerns of coordination chemists since the original discoveries using platinum and rhodium complexes has been to make and evaluate simple complex compounds of a variety of metals. In that connection, filamentous growth has been produced (Durig *et al.*, 1976) in *E. coli* by *cis*-$[Ru(NH_3)_3Cl_3]$ and, to a minor extent, by $[Pd(glycine)_2Cl_2]$. The tri-amine of ruthenium (III) had no effect on the morphology of *S. aureus*.

Platinum Blues

One of the most fascinating recent developments in the application of complex compounds of platinum in cancer therapy has been the observation that many of the so-called 'platinum blues' are very efficacious.

Blue compounds of platinum are relatively uncommon, although it is interesting that if a neutral solution of Rosenberg's complex *cis*-dichloro diammineplatinum(II) be left in the air for a few weeks, it becomes blue (Harrison, 1973). Also, although the compound bis-di-methylglyoximatoplatinum(II) is normally known as a red-brown substance, whose structure is analogous to that

of the well-known nickel(II) analogue, very finely divided samples of the platinum compound appear bright blue
in reflected light. This property is presumably somehow
related to the presence of a platinum-platinum contact,
which is a feature shared by most of the blue systems
which are currently being studied.

The original 'platinblau' was discovered by Hoffmann
and Bugge, by treating dichlorobis-acetonitrileplatinum
(II), in aqueous medium, with soluble silver salts. As
the silver chloride formed precipitates, so the solution
becomes blue, and it is possible to isolate blue glassy
solids. Despite a number of attempts, the nature of this
blue species has not fully been settled, although it is
fairly clear (Gillard and Wilkinson, 1964) that it contains ligands which are derived from acetamide.

Without being able to describe in detail all the blue
compounds now known, some of the more general aspects will
be covered. Mansy and Rosenberg discovered (Davidson *et*
al., 1975) the 'platinum pyrimidine blues'. Here, a
typical reaction scheme (Lippert, 1977) is:

$$\text{cis}[Pt(NH_3)_2Cl_2] + 2AgNO_3 \rightarrow \text{cis}[Pt(NH_3)_2(H_2O)_2](NO_3)_2$$

$$+ \text{ 2,4-dihydroxypyrimidine}$$

'blue'

Conditions:

$$2 < pH < 7$$

The present indication is that most of these 'blues' are
at least oligomeric, and many of them may be mixtures.

It is perhaps a pity that the generic name 'platinum
blues' has been applied to more or less any compound of
platinum which is blue. It seems likely that there are a
large number of classes of these compounds, ranging from

the relatively simple to the complicated polymeric mixtures. The fact that most of them have an obvious strong
absorption between 650 and 750 nm is not necessarily an
indication that they all contain similar structural
features.

Rosenberg's Compound and its Analogues

In order to give some idea of the spectacular results
obtained with the *cis*-dichloro complexes, Tables are presented, relating to effects on the experimental animals,
and then to clinical trials. The animal tests are based
on the following logic. It is more economical to use
transplanted tumours, since the animals may be injected
with the test drug shortly after transplantation. Several
of these transplanted tumours have been favoured as models
for chemotherapy, often because of the fact that some
particular class of chemical has affected them, and that
same class has subsequently proved to be useful in clinical work. Among such tumours are the Walker carcinoma,
and the sarcoma 180, and also the L1210 leukaemia, which
was originally induced by applying 20-methyl-cholanthrene
to the skin. This last L1210 tumour is the basis of the
National Cancer Institute's evaluation programme in the
testing of drugs.

Table II shows results obtained with *cis*-dichloro-
diammine-platinum(II) on a range of tumours of different
types, by comparison with the most extensively used alkylating agent in cancer chemotherapy, cyclophosphamide, and
with methotrexate, which is an anti-folate, used in clinical practice. The results in Table II are taken chiefly
from the work of Connors (1974).

From the results in Table II it will be clear that the
range of activity of the platinum-dichloro complex is
similar to that of the alkylating agent, rather than that
of methotrexate.

The effect of replacing the coordinated ammonia by

TABLE II

Results[a] of cis-$[Pt(NH_3)_2Cl_2]$ *on some transplanted tumours*

Transplanted	cis-[Pt(NH₃)₂Cl₂]	C[b]	M[c]
L 1210 leukemia	++	+++	++
Sarcoma 180	+++	++	+
Walker 256	++	+++	+
Lewis lung tumour	++	++	0
PC6 plasma cell tumours	+++	+++	0
TLX5 lymphoma	0	0	++
B16	++	+++	0

[a]Results are expressed as: 0, ineffective; + slight anti-tumour effect; ++, good (though not complete) regression of tumour, or increase in survival time from 50–100%; +++, complete regression or survival time more than tripled.
[b]Cyclophosphamide.
[c]Methotrexate.

TABLE III

Results obtained[a] on the PC6 plasma cell tumour, using cis-$[Pt(L)_2Cl_2]$

L	LD₅₀[bc]	ID₉₀[bd]	T.I.[e]
NH₃	13.0	1.6	8.1
CH₃NH₂	18.5	18.5	1.0
CH₃CN	27.0	27.0	1.0
cyclopropylamine	56.6	2.3	24.6
cyclobutylamine	90.0	2.9	31.0
cyclopentylamine	480	2.4	200
cyclohexylamine	3200	12.0	267
cyclo-octylamine	135	59	2.3

[a]Taken from T. Connors (1974).
[b]Doses are in mg per kg body weight.
[c]The LD₅₀ is the dose needed to kill 50% of the mice.
[d]The ID₉₀ is the minimum dose that gives complete regression of tumour.
[e]The T.I. (therapeutic index) is the ratio of LD₅₀ to ID₉₀. The larger the LD₅₀ the less toxic the drug. The smaller the ID₉₀ the more effective the drug. The larger their ratio, the better for utility.

other primary amines is shown in Table III.

As a result of a very large effort in modifying the structure of the original active platinum complexes, a number of very promising new substances have emerged. One of these is sulphato-1,2-diamino-cyclo-hexaneplatinum(II). It has been said (Owens *et al*., 1976) that 'its potency against leukaemia L1210 in mice, its excellent solubility in water, and its synergism when used in combination with cyclophosphamide and other anti tumour agents make this the most exciting platinum complex since the discovery of the cis-dichlorodiammineplatinum(II)'.

Utilizing a 'figure of merit', which is calculated as follows,

$$M = \frac{LD_{50} \times \%ILS}{ID_{99.9} \times 100}$$

(where LD_{50} is dose needed to kill half the mice; $ID_{99.9}$ is dose needed to kill 99.9% of tumour cells; %ILS is best obtainable percentage increased life span: the upper limit is 200, arbitrarily in these experiments), in experiments on leukaemia L1210, the shay tumour in rats, and the Gardner lymphoma OG in mice, the figure of merit (related to, but different from the therapeutic index) for the sulphato complex is 100, as against that for the dichlorodiammine complex of 3.3. Even when the sulphate, in sulphato-1,2-diaminocyclo-hexaneplatinum(II) is replaced with other 'leaving ligands', such as hydroxide, tetraborate $(B_4O_7^{2-})$, and bromate, the figures of merit remain very high (Ridgway *et al*., 1977): they are, for the leaving ligands named, 80, 70, and 32, respectively.

In order to show the wide potential for variants of structures of these active compounds, the results of Kidani and his collaborators (1977) are of interest. They synthesised the various isomers of dichlorocyclohexane-

1,2-diamineplatinum(II), where the isomerism depended on the configuration of the ligand. The three isomers are *cis*-, *trans*-(+), and *trans*-(-). Working with L1210 leukaemia, both *trans*-isomers showed stronger anti-tumour activities, and in studying the ILS (increase of life span), the (-)-isomer showed an ILS of 292%, whereas its enantiomer had a value of 191%. The compound of the *cis*-ligand had a corresponding value 111%. This is one of the very few cases where some evidence is available on the importance of the detailed stereochemical configuration about the platinum, and further work of this kind will undoubtedly be of the utmost value in understanding the detailed mechanism of action.

Central to the problem of oncogenesis is an understanding of the inter-relationship between cell growth and cell division and what often appears to be an interruption of this simple cycle when cells differentiate. In some cases (for instance nerve cells, erythrocytes) differentiation puts an end to further growth and reproduction. In other so-called mitotic tissues, such as epithelium, division of differentiated cells does occur. Tumour formation is marked by an interruption in the process of normal differentiation and a reversion to the simpler cell cycle of growth followed by reproduction similar to that occurring in embryonic tissues and prokaryotic microbes. Prokaryotic and eukaryotic microbes offer a wide range of experimental material in which these processes of growth, reproduction and differentiation can be studied under controlled conditions impossible to achieve as yet in the cells of higher organisms.

Most cytotoxic agents are aimed at inhibiting or slowing this simpler cell cycle. Coordination complexes of the platinum family of metals have been shown to be effective cytotoxic agents and have been studied as chemotherapeutic agents against a variety of malignant tumours in animals. As would be expected, these complexes have also

been shown to be inhibitors of the growth reproduction
of bacteria. In high concentrations growth is halted but
in low concentrations growth continues but cell division
is stopped; thus in rod-shaped bacteria long filaments
result. Complexes of platinum and rhodium are apparently
the most effective of this family of metals in this respect. The detailed differences in mode of action between
platinum and rhodium (Gillard, 1978) may be rather useful
in understanding cell division.

In prokaryotes such as the enteerobacteria, platinum
compounds appear to interrupt cell division by combining
with DNA and stopping effective replication. With rhodium
complexes of the type $trans-[Rh(C_5H_5N)_4X_2]Y$, X being
chloride or bromide, and Y a suitable anion, in marked
contrast, and under conditions in which filaments are produced, DNA replication is not affected but cell division
is. The filaments therefore contain DNA, probably as
individual genomes distributed along the length of each
filament. The effect is thus similar to that in the
filaments produced after recovery from radiation injury
and quite unlike those produced by treatment with other
cytotoxic agents, such as the flavins, which appear to act
like the platinum complexes.

One difficulty in studying such effects on growth and
cell division is that in a normal culture of bacteria and
other microbes, as well as in tissue cultures, both growth
and cell division are occurring randomly. Ideally, therefore, single cells only ought to be used and there have
been such attempts by exploiting ultra-micro methods,
such as those developed by Zeuthen. However, recently
more attention has been paid to developing cultures in
which cells divide at the same time, i.e. synchronous
cultures. By this method, massive cultures of a wide range
of microbes can be made to undergo at least two or three
divisions synchronously and a range of biochemical, genetic and morphological studies can be carried out through-

out the growth phase. Those processes connected with
growth and those with cell division can thus be more
clearly distinguished than in the randomly dividing cultures.

There are several methods for obtaining synchronous
cultures which can be placed in two main categories:

(1) *Induction Synchrony*. The cell population undergoes
either shock treatment (heat or cold) or some metabolic
block (DNA synthesis inhibitors or amino acid starvation)
in order to align the cells at some particular point in
the cell cycle. When the block is removed, the outgrowth
of cells is synchronous and the culture is said to have
been synchronized. However, using any of these methods,
the resulting synchronized cells have been shown to be
metabolically disturbed and cannot with any truth be
regarded as normal cells.

(2) *Selection Synchrony*. Cells that are selected from
a normal exponentially growing culture on either a size or
age basis show a synchronous outgrowth with little or no
metabolic disturbance. Selection synchrony should therefore be preferred to induction synchrony.

In this laboratory the effects of rhodium have been
studied in large scale, metabolically undisturbed synchronously and randomly dividing cultures of three genetically different strains of *Escherichia coli*. We have confirmed the morphological changes (filamentation) and also
shown that DNA replication is unaffected in the presence
of rhodium. The effects in synchronous cultures have
shown that sensitivity towards rhodium varies throughout
the cell cycle. Preliminary genetic studies suggest that
the genes determining rhodium sensitivity map at 10 minutes on the genome, close to the *lon* marker associated
with cell elongation. More biochemical studies are planned
with these and other mutants. Thus already, the combined
use of synchronous cultures with genetic and biochemical
approaches are yielding information not only about the

effects of rhodium on cell functions, but also the fundamental processes which determine the timing of events
during growth and cell division.

In prokaryotic microbes there now exist several fairly
detailed models of the inter-relationship between growth
and cell division. For instance, in some bacteria it has
been shown that reproduction and the initiation and inhibition of DNA replication are closely inter-related. Both
the increase in cell size (elongation) and cell division
(septation) appear to be regulated by specific proteins
controlled by defined genes such as the *Ion* locus already
referred to. In eukaryotes, though the situation is less
clear, similar processes can be seen to take place but
complications associated with multiple chromosomes,
spindle formation etc., are as yet ill understood. In
tissues, the situation is even less well defined, but the
added complication of the inter-relation between cell
surfaces must also be taken into account. Thus although
prokaryotes allow an understanding of the simplest and
most basic processes, it will be necessary to extend such
studies to the more complicated systems, a problem of far
greater difficulty than making new coordination compounds
for biological testing.

Conclusion

In view of the startling range of potentially useful
chemotherapeutic results already revealed by so small a
commitment of the global effort of coordination chemists,
it is almost superfluous to comment upon the crying need
for systematic testing of coordination compounds in
biological contexts. Recently, among the sporadic studies
which have naturally begun, acetylacetonato-1,5-cyclo-
octadiene-rhodium(I) was shown (Giraldi *et al.*, 1977) to
have an activity roughly comparable with that of *cis*-
$[Pt(NH_3)_2Cl_2]$ against Ehrlich ascites carcinoma, whereas
cis-$[Ru(DMSO)_4Cl_2]$ was less active. The hydrophobicity of

a series of dimeric dirhodium(II,II) tetra-carboxylates correlated (Howard *et al.*, 1977) with their activity against the same tumour.

References

Albert, A.A. (1968). 'Selective Toxicity', 4th edn. p. 331, Methuen.

Beck, D.J. and Brubecker, R.R. (1973). *J. Bact.* **116**, 1247.

Beck, D.J. and Brubecker, R.R. (1975). *Mut. Res.* **27**, 181.

Bielli, E., Gidney, P.M., Gillard R.D. and Heaton, B.T. (1974). *J. Chem. Soc.* (Dalton), 2133.

Bowen, H.J.M. (1966). 'Trace Elements in Biochemistry', Academic Press, 63.

Butler, H.M., Hurse, A., Thursky, E. and Shulman, A. (1969). *Austral. J. Exp. Biol. Med. Sci.* **47**, 541.

Butler, H.M., Laver, J.C., Shulman, A. and Wright, R.D. (1970). *Med. J. of Australia* **57**, 309.

Cade, G., Cohen, M., Shulman, A. (1970).*Austral. Vet. J.* **46**, 387.

Cade, G., Shankley, K.H., Shulman, A., Wright, R.D., Stahle, I.O., Macgibbon, C.B. and Lew-Song, F. (1970). *Med. J. of Australia* **57**, 304.

Chayen, J. and Bitensky, L. (1971). *Annals Rheumatic Diseases* **30**, 522.

Chisolm, J.J., Jr. (1968). *J. Pediatrics* **73**, 1.

Clusky, K.L. and Eichelberger, L. (1926). *J. Amer. Chem. Soc.* **48**, 136.

Connors, T.A. (1974). *In* 'Platinum Coordination Complexes in Cancer Chemotherapy' which is volume 48 of the series 'Recent Results in Cancer Research', ed. T.A. Connors and J.J. Roberts, Springer Verlag, 113. (This volume contains the Proceedings of the 2nd International Symposium on Platinum Coordination Complexes in Cancer Chemotherapy.)

Davidson, J.P., Faber, P.J., Fischer, R.G., Mansy, S., Peresie, H.J., Rosenberg, B. and van Camp, L. (1975). *Cancer Chemotherap. Rep.* **59**, 287.

Drew, H.D.K. (1934). *J. Chem. Soc.* 1790.

Drobnik, J., Urbankova, M. and Krekulova, A. (1973). *Mut. Res.* **17**, 13.

Durig, J.R., Danneman, J., Behnke, W.D. and Mercer, E.E. (1976). *Chem.-Biol. Interactions* **13**, 287.

Dwyer, F.P., Reid, I.K., Shulman, A., Laycock, G.M. and Dixson, S. (1969). *Australian J. Exp. Biol. Med. Sci.* **47**, 203.

Fibardi, L.S., Leon, W., Cruz, F.S. and Frausto da Silva, J.J.R. (1978). *J. Paracytology* (in press).

Freyberg, R.H., Block, W.D. and Levey, S. (1941). *J. Clinical Investigations* **20**, 401.

Freyberg, R.H. (1966). *In* 'Arthritis and Allied Conditions', ed. J.L. Hollander, published by Lea and Febiger, Philadelphia, 7th edn. 322.

Garner, R.J. (1959). *Nature* **184**, 733.

Gillard, R.D. (1976). *Educ. in Chemistry* **13**, 1.

Gillard, R.D. and Williams, P.A. (1977). *Transition Metal Chem.* **2**,14.

Gillard, R.D. and Hughes, C.T. (1977). *Chem. Comm.*

Gillard, R.D., Hughes, C.T. and Williams, P.A. (1976). *Transition Metal Chem.* **1**, 51.

Gillard, R.D. and Wilkinson, G. (1964). *J. Chem. Soc.* 2835.
Gillard, R.D. and Wilkinson, G. (1964). *J. Chem. Soc.* 1620.
Gillard, R.D. (1978). The work in my own laboratories, supported by
   the M.R.C., on the interference with cell division by compounds
   of the family *trans*-[Rh(py₄X₂]Y is largely unpublished, though
   parts have been described in: (a) Harrison, 1973, entitled
   'Biological Activity of Compounds of Rhodium'; (b) Bromfield, R.J.,
   Dainty, R.H., Gillard, R.D. and Heaton, B.T. (1969) *Nature*, **223**,
   735; (c) Gillard, R.D. (1977) *Kem. Kozlemenyek*, **47**, 107.
   (d) Gillard, R.D., Harrison, K. and Mather, I.H. in Connors (1974),
   pp. 29-33.
Giraldi, T., Sava, G., Bertoli, G., Mestroni, G. and Zassinovich, G.
   (1977). *Cancer Research* 37, 2662.
Goodgame, D.M.L., Jeeves, I., Phillips, F.L. and Shabski, A.C.
   (1974). *Biochim. Biophys. Acta* **378**, 153.
Haddow, A. (1974). *In* 'Platinum Coordination Complexes in Cancer
   Chemotherapy', ed. Connors, T.A. and Roberts, J.J., Springer
   Verlag, Berlin.
Harder, H.C. and Rosenberg, B. (1970). *Int. J. Cancer* **6**, 270.
Harder, H.C. (1975). *Chem.-Biol. Interactions* **10**, 27.
Harrison, K. (1973). Ph.D. thesis, Univ. of Kent.
Henry, M.S. and Hoffman, M.Z. (1977). *J. Amer. Chem. Soc.*
Herschel, J.F.W. (1819). *Edinburgh Phil. J.* **1**, 26; **2**, 154.
Howard, R.A., Sherwood, E., Erck, A., Kimball, A.P. and Bear, J.L.
   (1977). *J. Med. Chem.* 20, 943.
How Medica Inc. (1977). U.S. Patent, 4,012,228.
Howle, J.A. and Gale, G.R. (1970). *Biochem. Pharmacol.* **19**, 2757.
Howle, J.A. and Gale, G.R. (1970). *J. Bact.* **103**, 258.
Hughes, M.N. (1972). 'The Inorganic Chemistry of Biological Pro-
   cesses', John Wiley, London. 292.
Jessop, F.D., Vernon-Roberts, B. and Harris, J. (1973). *Annals
   Rheumatic Diseases* 32, 294.
Jessop, J.D. and Johns, R.G.S. (1973). *Annals Rheumatic Diseases* 32,
   228.
Johnson Matthey Co. Ltd., French Patent application, 2,303,087, 1977.
Kidani, Y., Inagaki, K., Sieto, R. and Tsukagoshi, S. (1977). *J. Clin.
   Hematology and Oncology* 7, 197.
Kirschner, S., Wei, Y.-K., Francis, D. and Bergman, J.G. (1966).
   J. Med. Chem. 9, 369.
Koch, J.H., Rogers, W.P., Dwyer, F.P. and Gyarfas, E.C. (1957).
   Austral. J. Biol. Sci. 10, 342.
Lippert, B. (1977). *J. Clin. Hematology and Oncology* 7, 26. (This
   contains the Proceedings of the 3rd International Symposium on
   Platinum Coordination Complexes in Cancer Chemotherapy, held at
   Wadley Institute of Molecular Medicine, Dallas, October 1976.)
Munchausen, L.L. (1974). *Proc. Nat. Acad. Sci.* 71, 4519.
Mylroie, R.L. and Hungate, R.E. (1954). *Canadian J. Microbiol.* 1, 55.
Möllgaard, H. (1924). "Chemotherapy of Tuberculosis", Copenhagen.
Odum, H.T. (1951). *Science* 114, 211.
Osler, G.K. (1971). *Nature* 234, 153.
Owens, D.R., Rolfe, R.D. and Hentges, D.J. (1976). *J. Clin. Micro-
   biol.* 3, 218.
Pascoe, J.M. and Roberts, J.J. (1974). *Biochem. Pharm.* 23, 1345.
Reslova, S. (1972). *Chem. Biol. Interactions* 4, 66.
Ridgway, H.J., Spear, R.J., Hall, L.M., Stewart, D.P., Newman, A.D.

and Hill, J.M. (1977). *J. Clin. Hematol. and Oncol.* **7**, 220.

Roberts, J.J. and Pascoe, J.M. (1972). *Nature* **235**, 282.

Rosenberg, B. (1971). *Platinum Metals Rev.* **15**, 42.

Rosenberg, B., van Camp, L. and Krigas, T. (1965). *Nature* **205**, 698

Rosenberg, B., van Camp, L., Grimley, E.B. and Thomson, A.J. (1967). *J. Biol. Chem.* **242**, 1347.

Rosenberg, B., van Camp, L., Trosko, J.E. and Mansour, V.H. (1969). *Nature* **222**, 385.

Rygh, O. (1949). *Bull. Soc. Chim. Biol.* **31**, 1052 and 1403.

Scharer, K. (1955). 'Biochemie der Spurenelemente', Parey, Berlin.

Shooter, K.V., Howse, R., Merrifield, R.K. and Robins, A.B. (1972). *Chem.-Biol. Interactions* **5**, 289.

Sowden, E.M. and Price, A. (1958). *Biochem. J.* **70**, 716.

Spear, R.J., Ridgway, H., Stewart, D.P., Hall, L.M., Zapata, A. and Hill, J.M. (1977). *J. Clinical Hematol. and Oncol.* **7**, 210.

Tchugaev, J. (1915). *Russ. Phys. Chem. Soc.* **47**, 213.

Vernon-Roberts, B., Dore, J.L., Jessop, J.D. and Henderson, W.J. (1976). *Ann. Rheum. Disease* **35**, 477.

Vinogradov, A.P. 'The elementary Chemical Composition of Marine Organisms', Sears Foundation, New Haven, Connecticut, 1953.

11 LITHIUM IN MEDICINE

N.J. Birch

Department of Biochemistry, University of Leeds, 9 Hyde Terrace, Leeds, LS2 9LS.

Introduction

Interest has been aroused in the use of lithium in medicine because of its apparent simplicity; it is after all the lightest solid element, and because fundamental aspects of drug action might be illuminated. I have based this review on two sets of questions: (1) what do chemists, biochemists and clinicians want to know about lithium? (2) what tools can they respectively offer for its study?

Lithium was introduced into psychiatric practice by Cade (1949) who has reviewed recently the early history of lithium and described in detail the studies which led to the first use in manic patients and the subsequent fall and rise in its fortunes (Cade, 1978). A measure of its current status may be determined from the report by Glen (1978) that more than one person in every thousand of the population of Edinburgh receives lithium prophylactically for treatment of affective disorder and the confirmation by Hullin (1978) that 6,700 kg of lithium carbonate were dispensed in 1976 in the United Kingdom alone, to an estimated 20 to 25,000 patients or approximately 1 in 2,000 of the population. The psychiatric research register maintained by the Maudsley Hospital in London suggests that 50,000 to 100,000 people might benefit from lithium treatment in Great Britain.

Lithium, therefore, is not merely a pharmacological oddity but has wide ranging social and economic significance. Dr. Hullin has recently calculated that at 1975 prices and based on the data of Hullin, McDonald and Allsopp (1975), the notional saving in hospital beds alone due to out-patient treatment with lithium in a group of 100 patients was £30,000 per year. It is impossible to measure the value of the social benefits since many of these patients would have been unmanageable without periodic or even long-term treatment in hospital. However, lithium is an oddity, its chemistry is anomalous and it has wide ranging pharmacological effects which cannot be explained yet in theoretical terms.

Historical

Lithium was introduced in 1859 for the treatment of 'gout and rheumatics' (Garrod, 1859) following the demonstration that of all urates the lithium salt was the most soluble and the suggestion that lithium could therefore be used as a uricosuric agent. Lithium was later detected in Spa waters in England and Germany and it was commonly held that the effectiveness of these waters was due to the lithium content, which was about 1 mM per litre (Thilenius, 1882). Over the next 90 years lithium was advocated for various disorders and then abandoned. In the light of current knowledge the most notable use was as a sedative. Lithium bromide was said to be the most effective of the bromides, though once the anion had been shown to have hypnotic properties of its own the role of lithium ion was not considered. In 1916 Squire reviewed the medicinal properties of lithium.

Throughout this early period of lithium treatment the dose of lithium involved was rather small. In common with many other mineral and spa treatments the element was given as one of a range of salts usually with 'copious draughts of water'. This pattern of treatment however was dramatically broken when in the late 1940s lithium chloride was promoted in the United States as a salt substitute, Westral, a taste corrective for patients on low salt diets for coronary and vascular disease. Liberal use of Westral resulted in several deaths from lithium poisoning and the drug was rapidly removed from the U.S. market. Reports of these toxic effects appeared at the same time as Cade's description of the use in treatment of mania. Without the persistence of a number of European psychiatrists, notably Schou and Baastrup, lithium treatment undoubtedly would have disappeared from normal medical practice. It is well recognised now that the doses of lithium ingested by the subjects who suffered the toxicities were in fact many times larger than those used in psychiatric practice and furthermore that these subjects, with cardiovascular disorders and low sodium intakes were particularly liable to lithium toxicity.

Lithium, the first of the modern generation of psychotropic drugs, was used initially in the treatment of mania. In this context it was largely superseded in the mid 1950s by the advent of phenothiazine tranquilisers which acted more quickly and were less open to doubt in terms of toxicity. However it had been observed that lithium appeared to have a preventative, or prophylactic, effect against recurrent attacks of mania and depression in those patients whose illness was characterized by a periodic course (Schou, 1968).

The veracity of the prophylactic effects of lithium was strongly challenged by Blackwell and Shepherd (1968) on statistical grounds. Enthusiastic clinicians who had seen sometimes spectacular effects of lithium on manic depressive patients were now forced to move from the realm of 'clinical impression' to designed clinical trials. A number of such trials were reported in the early 1970s (Baastrup *et al.*, 1970; Schou *et al.*, 1970;

Coppen *et al.*, 1971, Hullin *et al.*, 1972). Schou and
Thomsen (1975) reviewed the 'double blind' trials which
had been published and concluded that there could be very
little doubt of the existence of the prophylactic effect
of lithium against bipolar (manic-depressive) psychoses.
However there is some doubt of the efficacy of lithium
against unipolar (depressive) affective disorder and
trials of this are in progress (Fieve *et al.*, 1976;
Dunner *et al.*, 1976).

Clinical Aspects

Manic depressive psychoses (periodic affective disorder)

The major indication for the use of lithium is in
prophylactic treatment of manic depressive disorder. It
should be understood that psychiatric classification of
diseases can be carried out only in terms of symptoms
presented by the patient. It is likely therefore that
under any one diagnostic category we are not dealing with
a single nosological entity but that a number of different disease processes may be represented.

The major division in psychiatric terminology is between neuroses and psychoses, the latter being the states
in which the patient has no insight into his condition
and is unable to recognise that his behaviour and thought
processes are abnormal. The psychoses are further divided.
Toxic psychoses are those where the toxic agent has been
exogenously administered (for instance in cases of heavy
metal poisoning). The organic psychoses occur where there
is brain damage as a result of injury, degenerative disease or neurological injury as a result of somatic disease. The functional psychoses are the schizophrenias and
the affective disorders, both of unknown aetiology.

In the affective disorders the major disturbance is in
mood (affect) and this is reflected in a number of features such as appearance, speech and motor activity,

speed of thought, thought content, self esteem, response
to outside stimuli and energy. In the case of mania or
hypomania some or all of these features may be elevated
and the patient is cheerful, hyperactive, interfering and
talkative. In depression the patient looks pale and de-
jected and is difficult to provoke into action or speech,
has little energy and low self esteem.

In contrast to the affective disorders, in schizophrenia there is an inability to interact emotionally,
the so-called 'glass wall' phenomenon. (Schizophrenia
does not mean 'split personality'.) The schizophrenic
frequently has auditory, visual and tactile hallucina-
tions and the resultant bizarre behaviour leads to the
'rule of thumb' distinction that one laughs *at* a schizophrenic but *with* a manic patient. In fact periodic psy-
chotic patients in mania often have a repertoire of
amusing but grandiose and improbable stories worthy of
Baron Munchausen.

The course of the disease may be one in which both
manic and depressive episodes occur (the bipolar or manic-
depressive type) or in which depression alone is seen
(unipolar or recurrent depressive type). A third sub-type
is now recognized in which schizophrenic symptoms may be
seen, though the underlying primary disorder is in mood
(schizoaffective disorder).

For a patient the most distressing part of the disease
is during depression when not only is the thought content
self critical and morbidly preoccupied but there is difficulty in formulating ideas and in acting upon them. The
inability of depressed patients to carry out simple tasks
without an agony of decision is frequently misunderstood
by those around and this leads to further anxiety and
self-recrimination in the patient. Indeed, this 'anergia'
is characteristic, so that suicidal attempts, commonly
thought of as occurring in the depths of despair, usually
arise, not during the most severe period of depression,

but during recovery.

In mania the relatives and close companions of the
patient suffer because of the excesses and constant high
level of activity and pressure of speech. Social problems
may arise as a result of profligate spending, excessive
alcoholic intake, petty pilfering, intrusive, aggressive
and inconsiderate behaviour. The patient on the other
hand enjoys this phase of his disease and provided it is
not too exaggerated this may be a period of intense prod-
uctivity in a well-motivated subject. A very clear description of one such patient in the various stages of
his disease has been provided by Jenner *et al.*, 1967.

In many patients single episodes of depression, trig-
gered by an external event such as death of a loved one,
or severe stress, may not recur within the next twenty
years. However, in a large number of patients the first
episode of depression is followed by others at gradually
increasing frequency. If episodes of mania appear, the
course is almost invariably cyclic. Cycles may vary in
length from 48 hours to many months. The patient described
by Jenner *et al.* (1967) had a disease course in which one
day of mania was followed by one day of depression con-
tinuously for 11 years until the cycle was stopped by
administration of lithium (Hanna *et al.*, 1972). This very
rapid cycle is however unusual. A particularly clear example of a more usual short cycle manic depressive subject has been presented by Lee and Paschalis (1978) and
this is reproduced in Fig. 1.

Fig. 1 demonstrates some of the more obvious features
of the manic depressive syndrome. There is a close
correlation between mood, activity and speech. The magnitude of the hyperactivity may be seen from the distance
recorded on the pedometer at the very height of mania
(22 miles in one day).

This, then, is the clinical problem against which
lithium therapy, with all its disadvantages and limita-

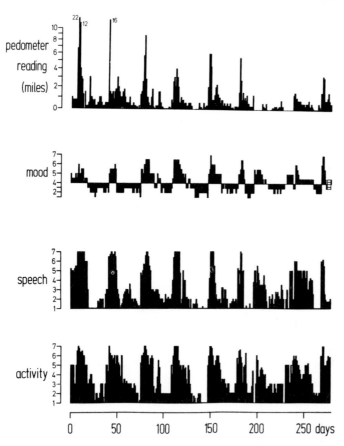

Fig. 1. Taken from Lee and Paschalis (1978) with permission of the copyright owners. The patient was under the clinical care of Professor F.A. Jenner whose help and co-operation is gratefully acknowledged by the authors.

tions must be judged.

Practical Aspects of Lithium Therapy

The indications for lithium therapy, the method of initiating treatment, the monitoring of the treatment and various difficulties which may arise have been discussed in great detail by Schou and Thomsen (1975), Schou (1973a, 1976c) and Kerry (1978). Fieve (1975) has described the organization and running of a special lithium clinic.

Lithium is accepted to be effective in the acute treatment of mania and in prophylaxis against recurrent affective disorders although the efficacy of lithium against
periodic depression is unproven. Other suggested psychiatric uses, discussed by Schou (1978), are in the treatment
of recurrent schizo-affective disorder, emotional instability in children and adolescence, pathological impulsive
aggression, pre-menstrual tension syndrome, alcoholism,
drug addiction, and affective psychoses in organic brain
syndromes. Non-psychiatric uses have been suggested as
follows; movement disorders (Huntington's Chorea, Tardive
Dyskinesia, Parkinsonian Hypokinesis), migraine and
cluster headache, inappropriate secretion of antidiuretic
hormone, hyperthyroidism, thyroid cancer and granulo-
cytopenia.

By its very nature, prophylactic treatment with lithium
is a long-term process and may be compared with the insti-
tution of insulin therapy in diabetic patients. Because
the therapeutic index of lithium is low a number of precautions must be taken prior to starting a patient on
lithium treatment and regular monitoring and recording
of plasma lithium concentration together with measures of
thyroid and renal function are essential. The objective
is to maintain a plasma lithium within the range 0.6—1.2
m.mol 1^{-1}. Toxic symptoms are often seen if the plasma
lithium rises to above 2 m.mol 1^{-1}. The lower limit was
set arbitrarily as a result of experience in the acute
treatment of mania and has recently been questioned by
Jerram and McDonald (1978). Most patients require between
1 and 2 g lithium carbonate each day in divided doses to
maintain their plasma lithium within the therapeutic
'window'. A large measure of individual variation occurs,
however, depending on volume of distribution (that is,
body size) and renal function. In elderly patients, whose
kidneys have reduced function, doses as low as 400 mg
lithium carbonate per day may maintain the therapeutic

levels while in younger patients much higher doses may be required to maintain the therapeutic effect (Hewick *et al.*, 1977). Therefore it is essential, when commencing therapy, either to institute a gradually increasing dosage regimen with frequent serum lithium analysis or to perform some test of renal clearance to determine the dose required (for example Schou, 1973a, Cooper *et al.*, 1973, 1976).

During the early stages of lithium treatment blood samples should be taken at fairly frequent intervals and for this reason it is often considered desirable that the treatment should be started on an in-patient basis and indeed if the patient is single and living alone this is almost essential. However if there are no other contra-indications and there is a fairly responsible person living with or near the patient, it is possible to institute treatment as an out-patient provided both patient and observer are primed to look for toxic signs which might arise to indicate impending toxicity. Under these conditions weekly blood samples are taken for the first month and the interval gradually increased until the normal frequency of visits is between six and eight weeks. In very old patients, in patients with cardiovascular or renal disease and in patients who are erratic in taking their medication the interval should obviously be shorter. It is important that records are kept of plasma or serum lithium at each visit together with any additional information on renal function, such as urea and creatinine excretion and thyroid function. In this way gradually developing intoxication may be prevented and abnormalities in for instance thyroid function may be treated before they become too severe.

Side Effects and Toxicity

Minor side effects are frequent in the early stages of lithium treatment and most of them disappear within

the first few weeks. The most frequent side effect is a
slight tremor of the hand which is particularly marked
when the patient is tired or tense. This tremor is not
Parkinsonian but may be treated by the β adrenergic receptor blocker, propranolol. In some patients there is a
tendency to gain weight during lithium therapy. A side
effect which develops in about 10—15 per cent of patients
is hypothyroidism during lithium treatment and this may
be treated by small doses of thyroxine. A late side
effect in a few cases is polydipsia-polyuria. Signs of
impending intoxication are vomiting and diarrhoea, development of coarse tremor of the hands, sluggishness,
drowsiness, vertigo and slurred speech (Schou, 1973a).

If intoxication is allowed to develop the effects are
mainly on the central nervous system with asymmetrical
neurological signs and ultimately coma and death. Kidney
damage may occur leading to further exacerbation of the
toxicity. The treatment of choice is by haemodialysis
to remove as much of the body load of lithium as possible.

According to Schou there may be three sets of circumstances for development of lithium poisoning. The patient
may be receiving too high a dose, the renal clearance of
lithium may fail due to, for instance, kidney disease,
decreased sodium intake, diuretic therapy or to inter-
current infection. Finally lithium toxicity may occur as
a result of overdosage with suicidal intent and this is
of course a particular problem with depressed patients.

Chemical Aspects

The chemical aspects of lithium in relation to its
use in manic depressive psychoses have been reviewed in
detail by Williams (1973) and discussed by the partici-
pants at a recent symposium (Bunney and Murphy, 1976).

The physical properties of lithium are summarized in
Table I. Lithium is a member of group I in the periodic
table of elements which includes the other alkali metals

TABLE I

Physical properties of lithium

Atomic number		3	Atomic Weight	6.94
Isotopes	Atomic mass 5		Half life	10^{-21} s
	Atomic mass 6		Stable. Abundance	7%
	Atomic mass 7		Stable. Abundance	93%
	Atomic mass 8		Half life	0.8 s
	Atomic mass 9		Half life	0.2 s
Melting point (°C)		180.5°	Boiling point	1336°
Specific Heat at 25°C		9.85		
Heat of fusion (cal/g)		103.2	Heat of vaporization	5100 (cal/g)

sodium, potassium, rubidium and caesium. Lithium, a first
row element, is typically anomalous in that its chemistry
may be likened to that of magnesium, a member of group II.
These diagonal similarities (Heslop and Jones, 1976) result from the relatively large increment in charge and
size with successive members of the first row of elements
and the relatively large decrement in size between the
first and second members of any group. The polarising
powers of diagonally related elements may be very similar.
This similarity is reflected in the physical properties
of the elements (Table II).

Lithium has a greater tendency to form covalent bonds
than the other alkali metals and this and its other anomalous properties result mainly from the small size of its
atom and ion and hence high charge concentration leading
to a high degree of solvation in solution. It has been
suggested that 'lithium bonds' exist which are similar in
character to hydrogen bonds (Kollman *et al.*, 1970).

Anomalous properties of lithium compounds also result
from the small size of the ion. This is due to the exceptional stability of salts of lithium with small anions
since they have very high lattice energies and the relatively unstable salts with large anions due to poor

TABLE II

*Physical and chemical properties of
alkali and alkaline earth metals*

		K	Na	Li	Mg	Ca
Atomic radius (Å)		2.03	1.57	1.33	1.36	1.74
Crystal ionic radius (Å)		1.33	0.95	0.60	0.65	0.99
Hydrated radius* (Å)		2.32	2.76	3.40	4.65	3.21
Ionisation potential	1st	4.34	5.14	5.39	7.64	6.11
(eV)	2nd	31.8	47.3	75.6	15.0	11.9
Standard Electrode Potential E° (V)		-2.92	-2.71	-3.02	-2.37	-2.87
Electronegativity		0.8	0.9	1.0	1.2	1.0
Polarizing power (z/r^2)		0.56	1.12	2.8	4.7	2.05

*Stern and Amis (1959)

packing of large and small ions (Cotton and Wilkinson,
1972). As with the other alkali metals, lithium salts
have high melting points, are electrically conductive in
the molten state and readily soluble in water. Lithium
salts are often hydrated when the salts of other alkali
metals with the same acids are not.

Solution Chemistry of Lithium

The solution chemistry of lithium is again anomalous
compared with its congeners. Hydrates of lithium salts
rarely contain more than 4 water molecules unless the
anions themselves are hydrated (Cotton and Wilkinson,
1972). However, because of the small crystal radius and
high charge density, the lithium ion has the highest
hydrated radius since it binds additional water molecules
in outer hydration spheres by electro-static forces. The
exact size is open to dispute. Because of the small size
of the naked ion compared with solvent molecules, devia-
tion occurs from Stokes' law and mobility data are not
accurate (Stern and Amis, 1959). For similar reasons
lithium salts deviate from ideal solution behaviour and
hence have abnormal colligative properties.

Complexes

The alkali metal ions have a noble gas structure and a single positive charge and therefore generally behave indifferently towards ligands (Winkler, 1976). Selectivity usually occurs by variation in size of site available. Lithium has the smallest crystal radius but the largest hydrated radius of the alkali metals and it is possible to envisage two types of structural site which might have differing properties and which might accommodate lithium either when hydrated or alternatively when stripped of its hydration shells. However, because of its relatively high polarizing power, and hence its ability to distort electron clouds of ligands, lithium has a higher affinity than the other alkali metals for oxygen and nitrogen binding sites on ligands and the latter is reminiscent of the chemistry of magnesium (Williams, 1973). The range of ligands to which lithium might 'bind', as distinct from being enclosed, may therefore be greater than the other members of Group I.

Until recently, comparatively few substances were known to form complexes with alkali metals and it was difficult to imagine how the theory of receptor site occupation could be translated into biochemical terms for the active sites of alkali metal dependent enzymes, transport processes and neurohumoral receptors. Recent developments in chemistry have provided models from which it may be eventually possible to extrapolate to the biological systems (Fenton, 1977).

The first advance came when it was demonstrated that one action of a number of naturally occurring antibiotics was to complex selectively particular metal ions and facilitate their transport across living membranes (Moore and Pressman, 1964). Thus, for instance, valino-mycin binds potassium in preference to sodium; actino-mycin on the other hand has the opposite selectivity (Williams, 1970). Series of antibiotic ligands ('Iono-

phores' as they are now called) may be produced, each
with varying metal specificity. Examples are the actin
series (Fig. 2) and the enniatin series (Fig. 3).

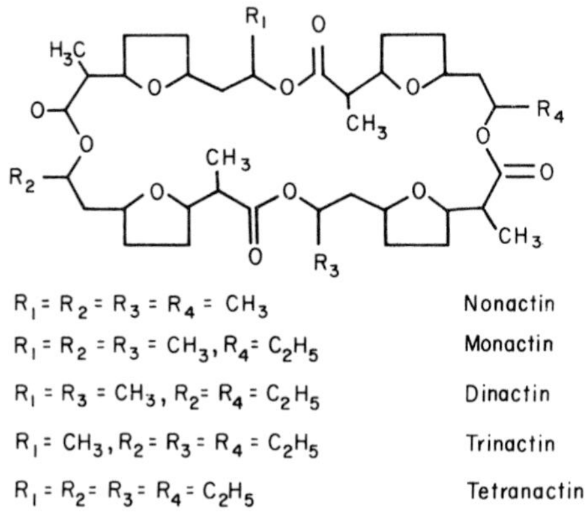

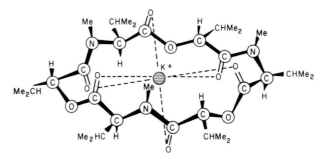

$$R_1 = R_2 = R_3 = R_4 = CH_3 \qquad \text{Nonactin}$$
$$R_1 = R_2 = R_3 = CH_3, R_4 = C_2H_5 \qquad \text{Monactin}$$
$$R_1 = R_3 = CH_3, R_2 = R_4 = C_2H_5 \qquad \text{Dinactin}$$
$$R_1 = CH_3, R_2 = R_3 = R_4 = C_2H_5 \qquad \text{Trinactin}$$
$$R_1 = R_2 = R_3 = R_4 = C_2H_5 \qquad \text{Tetranactin}$$

Fig. 2. The actins. Taken from Fenton (1977) with permission of the
copyright owner.

Fig. 3. Taken from Fenton (1977) with permission of the copyright
owner.

The next major step was the synthesis of a series of
'crown' ligands with a cyclic polyether structure and
with varying sizes of enclosed hole to match a variety of
metal ions (Pedersen, 1967) (Fig. 4). The ability to
design and synthesise ionophores for specific ions has
been a great achievement, though its significance probably has not been fully appreciated in the biological
sciences. The cyclic polyethers recently have been supplemented by the 'cryptands' which are three-dimensional

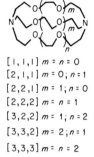

Fig. 4. A crown polyether, dibenzo-18-crown-6 (dibenzo refers to the non-ethyleneoxy portion, 18 to the number of atoms in the crown ring and 6 to the number of heteroatoms in the ring). Taken from Fenton (1977) with permission of the copyright owners.

[1,1,1] $m = n = 0$
[2,1,1] $m = 0; n = 1$
[2,2,1] $m = 1; n = 0$
[2,2,2] $m = n = 1$
[3,2,2] $m = 1; n = 2$
[3,3,2] $m = 2; n = 1$
[3,3,3] $m = n = 2$

Fig. 5. A series of cryptands. In the cryptand discussed by Lehn (1976) as a possible lithium carrier ('Ligand 4') one of the 'm' bridges is replaced by a hydrocarbon chain and the structures is: n = 1, 2nd bridge m = 0, 3rd bridge is $-(CH_2)_5-$. Figure taken from Fenton (1977) with permission of the copyright owner.

structures with a closed-cage (Lehn, 1973), for example Fig. 5. Cryptands provide a very accurately determined space for the ion since their structure is rather more rigid. Both the crown ligands and the cryptands are able to transport highly charged ions through lipid membranes by enclosing the ion and presenting to the outside a lypophilic surface. Lehn (1976) has described how cryptands can be adapted to make a specific lithium binding agent. The best result reported was that of a cryptand (Fig. 5) in which the relative transport of lithium:sodium:potassium was 1:1.8:0. It was suggested that further modification might make this compound selective for lithium alone.

One of the problems associated with the modelling of transport systems or switching devices is the problem of

stability versus release properties. To be an effective
model for a biological system the stability at the point
of uptake of the metal must allow complex formation yet
this stability must not be so great that release of metal
does not occur readily at the release site or on the
release stimulus. Furthermore, the stability of the specific metal-ligand complex under the conditions obtaining
at the point of uptake must be much greater than that of
likely competitor cations though differing stability
under other conditions might be an adequate release mechanism. In addition the energy required to strip off hydration shells must closely match that gained by formation of
a stable complex. Winkler (1976) has pointed out that in
the case of lithium, which has a very high field density
at the surface, the energy required to strip off the inner
hydration shell is 122 Kcal Mol^{-1} and that the interaction
of any acceptor ligand with the cation must be very favourable to compete with this hydration energy. However individual water molecules in this shell have a very high
rate of substitution by ligands and Winkler suggests that
the optimal acceptor would be one which substitutes water
molecules in a step-wise fashion and was itself flexible
enough to change conformation sequentially to a more compact structure. A more rigid ('non-breathing') structure
would have a much slower rate of formation.

The various families of ionophores have not been
applied extensively to lithium chemistry and there is the
exciting possibility that a 'lithium model receptor'
might be identified. Each of the groups of ionophores has
its own characteristics and their use in pharmacological
systems and in model membranes might provide much information on lithium 'binding' and action. Furthermore some
clue might be given to the structural characteristics of
molecules which bind lithium *in vivo*.

Pharmacology and Biochemistry of Lithium

The biological effects of lithium are diverse and affect many systems of the body. In this section I hope to review the basic knowledge of lithium pharmacology and to provide signposts to areas of current interest. I shall discuss therefore only a few topics in great detail and this will necessarily represent a personal view of the likely points of growth. Recent reviews of all aspects of lithium have been published in the volumes edited by Gershon and Shopsin (1973), Johnson (1975), Bunney and Murphy (1976) and Johnson and Johnson (1978). Comprehensive reviews of the pharmacology of lithium have been published by Schou (1957, 1976a), Davis and Fann (1971) and the biochemical aspects of lithium by Schou (1973b), Birch (1978). A cumulative lithium bibliography has been published by Schou (1969, 1972, 1976b,d).

Absorbtion and Distribution of Lithium

Lithium is administered orally, its absorption is rapid and occurs in the upper small intestine. Once it appears in the blood stream its distribution in the extracellular fluids occurs fairly quickly. The rate of transfer into the various tissues however varies. In rats, following acute intravenous administration of 5 mM/kg body weight, lithium soon appears in kidney but appears rather more slowly in liver, muscle and bone. The transfer of lithium into brain is rather slow (Schou, 1958). The efflux from the tissues has similar rate characteristics. However, in therapeutic practice lithium is used as a long-term treatment and such short-term pharmacokinetic data may not reflect the long-term distribution. A number of studies have been carried out on the distribution of lithium in various tissues, notably the brain. Bond (1978) reviews the published data and reports a number of areas of discrepancy. It is difficult to assess the significance of many of the results since the various brain

areas studied contain variable quantities of lipids and
it may be that the lithium concentrations observed reflect merely the distribution of aqueous phase. A number
of reports suggest that there is an increased lithium
content in the hypothalamus, which is particularly interesting since this is the major area of convergence of
the two divisions of control of bodily function, the nervous system and the endocrine system. A number of the
endocrine changes to be described later could result from
effects on the hypothalamus. However, Bond (1978) is careful to emphasise that the variety of dissection techniques used makes comparison very difficult and these
results should not be considered to be definitive.

The distribution of lithium in most tissues, including
brain, is rather similar to the plasma concentration from
0.5 to 1.5 m.mol.kg wet weight of tissue^{-1}. However,
Désgrez and Meunier (1927) had already shown that lithium
occurred in bone and teeth and it should not have been
surprising, therefore, that we were able to demonstrate
a higher concentration of lithium in bone than in other
tissues of rats following long term treatment (Birch and
Jenner, 1973, Birch and Hullin, 1972) (Fig. 6). It was
also shown that much of the lithium found in bone was
tenaciously held after discontinuation of lithium, for
instance, as shown by the hatched histogram in Fig. 6.

Subsequent experiments in rats which had received
lithium in the drinking water at a dose of 1 mM/kg body
weight/day for 28 days, followed by discontinuation of
the lithium and sampling at various intervals showed
that one fraction of bone lithium, corresponding to bone
water, was lost rapidly. About 50 per cent of the original
lithium concentration was still present after 42 days
(Birch 1974) (Fig. 7). Furthermore the concentration of
lithium attained in the bone after 28 days lithium treatment bore a strong inverse relationship with the age at
which lithium treatment was commenced. This suggests that

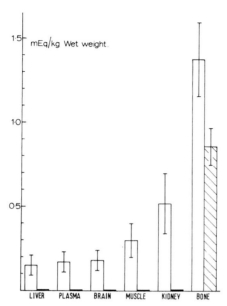

Fig. 6. Distribution of lithium in tissues of rats after long-term administration of LiCl in drinking fluid (approx. 1 mmol/kg body weight/day). Open columns indicate lithium concentrations in animals killed within 12 hours of last lithium intake, hatched columns show concentrations in tissues after six weeks during which the lithium drinking fluid was replaced by distilled water. Reproduced from Birch and Hullin (1972) with permission.

the lithium was laid down in the bone mineral during the mineralisation process (Birch, 1974).

Lithium has been known for some time to accumulate in the thyroid gland (Berens and Wolff, 1975). This accumulation together with that in a number of other endocrine organs has recently been studied by Stern *et al.* (1977) and these workers have demonstrated a rapid accumulation not only in the thyroid but also in the pituitary (tissue lithium/plasma lithium is 3.21 in each case after 2 hours). The adrenal gland is also relatively rich in lithium (tissue lighium/plasma lithium = 1.7 after 2 hours). Taken in conjunction with the report of high

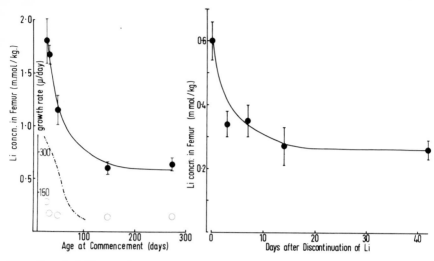

Fig. 7. a) Effect of age at which lithium treatment was commenced
on lithium content of rat femur after 28 days' administration
(approx. 1 mmol/kg body weight/day) in drinking fluid. b) Effect
of discontinuation of lithium after 28 days' treatment on lithium
content of rat femur. Open circles indicate water treated control
group. The graph of longitudinal bone growth is redrawn from data
of Hansson, L.I., Menander-Sellman, K., Senström, A., Thorngren,
K.-G. (1972) *Calcified Tissue Research* **10**, 238-251. Figure taken,
with modification, from Birch (1974) by kind permission of the
Editors of Clinical Science and Molecular Medicine and Blackwell
Scientific Publications Ltd.

lithium concentration in the hypothalamus, this suggests
the possibility of lithium pooling at critical sites of
neuro-endocrine control.

An elegant technique has been reported recently by
Thellier and his colleagues (Thellier *et al*., 1976a,b;
Wissocq *et al*., 1976) in which lithium is directly localised in histological specimens by auto-radiography following neutron activation. Irradiation of the stable
isotope ^{6}Li with thermal neutrons in a nuclear pile leads
to the fission reaction: $^6_3\text{Li}(^1_0\text{n}, ^4_2\alpha)^3_1\text{H}$. By the use of
cellulose nitrate detectors which are penetrated by α
particles but not affected by neutrons or gamma rays,
which fog normal photographic detectors, autoradiographs

may be made of samples in contact with the detector during
neutron activation. The cellulose nitrate detector is
developed by treatment with sodium hydroxide, which causes
enlargement of the tracks left by α-particles. Subsequent
viewing with transmitted green light allows the observation of the track distribution. The holes appear as
bright points on a dark background.

This technique has allowed estimation of ^{6}Li in the
tissues of mice previously treated with the stable isotope. Fig. 8a shows the distribution of ^{6}Li in the brain
of a mouse which had received 12 m.mol/kg body weight
^{6}LiCl daily for 3 days by intraperitoneal injection. The
comparison of the histological section (Fig. 8b) and the
etched detector (Fig. 8a) shows clearly that lithium is
not distributed uniformly throughout the brain. Wissocq
et al. (1976) identified particularly the following areas
of high lithium content: cerebral cortex, olfactory lobe
and hippocampus. They comment that the thalamus has a
particularly low lithium content.

In addition, one can see from the etched detector that
lithium is accumulated in selected areas of the corpus
striatum, emphasising the striations seen. This suggests
that there might be localised areas of high lithium content within specific tracts of the basal ganglia and
these might have functional significance. This finding
confirms the earlier analytical studies of Ebadi *et al.*
(1974) and Bond *et al.* (1975) who both showed high concentrations of lithium in the basal ganglia. It is interesting to note that lithium has been suggested in the
treatment of Huntington's Chorea and Tardive Diskinesia
(Schou 1978) both of which are associated with disorders
of the basal ganglia. Similarly the high concentration of
lithium in the hippocampus may relate to the role of the
limbic system in the control of feeding behaviour and in
the expression of rage. One well known side effect of
lithium is weight gain (for example see Birch, 1978) and

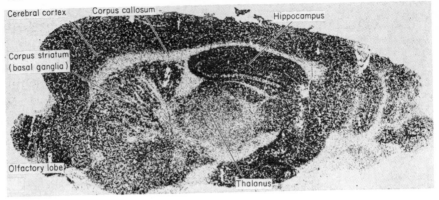

Fig. 8a

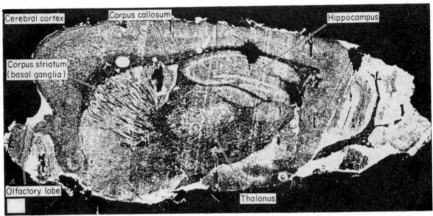

Fig. 8b

Fig. 8. Microlocalisation of ^{6}Li in mouse brain by neutron activation using cellulose nitrate detector, following three days' treatment by intraperitoneal ^{6}LiCl injection (12 mmol.kg body weight^{-1} once daily). (a) Cellulose nitrate detector after etching with 10% NaOH and subsequent viewing with transmitted green light. Areas of lithium accumulation appear dark. (b) Histological section of mouse brain corresponding to detector. Parasagittal section. Reproduced from Thellier, M., Steltz, T., Wissocq, J.-C. (1976) *Journal de Microscopie et de Biologie Cellulaire* 27, 157-168, by kind permission of authors and publishers.

lithium is currently under investigation for the treatment of pathological aggression (Worrall, 1978).

Thellier *et al.* (1976b) have also reported the quantitative evaluation of lithium content in different organs of the mouse. Their finding of high concentrations of lithium in hypophysis (pituitary) corroborates the work of Stern *et al.* (1977). It is particularly of note that they find considerably higher concentration of the element in the posterior hypophysis since this is the site of production of the antidiuretic hormone (ADH) which controls the reabsorbtion of water by the kidney. Polyuria (excessive urine volume) is a well known side effect of lithium treatment which will be discussed later.

One other notable feature of the lithium distribution demonstrated by Thellier *et al.* (1976b) is that despite administration by intra peritoneal injection there appears to be accumulation of lithium in all sites where significant membrane transport occurs of alkali and alkaline earth metals. Thus autoradiographs demonstrate high concentrations of lithium in tongue, salivary gland, intestine, stomach, kidney and bladder, suggesting that lithium is either freely diffusable or actively transported across epithelia into the lumens of these organs. Further studies with this exciting new technique particularly in the different areas of the brain may throw more light on the mechanisms of lithium action.

Pharmacokinetics of Lithium

Clinically, lithium is usually administered in the form of lithium carbonate tablets, the dose range being between 500 and 2000 mg/day. It has been suggested that a number of side effects are related to the peak in plasma lithium following the absorbtion of a single dose and a number of slow release preparations have been marketed. The administration of lithium to animals during experimental studies has been by intravenous injection, subcutaneous and intra-peritoneal injections, in the drinking fluid and in the food. Each of these methods has its advantages and these factors have been discussed (Birch 1978).

However, it is clear that one should be wary of extrapolating to the clinical situation results from animal

experiments in which single acute injections of lithium
salts were given since it might be argued that the effects
observed were due to high lithium concentrations in the
tissues immediately following injections rather than to
a prolonged effect. The pharmacokinetic profile following intra peritoneal injection of 3 mM/kg body weight of
lithium chloride in various tissues has been reported by
Mukherjee *et al*. (1976) and these results confirm the
earlier results of Schou (1957) that the uptake of lithium
into brain tissue was considerably slower than kidney,
heart, muscle or liver.

We have studied in patients and volunteers the pharma-
codynamic aspects of the use of 'slow release' preparations of lithium compared with conventional lithium carbonate B.P. We conclude that there is no significant
advantage in the use of the slow release preparations.
The profile of serum lithium obtained with the slow release preparations available in Great Britain shows no
significant difference from conventional lithium carbonate in the time of peak concentration and any difference in the height of the peak is related to the degree
of absorbtion of the lithium (Tyrer *et al*., 1976). In one
preparation as much as 50% of the lithium dose could be
recovered from the faeces of certain patients (Tyrer *et
al*., 1976). Not only is this wasteful but should gastro
intestinal function be altered for any reason, such as
infection or change in dietary habits, there a possibility of changed rate of absorbtion and concomitant
danger of slowly developing lithium intoxication. These
results have received independent confirmation from a
trial carried out by different methods (Thornhill, 1978).

Use of the Stable Isotope ^6Li in Pharmacokinetic Studies

One of the particular problems in the study of lithium
pharmacology is the lack of a suitable radioactive isotope. The isotopes ^5Li, ^8Li and ^9Li have half lives of

10^{-21}, 0.8 and 0.2 seconds and clearly are not useful in conventional radiotracer technology. We have therefore developed a method for the determination of the stable isotope ^{6}Li by atomic absorption spectroscopy (A.A.S.) (Birch *et al.*, 1978). Conventional stable isotope techniques involve the use of mass spectrometry. Indeed this technique and that of neutron activation analysis provide the basis of the work of Thellier *et al.* outlined in a previous section, which had its origins in the study of the effects of lithium on development in growing plants (Desbiez and Thellier, 1975).

Mass spectrometry requires expensive apparatus which is not available in most clinical laboratories and the lengthy preparative and analytical time places a severe restriction on the number of samples which may be readily handled. We modified, therefore, the technique of Wheat (1971) for the determination of the lithium isotope ratio in nuclear reactor materials. This method, which may be performed on simple, unmodified, AAS equipment, takes advantage of the coincidence of the spectral shift of the isotopes ^{6}Li and ^{7}Li so that by determination of the absorption by single samples of a range of isotopic enrichments (^{6}Li/^{7}Li) both of the light emitted by a ^{6}Li hollow cathode lamp (A_6) and that from one made from ^{7}Li (A_7). An exponential calibration curve (^{6}Li/^{7}Li vs. A_6/A_7) may be obtained for each of a range of concentrations of total lithium and unknown solutions may thus be estimated (Birch *et al.*, 1978).

We have applied this technique to the study of the pharmacokinetics of a single dose of lithium carbonate given to subjects who have received loading doses for several days and it is intended that we should determine whether or not there has been any change in pharmacokinetic response to a lithium dose following long-term lithium administration in patients. An example of the results obtained is seen in Fig. 9. A male volunteer

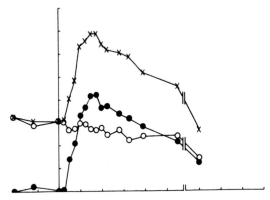

Fig. 9. Serum concentrations of total Li (X), ^{6}Li ($\bullet$), ^{7}Li ($\circ$), after
a single dose (750 mg) ^{6}Li$_2$CO$_3$ at 11.00 h following four days'
treatment with ^{7}Li$_2$CO$_3$ (250 mg, thrice daily). Normal, male volunteer 33 years. Redrawn from data to be published in British Journal
of Clinical Pharmacology (Birch *et al.*, 1978) by kind permission of
the Editorial Board.

received 750 mg ^{6}Li$_2$CO$_3$ in a single dose after 4 days pre-
treatment with 250 mg ^{7}Li$_2$CO$_3$ three times daily. It is
clear that the increase in total serum lithium closely
parallels that of serum ^{6}Li and furthermore that the
plasma concentration of ^{7}Li continues to decline after
the administration of the ^{6}Li$_2$CO$_3$.

In chronic toxicity studies we have maintained rats on
^{6}LiCl (1 mM/kg body weight/day) for a period of 18 months
with no evidence of increased toxicity over controls
treated with a similar dose of ^{7}LiCl. The use of the
stable isotope ^{6}Li should have wide applications in the
study of lithium fluxes across membranes as well as in
whole-body kinetics.

Lithium Intoxication

Toxic effects of lithium may be divided into two main
varieties, those which are due to a rapid increase in the
quantity of lithium retained by the body, lithium intoxication, and those which are usually considered to be side
effects and occur after long-term stabilisation on lithium

with no increase in plasma lithium. The latter are often
considered to be either unavoidable sequelae in a proportion of cases or may be treated by the administration of
other drugs.

Much of the work of elucidating the mechanisms of
initiation of lithium intoxication was done by Thomsen in
Aarhus and is described in detail in two recent papers
(Thomsen, 1976; Thomsen *et al.*, 1976).

Although the final stages of lithium intoxication
affect mainly the nervous system, the rapid rise in plasma
lithium which causes these neurological effects is due to
the effects of lithium in renal function.

Lithium is almost exclusively excreted through the
kidneys. In the normal state serum lithium is regulated
by the excretion of an amount which is approximately
equivalent to the daily intake. Lithium is freely filtered
through the glomerulus of the kidney and 80% of that
filtered is reabsorbed along with water and sodium in the
proximal tubule. The remainder passes to the distal parts
of the renal tubule where very little is absorbed and it
is therefore excreted. This contrasts with the absorbtion
of sodium where 80% of the filtrate is reabsorbed in the
proximal tubule and a further 18 or 19% is reabsorbed in
the distal tubule. Much of the fine control of sodium
excretion is exerted on the fraction reabsorbed in the
distal tubule. If therefore a fall in dietary sodium intake occurs, more sodium is reabsorbed in the distal
tubule under the influence of the mineralocorticoid hormones. If, however, sodium deprivation is so severe that
the changes in distal reabsorbtion can no longer cope
with the need to conserve sodium, an increase occurs in
the proximal reabsorbtion and, in lithium treated subjects, this is accompanied by an increased reabsorbtion
of lithium. However lithium inhibits reabsorbtion of
sodium in the distal tubule due to its effects on the
response to aldosterone. It therefore causes sodium

depletion and hence increased proximal sodium reabsorbtion and further lithium reabsorbtion which itself exacer-
bates the distal tubular effects. Once the critical stage
has been reached a vicious circle is initiated. This
model of lithium intoxication elucidated in the rat is
applicable in general terms to man (Thomsen *et al.*, 1976).

Since a number of diuretics, notably the thiazides,
and also adrenalectomy, lead to the inhibition of distal
sodium reabsorbtion and consequent natriuresis, these
may also lead to lithium intoxication. The risk of intoxication is also increased when sodium is lost through
heavy sweating, vomiting or diarrhoea. Conversely if sodium intake is reduced as in the case of self administered
weight reducing diets, there is also the risk of intoxication. From the point of view of patient management it
is therefore important that patients should be warned,
especially in hot weather, to maintain their sodium intake and that their general practitioners should be
aware of the potential dangers of diuretic therapy whilst
receiving lithium. However, a recent report has shown that
with careful monitoring of lithium status, diuretic therapy may be maintained with a lowered lithium dose (Kerry,
1978).

Metabolic Effects of Lithium

The range of metabolic effects of lithium is large. It
extends from those effects which are manifested at the
level of the whole organism, for instance effects on body
weight, and encompasses effects on individual organs or
tissues, such as the well known thyroid effects, and
finally extends to the molecular level where individual
enzyme systems may be disturbed by addition of lithium.

One of the difficulties in evaluating reports of metabolic effects of lithium is that the majority of the
literature has arisen, not as a direct attempt to determine a mechanism of lithium action, but incidental to

other studies. Thus lithium is frequently used as a substitute for sodium in physiological fluids and may be
present in concentrations as high as 150 mM/l. Effects of
such solutions are often found though it is not common to
ascribe any part of the effect to the pharmacological
action of lithium. The false logic behind these experiments seems to be: 1) Lithium is similar to sodium
therefore one can replace sodium by lithium in solution.
2) Lithium is not sodium therefore any effects seen must
be due to lack of sodium. A further common experimental
use of lithium is as one of a series of alkaline metals
of differing sizes used to test binding sites, transport
mechanisms, and membrane 'pores'. I have discussed in an
earlier section the anomalous chemistry of lithium and
the likelihood that its chemistry in complexes may result
not only from considerations of size but also from its
high polarizing power. Finally, lithium salts are used
as reagents for example in the disaggregation of ribosomes at concentrations of 2 mole 1^{-1}.

Further problems arise in the interpretation of animal
experiments on lithium. Quite large doses of lithium
salts may be given by injection over a short period. The
biochemical or behavioural changes seen after such treatment may bear no relation to the effects of lithium but
rather to the stress of the pain and discomfort induced
by injection of very hypertonic solutions. A very acrid
exchange in the literature illuminates this point since
one of the items of dispute was the use of acute intraperitoneal injection of 2.5 mole L^{-1} LiCl in rats which
were immediately used in a behavioural experiment. According to one author injection of such a solution causes
severe pain, haemorrhage, and inflamation of the intestine
(Smith, 1977, Johnson, 1977).

A large number of pharmacological experiments have used
the intraperitoneal route of lithium administration and
this may be legitimate with dilute solutions if a divided

dosage schedule is used and the experiment is essentially
long term. The pharmacological situation attained in
patients receiving prophylactic lithium, however, must
differ from that obtaining during a single dose acute
study. Finally, one must beware of extrapolating results
from one species to another.

Whole body effects

   One side effect of lithium which is of concern to
clinicians is that certain patients, probably less than
10%, appear to gain significant amounts of body weight
(sometimes over 10 kg). The mechanism for this weight
gain is largely unknown though it has been suggested that
it is a reflection of the general improvement in health
accompanying the mood stabilisation (Kerry *et al.*, 1970).
Lithium is also known to have effects on lipid and carbohydrate metabolism (Mellerup and Rafaelsen, 1975), water
and electrolyte metabolism (Jenner, 1973) and thyroid
function (Wolff, 1976), all of which may contribute.
Factors involved in the control of body weight and in
lithium's effects have been reviewed by Birch (1978).

   Dempsey *et al.* (1976) have reported successful treatment of excessive weight gain during lithium therapy by
using individually designed low caloric diets with strong
emphasis on maintaining sodium and potassium intake. The
success of this treatment lends support to the view that
lithium must be acting centrally on brain areas which
control feeding behaviour. Plenge *et al.* (1973) have
shown that environmental factors are important in the
development of weight gain in rats during lithium treatment.

   Nevertheless, changes have been demonstrated in carbohydrate metabolism in rats (Mellerup *et al.*, 1974) and
studies *in vitro* have shown changes in glucose utilisation
and glycogen synthesis (Haugaard *et al.*, 1975). Clinical
studies have been, so far, equivocal, with demonstration

of initial changes in glucose tolerance, a measure of
release and action of the hormone insulin, during the
early stages of lithium prophylaxis (Van der Velde and
Gordon, 1969; Shopsin *et al.*, 1972). However after six
months treatment with lithium Vendsborg and Prytz (1976)
were unable to show abnormality in response to a glucose
load, and this is in agreement with Gordon and Van der
Velde (1974).

Fig. 10 shows the various factors which influence the
control of body weight. It is clear that the two main

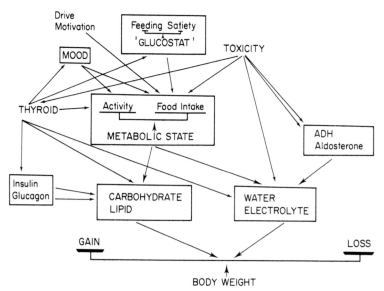

Fig. 10. Reproduced from Birch (1977) by permission of the copyright
owner.

effector domains are those of lipid and carbohydrate
metabolism and the control of body fluids and electrolytes. These effector systems are integrated by endocrine
functions directed largely from the hypothalamus and the
pituitary gland. We will now consider the effects of
lithium on the various endocrine glands and their hormones, then in a later section we will consider the control of body fluids and electrolytes by the action of
these hormones on the kidney.

Effects of Lithium on Endocrine Systems

Lithium has marked effects on a number of endocrine organs though the mechanism of this action is sometimes obscure. Endocrine glands, under the control of the nervous system and neurosecretory cells of the hypothalamo-hypophyseal nuclei and tracts, secrete hormones which are transported in the blood to act at sites distant from the endocrine organ. The regulation of this system is by inter-related feedback systems, where changes in the blood concentration of hormone act on the centres in the nervous system which regulate the release of that hormone. The major sites at which lithium might therefore effect a single endocrine function are as follows: (a) in the sensitivity of neuro-receptor to circulating hormone, (b) on the integration of response from this receptor with that of other receptors monitoring other aspects of the controlled function, (c) on the release of local hormone from the hypothalamus to the pituitary, (d) on the sensitivity of the pituitary to this local releasing factor. These points of influence are all within the central nervous system. At a slightly lower level of organisation lithium might influence (e) the synthesis and storage of the trophic hormone by the pituitary gland, (f) the release of this trophic hormone by the pituitary and (g) the sensitivity of the final effector endocrine gland to the circulating trophic hormone to the pituitary. Finally, at the level of the endocrine gland and final target organ lithium might influence (h) synthesis and storage of the effector hormone, (i) the release of the effector hormone by the endocrine gland, (j) the sensitivity of the target organ to the circulating effector hormone and (k) the transduction of the endocrine message by means of cyclic 3'5'AMP into biochemical action by the target organ.

This very long chain has a number of links which are accessible and have been studied in some detail. Thus,

effects of lithium on cyclic 3'5'AMP have been investigated, since this is the common final transducing mechanism
for many hormones. The plasma concentrations of various
hormones has been investigated following lithium and the
potential of this research area has rapidly expanded
following the advent of radioimmuno assay techniques for
small peptide hormones. Berens and Wolff (1975) surveyed
the endocrine effects of lithium and Forn (1975) discussed effects on cyclic AMP and adenylcyclase.

In general it appears that lithium inhibits hormone
stimulated adenyl cyclase activity and this occurs in the
thyroid, in the kidney, where both antidiuretic hormone
sensitive and parathormone sensitive adenyl cyclase are
inhibited, and in various sites under the influence of
noradrenaline (norepinephrine). (See reviews: Forn,
1975; Geisler *et al.*, 1978; Birch, 1978.)

One of the well-known side effects of lithium is a
tendency for lithium treated patients to develop hypothyroidism. The biochemical basis of the inhibition of
thyroid function has been discussed by Hullin (1978), and
more extensively by Berens and Wolff (1975). Lithium has
been shown to block thyroid adenyl cyclase sensitive to
thyroid stimulating hormone (released from the pituitary)
and to inhibit the coupling of iodotyrosines to form the
thyroid hormones triiodothyronine (T_3) and tetraiodo-
thyronine (T_4, thyroxine). It also interferes with the
release of T_3 and T_4 by the disruption of the secretory
micro-tubules (Bhattacharyya and Wolff, 1976). These
studies therefore suggest that lithium has actions at
multiple sites in the synthesis and the release of thyroid hormones. The influence of thyroid hormones on various controlling factors of body weight has been shown in
Fig. 10 though the effects of lithium on the response to
thyroid hormones of various target organs has not been
investigated.

Control of Fluid and Electrolyte Metabolism and Renal Effects of Lithium

The second major area of endocrinology which has been investigated with respect to lithium action is that of the control of fluid and electrolyte metabolism. Early studies showed that water and electrolyte metabolism in the manic depressive patient was abnormal, there being retention of sodium and water during depression (see Hullin, 1975; Baer, 1973).

The control of fluid balance in the body is carried out mainly by the action of the antidiuretic hormone (ADH) which is secreted by the posterior pituitary and acts on the renal tubule to cause reabsorbtion of water. The control of electrolyte metabolism is by a range of hormones, the mineralocorticoids, secreted by the adrenal cortex. The most potent, aldosterone, is secreted by the adrenal cortex in response to the circulating peptide angiotensin which in turn is converted from its inactive form by the action of renin, secreted by the juxta-glomerular apparatus (J.G.A.) of the kidney. The ADH and Renin-Aldosterone systems are intimately related since increased reabsorbtion of sodium also causes concomitant fluid retention and similarly diuresis necessarily in-volves loss of electrolytes. Jenner and Eastwood (1978) have reviewed the renal effects of lithium and MacNeil and Jenner (1975) and Singer (1976) have discussed in detail the effects of lithium on fluid balance and the antidiuretic hormone.

Allsopp *et al.* (1972) and Hullin *et al.* (1977) have shown abnormalities in renin and aldosterone relationships in manic depressive psychoses, there being some dissocia-tion between the normal relationships of renin, aldo-sterone and electrolyte excretion. The response of manic depressive patients to changes in posture, a potent stimulus to renin secretion, is blunted. Recent studies by O'Brien (reported by Hullin, 1978) have suggested that in manic depressive subjects there is either an inhibitor present of aldosterone synthesis or conversely the absence of a stimulator leading to a reduced response to circulating renin of aldosterone synthesis. These studies have not systematically investigated effects of lithium but it is notable that the one subject who was receiving lithium had an abnormally high resting plasma renin activity and an exceptionally high response to change in posture.

Lithium, however, has been shown to inhibit the action of antidiuretic hormone in the reabsorption of water (Harris and Jenner, 1972) and it is possible that this may be the basis of the polyuria-polydipsia seen in a number of patients, who are reported also to be vasopressin (ADH) resistant. Recent reports suggest that a compensatory increase in ADH production occurs since both excretion of ADH and plasma ADH have been shown to be increased in human subjects receiving lithium (Jenner and Eastwood, 1978). Lithium has been reported to inhibit ADH sensitive adenyl cyclase (Forn, 1975) though this has been critically reviewed by Jenner and Eastwood (1978) who suggest that prolonged treatment of an intact animal with lithium causes a depletion of the tissue concentration of renal adenyl cyclase and hence decreased measured enzyme activity.

Horrobin and his colleagues (1978) have made the interesting suggestion that prostaglandins are essential 'permissive' agents for the action of antidiuretic hormone on adenyl cyclase and that lithium, at low concentrations, is able to inhibit the effects of vasopressin on prostaglandin synthesis. These workers also propose a more general hypothesis that hormone stimulated prostaglandin synthesis is a factor in the aetiology of the manic-depressive psychoses.

We have previously considered the work of Thomsen on the development of renal toxicity of lithium and it is clear that even at non-toxic plasma concentrations lithium

has effects on the excretion of water and electrolytes.
Thomsen's view of lithium's effect on sodium transport
in the nephron is supported by the micropuncture studies
of Harris and Dirks (1973) which suggest that lithium
inhibits sodium reabsorbtion in the loop of Henle. It is
clear therefore that lithium has effects on water and
electrolyte excretion in the kidney though the molecular
mechanisms involved are not yet clear. It is difficult
to unravel the phsyicochemical effects on the various
exchanges in the renal tubule from the various hormone
and hormone-messenger systems which act there.

Subcellular Effects of Lithium

The range of effects reported of lithium at the sub-
cellular level is large. Much of this work however must
be neglected in the investigation of lithium in manic
depressive psychoses since non-pharmacological conditions
have been used. Johnson (1975) and Birch (1978) have reviewed this literature in some detail and a valuable new
viewpoint was presented by a symposium organised by the
Neuro Sciences Research Programme in the United States
when physical and inorganic chemists met with neuro
biologists to provide fresh perspectives (Bunney and
Murphy, 1976).

In much of the early work on lithium it was considered
to be a potential competitor for functions of alkali
metals. However the chemistry of lithium may be described
in terms of the 'diagonal relationship' in the periodic
table and perhaps one should rather consider the biochemistry of lithium in this light (Birch, 1970). Early
studies confirmed that lithium did indeed have effects on
magnesium (Birch, 1971, Birch and Jenner, 1973) and independent studies by Mellerup and his colleagues (1970)
supported these findings. Williams (1973), Bunney and
Murphy (1976), and Frausto da Silva and Williams (1976)
have provided theoretical chemical justification for the

lithium-magnesium relationship but biochemical and pharmacological studies, though encouraging, have failed so far
to provide definitive confirmation of the hypothesis.

Magnesium dependent enzymes

Very many biological processes are magnesium dependent
and include a large number of enzymes which require magnesium as a co-factor (Aikawa, 1976). The metabolic
significance of even a small change induced by lithium in
a series of such enzymes might be very large, particularly
in an organ as uniquely sensitive to energy availability
as the brain. As an example, one of the major pathways
regulating the energy supply in cells is of course glycolysis and it can be seen from Fig. 11 that many of these
enzymes are magnesium dependent and in fact two have
been shown to be affected by lithium.

Table III shows the range of enzymes which has been
shown to be affected by lithium in studies both *in vitro*

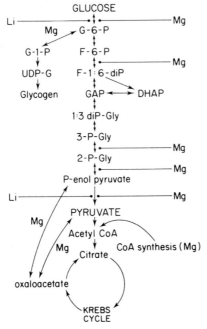

Fig. 11. Glycolysis, showing magnesium dependent processes.

TABLE III

*Effects of lithium have been reported on each
of the following enzyme systems*

(a) *In vitro*	(b) *In vivo*
ATPases, (Na^+K^+), Mg^{2+}, Ca^{2+}	Acetyl cholinesterase
Alkaline phosphatase	Acid phosphatase
DNA polymerase	Aconitase
Enolase	Alkaline phosphatase
Fructose 1,6 diphosphatase	Aryl sulphatase
Hexokinase	Cholinesterase
Pyruvate kinase	Glucokinase
RNA synthetase	Hexokinase
	Pyruvate kinase
	RNA synthesis
	Succinic dehydrogenase
	Tryptophan oxygenase
	Tyrosine aminotransferase

For bibliographic detail see Birch (1978). Taken from Birch (1977) with permission.

and *in vivo*. No common pattern may be seen in these enzymes and the molecular details of the lithium effect have rarely been investigated.

We have studied the interaction of lithium with pyruvate kinase (Birch, 1978) and have shown that lithium is not competitive with magnesium though its inhibition is similar to that of calcium ion. We have also investigated whether lithium might compete with magnesium at magnesium-ADP and magnesium-ATP complexes which are the substrates for some enzymes and were able to provide preliminary evidence of possible ternary complex of the type Li-Mg-ADP (Birch and Goulding, 1975).

In the previous section we discussed the effects of lithium on adenyl cyclase in the kidney. It is interesting to note that both adenyl cyclase the synthetic enzyme, and phosphodiesterase, the enzyme which inactivates cyclic 3'5' AMP, are magnesium dependent. If lithium were to compete for magnesium in either of these important enzymes this could provide a basis to explain the multiple and varied actions of lithium which so far bear no recognised relationship to each other.

Effects of lithium on various systems of neurotransmitter substances have produced a large number of studies.
Lithium inhibits various stages of acetylcholine synthesis and release and this has been reviewed by Vizi
(1975). Shaw (1978) has reviewed the field of trypto-
phan metabolism and its role in affective disorders and
he has also attempted the daunting task of sorting the
'hay-stack of needles', an apt description, which is the
literature of the indoleamine and catecholamine research
in psychopharmacology (Shaw, 1975).

The amine hypotheses have maintained considerable
currency though the interpretation of experimental results
rests on multiple pharmacological insults and high dosage
and the uncertainty of the tissue source of the various
metabolites estimated. Lithium effects have been reported on catecholamine release, reuptake, synthesis and
turnover and on catecholamine receptors (Murphy, 1976).
Murphy has also reviewed glutamate, GABA, and other transmitters affected by lithium. Mandell (1976) has discussed
the short term stimulation of serotonin synthesis by
lithium and its reversal with time.

Lithium and Membranes

One area of considerable controversy at the present is
the transport of lithium across the red blood cell (erythrocyte) membrane. It has been proposed that the concentration of lithium in the red blood cell might be used as
a predictive test for response to lithium, or to discriminate between differing subtypes of the underlying
affective psychoses (Mendels and Frazer, 1973; Frazer *et
al.*, 1973). Other workers were unable to agree with these
findings though the methods used and the psychiatric
criteria may have been different. Lee and Paschalis
(1978) have reviewed these various studies and they conclude that certain patients with affective or schizo-

affective disorders may have abnormal lithium metabolism
in the red cell but that predictive tests using red cell
lithium are unreliable.

Investigation of the kinetics of lithium uptake by the
red cell has been carried out since if membrane function
is abnormal in the affective disorders, this may be
reflected in abnormal red cell membranes. There is little
doubt that the most definitive work in this area is due
to Greil and his colleagues (Duhm *et al.*, 1976; Duhm and
Becker, 1977; Greil *et al.*, 1977). This series of experiments has been summarized by Greil and Eisenried (1978)
who report that lithium uptake by erythrocytes consists
of three different components: (a) Ouabain sensitive
uptake of lithium which results from the sodium-potassium
pump; (b) a phloretin sensitive uptake which is a result
of electrogenic sodium-lithium counter transport, and
(c) leak diffusion of lithium. Aspects of these postulated transport systems have been confirmed by other workers
(Meltzer *et al.*, 1976; Haas *et al.*, 1975; Frazer *et al.*,
1977). Meltzer *et al.* (1977) have reported that though
the lithium pump activity in manic-depressives is not
different from a normal population prior to lithium
treatment, after several days or weeks of lithium administration the pump activity is decreased. This suggests that
lithium has long-term effects on membrane structure and
function and this is further supported by the report of
an irreversible inhibition of choline uptake by erythrocytes following lithium (Martin, 1978; Lingsch and Martin,
1976).

Finally, lithium has effects on the sodium content of
red cells and on sodium-potassium-ATPases (Dick *et al.*,
1978; Glen, 1978). It has been proposed that abnormalities
in sodium transport are one of the basic metabolic disturbances in the manic-depressive psychoses (Hesketh *et
al.*, 1977; Naylor *et al.*, 1976; Naylor *et al.*, 1977) and
that lithium may reverse the defect. Glen (1978) has

emphasised particularly that ATPases have specific orientation within membranes and different effects of lithium are seen when presented at different sides of the membrane. This may be of particular importance in the action of lithium in nerve cells.

Mode of Action of Lithium: Theories and Speculations

The mode of action of lithium is unknown. We have, though, a number of interesting lines of approach which may provide some answer.

From the theoretical chemical point of view Frausto da Silva and Williams (1976) have suggested that despite apparently unfavourable conditions within cells for competition between lithium and other cations, if the *Milieu Intérieur* of the cell is considered as a whole rather than as individual complexes in isolation, various chelating ligands present, particularly nucleotides, may sequester ions such as magnesium and leave available for occupation sites which might otherwise have been occupied by magnesium. Under these conditions, lithium might compete effectively for sites despite its lower affinity. They therefore suggest in predictive work the use of 'conditional' association constants determined under conditions similar to the internal *milieu* of the cell rather than association constants which have been determined in simple aqueous solution. It is possible that the microenvironments of ligands in different parts of the cell might be sufficiently variable for Li-Ca-Mg interchanges to take place, changing the conformation of the ligands and acting as an effective switching device.

At the next level of organisation we have theories which suggest that lithium interferes with enzymes due to competition with magnesium or sodium. Unifying hypotheses have been put forward suggesting a major locus of effect either on the cyclic AMP system or on the ATPase of the cell membrane or by the modulating effects of prosta-

glandins. Furthermore, it has been suggested that lithium's action might be on the synthesis or release of
neurotransmitter substances and a whole host of sites
might thus be implicated.

Lithium action might be on the regulation of body
fluids and electrolytes or energy availability due to
endocrine effects and this in turn is related to cyclic
AMP action. One might also consider despite its wide distribution that the somatic effects are irrelevant to the
psychiatric effect and that lithium has a specific action
at a limited locus in the brain.

The variety of models, ranging from the biochemical to
the mathematical, has been reviewed by Johnson (1978)
who maintains that the function of a model is to provide
a springboard for further ideas rather than an explanation of a phenomenon. The biochemist, the chemist and
the psychiatrist have each their own models for lithium
action. It is important that each should test the others'
springboards so that more powerful and accurate versions
may be made.

Lithium is of great benefit to the 25,000 patients
who are receiving it in Britain. However, it may be of
even greater use if it can provide the clue to the aetiology of recurrent affective disorder.

Acknowledgements

I wish to thank all those who have contributed to this
review by discussion and publication of their various
approaches to lithium and particularly those colleagues
who have allowed me to reproduce their results here. I
am most indebted to Dr. D.E. Fenton and Dr. T.C. Jerram
who read parts of the text and provided valuable criti-
cism. I thank Dr. R.P. Hullin for his continuing encour-
agement and support. N.J.B. is supported by an M.R.C.
grant.

References

Aikawa, J.K. (1976). *In* 'Trace Elements in Human Health and Disease' (A.S. Prasad, ed.), Vol. II, pp. 47-48. Academic Press, New York.
Allsopp, M.N.E., Levell, M.J., Stitch, S.R., Hullin, R.P. (1972). *Brit. J. Psychiat.* **120**, 399-404.
Baastrup, P.C., Poulsen, J.C., Schou, M., Thomsen, K., Amdisen, A. (1970). *Lancet* (ii), 326-330.
Baer, L. (1973). *In* 'Lithium, its Role in Psychiatric Research and Treatment' (S. Gershon, B. Shopsin, eds.), pp. 33-49. Plenum Press, New York and London.
Berens, S.C., Wolff, J. (1975). *In* 'Lithium Research and Therapy' (F.N. Johnson, ed.), pp. 443-472. Academic Press, London.
Bhattacharyya, B., Wolff, J. (1976). *Biochem. Biophys. Res. Comm.* **73**, 383-390.
Birch, N.J. (1970). *Brit. J. Psychiat.* **116**, 461.
Birch, N.J. (1971). Ph.D. thesis, University of Sheffield (Microfilm D3225/73 British Library, Lending Division, Boston Spa, Yorks, U.K.).
Birch, N.J. (1974). *Clin. Sci. & Mol. Med.* **46**, 409-413.
Birch, N.J. (1977). *Trends in Biochemical Sciences* 2, 282-284.
Birch, N.J. (1978). *In* 'Lithium in Medical Practice' (F.M. Johnson, S. Johnson, eds.), pp. 89-114. M.T.P. Press, Lancaster.
Birch, N.J., Goulding, I. (1975). *Analyt. Biochem.* **66**, 293-297.
Birch, N.J., Hullin, R.P. (1972). *Life Sci.* **11**, 1095-1099.
Birch, N.J., Jenner, F.A. (1973). *Brit. J. Pharmacol.* **47**, 586-594.
Birch, N.J., Hullin, R.P., Inie, R.A., Robinson, D. (1978). *Brit. J. Clin. Pharmacol.* **5**, 351P-352P.
Blackwell, B., Shepherd, M. (1968). *Lancet* (i), 968-971.
Bond, P.A. (1978). *In* 'Lithium in Medical Practice' (F.M. Johnson, S. Johnson, eds.), pp. 215-223. M.T.P. Press, Lancaster.
Bond, P.A., Brooks, B.A., Judd, A. (1975). *Brit. J. Pharmacol.* 53, 235-239.
Bunney, W.E., Murphy, D.L. (eds.) (1976). 'The Neurobiology of Lithium', Neurosciences Research Program Bulletin Vol. 14, pp. 111-207, N.R.P., Boston.
Cade, J.F.J. (1949). *Med. J. Austral.* **36**, 349-352.
Cade, J.F.J. (1978). *In* 'Lithium Research and Therapy' (F.N. Johnson, ed.), pp. 5-16. Academic Press, London.
Cooper, T.B., Bergner, P.-E.E., Simpson, G.M. (1973). *Am. J. Psychiat.* **130**, 601-603.
Cooper, T.B., Simpson, G.M. (1976). *Am. J. Psychiat.* **133**, 440-443.
Coppen, A., Noguera, R., Bailey, J., Burns, B.H., Swani, M.S., Hare, E.H., Gardner, R., Maggs, R. (1971). *Lancet* (ii), 275-279.
Cotton, F.A., Wilkinson, G. (1972). 'Advanced Inorganic Chemistry'. Interscience Pub., New York.
Davis, J.M., Fann, W.E. (1971). *Ann. Rev. Pharmacol.* **11**, 285-302.
Dempsey, G.M., Dunner, D.L., Fieve, R.R., Farkas, T., Wong, J. (1976). *Am. J. Psychiat.* **133**, 1082-1084.
Desbiez, M.O., Thellier, M. (1975). *Plant Sci. Lett.* **4**, 315-321.
Desgrez, A., Meunier, J. (1927). *C.R. Acad. Sci. (Paris)* **185**, 160-163.
Dick, D.A.T., Naylor, G.J., Dick, E.G. (1978). *In* 'Lithium in Medical Practice' (F.N. Johnson, S. Johnson, eds.), pp. 173-182. M.T.P. Press, Lancaster.

Duhm, J., Becker, B.R. (1977). *Pflügers Archiv.* **368**, 203-208.
Duhm, J., Eisenried, F., Becker, B.F., Greil, W. (1976). *Pflügers Archiv.* **364**, 147-155.
Dunner, D.L., Stallone, F., Fieve, R.R. (1976). *Arch. Gen. Psychiat.* **33**, 117-120.
Ebadi, M.S., Simmons, V.J., Hendrickson, M.J., Lacy, P.S. (1974). *Eur. J. Pharmacol.* **27**, 324-329.
Fenton, D.E. (1977). *Chem. Soc. Rev.* **6**, 325-343.
Fieve, R.R. (1975). *Am. J. Psychiat.* **132**, 1018-1022.
Fieve, R.R., Kumbaraci, T., Dunner, D.L. (1976). *Am. J. Psychiat.* **133**, 925-929.
Forn, J. (1975). *In* 'Lithium Research and Therapy' (F.N. Johnson, ed.), pp. 485-497. Academic Press, London.
Frausto da Silva, J.J.R., Williams, R.J.P. (1976). *Nature* **263**, 237-239.
Frazer, A., Mendels, J., Secunda, S.K., Cochrane, C.M., Bianchi, C.P. (1973). *J. psychiat. Res.* **10**, 1-7.
Frazer, A., Mendels, J., Brunswick, D. (1977). *Commun. Psychopharmacol.* **1**, 255-269.
Garrod, A.B. (1859). 'Gout and Rheumatic Gout'. Walton and Maberly, London.
Geisler, A., Klysner, R., Thams, P. (1978). *In* 'Lithium in Medical Practice' (F.N. Johnson, S. Johnson, eds.), pp. 159-165. M.T.P. Press, Lancaster.
Gershon, S., Shopsin, B. (eds.) (1973). 'Lithium, Its Role in Psychiatric Research and Treatment'. Plenum Press, New York and London.
Glen, A.I.M. (1978). *In* 'Lithium in Medical Practice' (F.N. Johnson, S. Johnson, eds.), pp. 183-192. M.T.P. Press, Lancaster.
Gordon, M.W., Van der Velde, C.D. (1974). *Nature* **247**, 160-162.
Greil, W., Eisenried, F. (1978). *In* 'Lithium in Medical Practice' (F.N. Johnson, S. Johnson, eds.), pp. 415-420. M.T.P. Press, Lancaster.
Greil, W., Eisenried, F., Becker, B.F., Duhm, J. (1977). *Psychopharmacol.* **53**, 19-26.
Haas, M., Schooler, J., Tosteson, D.C. (1975). *Nature* **258**, 425-427.
Hanna, S.M., Jenner, F.A., Pearson, I.B., Sampson, C.A., Thompson, E.A. (1972). *Brit. J. Psychiat.* **121**, 271-280.
Harris, C.A., Dirks, J.H. (1973). *Fed. Proc.* **32**, 381.
Harris, C.A., Jenner, F.A. (1972). *Brit. J. Pharmacol.* **44**, 223-232.
Haugaard, E.S., Frazer, A., Mendels, J., Haugaard, N. (1975). *Biochem. Pharmacol.* **24**, 1187-1191.
Hesketh, J., Glen, A.I.M., Reading, H.W. (1977). *J. Neurochem.* **28**, 1401-1402.
Heslop, R.B., Jones, K. (1976). 'Inorganic Chemistry'. Elsevier, Amsterdam.
Hewick, D.S., Newbury, P., Hopwood, S., Naylor, G., Moody, J. (1977). *Brit. J. Clin. Pharmacol.* **4**, 201-205.
Horrobin, D.F., Mtabaji, J.P., Manku, M.S., Karmazyn, M. (1978). *In* 'Lithium in Medical Practice' (F.N. Johnson, S. Johnson, eds.), pp. 243-246. M.T.P. Press, Lancaster.
Hullin, R.P. (1975). *In* 'Lithium Research and Therapy' (F.N. Johnson, ed.), pp. 359-379. Academic Press, London.
Hullin, R.P. (1978). *In* 'Lithium in Medical Practice' (F.N. Johnson,

S. Johnson, eds.), pp. 433-454. M.T.P. Press, Lancaster.

Hullin, R.P., McDonald, R., Allsopp, M.N.E. (1972). *Lancet* (i), 1044-1046.

Hullin, R.P., McDonald, R., Allsopp, M.N.E. (1975). *Brit. J. Psychiat.* **126**, 281-284.

Hullin, R.P., Jerram, T.C., Lee, M.R., Levell, M.J., Tyrer, S.P. (1977). *Brit. J. Psychiat.* **131**, 575-581.

Jenner, F.A. (1973). *Biochem. Soc. Spec. Publ.* **1**, 101-111.

Jenner, F.A., Eastwood, P.R. (1978). *In* 'Lithium in Medical Practice' (F.N. Johnson, S. Johnson, eds.), pp. 247-263. M.T.P. Press, Lancaster.

Jenner, F.A., Gjessing, L.R., Cox, J.R., Davies-Jones, A., Hullin, R.P., Hanna, S.M. (1967). *Brit. J. Psychiat.* **113**, 895-910.

Jerram, T.C., McDonald, R. (1978). *In* 'Lithium in Medical Practice' (F.N. Johnson, S. Johnson, eds.), pp. 407-413. M.T.P. Press, Lancaster.

Johnson, F.N. (ed.) (1975). 'Lithium Research and Therapy'. Academic Press, London.

Johnson, F.N. (1977). *Compr. Psychiat.* **18**, 453-457.

Johnson, F.N. (1978). *In* 'Lithium in Medical Practice' (F.N. Johnson, S. Johnson, eds.), pp. 305-327. M.T.P. Press, Lancaster.

Johnson, F.N., Johnson, S. (eds.) (1978). 'Lithium in Medical Practice'. M.T.P. Press, Lancaster.

Johnson, S. (1975). *In* 'Lithium Research and Therapy' (F.N. Johnson, ed.), pp. 533-556. Academic Press, London.

Kerry, R.J. (1978). *In* 'Lithium in Medical Practice' (F.N. Johnson, S. Johnson, eds.), pp. 337-353. M.T.P. Press, Lancaster.

Kerry, R.J., Liebling, L.I., Owen, G. (1970). *Acta Psychiat. Scand.* **46**, 238-243.

Kollman, P.A., Liebman, J.F., Allen, L.C. (1970). *J. Am. Chem. Soc.* **92**, 1142-1150.

Lee, C.R., Paschalis, C. (1978). *In* 'Lithium in Medical Practice' (F.N. Johnson, S. Johnson, eds.), pp. 365-379. M.T.P. Press, Lancaster.

Lehn, J.M. (1973). *Structure and Bonding* **16**, 2-69.

Lehn, J.-M. (1976). *In* 'The Neurobiology of Lithium' (W.E. Bunney, D.L. Murphy, eds.), pp. 133-137. Neurosciences Research Program Bulletin, Vol. 14. N.R.P., Boston.

Lingsch, C., Martin, K. (1976). *Brit. J. Pharmacol.* **57**, 323-327.

MacNeil, S., Jenner, F.A. (1975). *In* 'Lithium Research and Therapy' (F.N. Johnson, ed.), pp. 473-484. Academic Press, London.

Mandell, A.J. (1976). *In* 'The Neurobiology of Lithium' (W.E. Bunney, D.L. Murphy, eds.), pp. 169-173. Neurosciences Research Program Bulletin, Vol. 14, N.R.P., Boston.

Martin, K. (1978). *In* 'Lithium in Medical Practice' (F.N. Johnson, S. Johnson, eds.), pp. 167-171. M.T.P. Press, Lancaster.

Mendels, J., Frazer, A. (1973). *J. Psychiat. Res.* **10**, 9-18.

Mellerup, E.T., Rafaelsen, O.J. (1975). *In* 'Lithium Research and Therapy' (F.N. Johnson, ed.), pp. 381-389. Academic Press, London.

Mellerup, E.T., Plenge, P.K., Rafaelsen, O.J. (1974). *Dan. Med. Bull.* **21**, 88-92.

Meltzer, H.L., Rosoff, C.J., Kassir, S., Fieve, R.R. (1976). *Life Sci.* **19**, 371-380.

Meltzer, H.L., Kassir, S., Dunner, D.L., Fieve, R.R. (1977). *Psycho-*

434N.J. BIRCH

pharmacol. **54**, 113-118.
Moore, C., Pressman, B.C. (1964). *Biochem. Biophys. Res. Comm.* **15**, 562-567.
Mukherjee, B.F., Bailey, P.T., Pradhan, S.N. (1976). *Psychopharmacol.* **48**, 119-121.
Murphy, D.L. (1976). *In* 'The Neurobiology of Lithium' (W.E. Bunney, D.L. Murphy, eds.), pp. 165-169. Neurosciences Research Program Bulletin, Vol. 14. N.R.P., Boston.
Naylor, G.J., Dick, D.A.T., Dick, E.G. (1976). *Psycholog. Med.* **6**, 257-263.
Naylor, G.J., Smith, A., Boardman, L.J., Dick, D.A.T., Dick, E.G., Dick, P. (1977). *Psycholog. Med.* **7**, 229-233.
Pedersen, C.J. (1967). *J. Am. Chem. Soc.* **89**, 7017-7036.
Plenge, P.K., Mellerup, E.T., Rafaelsen, O.T. (1973). *Int. Pharmacopsychiat.* **8**, 234-238.
Schou, M. (1957). *Pharmacological Rev.* **9**, 17-58.
Schou, M. (1958). *Acta Pharmacol. (Kbh).* **15**, 115-124.
Schou, M. (1968). *J. psychiat. Res.* **6**, 67-95.
Schou, M. (1969). *Psychopharmacol. Bull.* **5** (Pt. 4), 33-62.
Schou, M. (1972). *Psychopharmacol. Bull.* **8** (Pt. 4), 36-62.
Schou, M. (1973a). *Psychiat. Neurol. Neurochir. (Amst.)* **76**, 511-522.
Schou, M. (1973b). *Biochem. Soc. Trans.* **1**, 81-87.
Schou, M. (1976a). *Ann. Rev. Pharmacol.* **16**, 231-243.
Schou, M. (1976b). *Psychopharmacol. Bull.* **12** (Pt. 1), 49-74; **12** (Pt. 2), 69-83; **12** (Pt. 3), 86-99.
Schou, M. (1976c). *In* 'The Neurobiology of Lithium' (W.E. Bunney, D.L. Murphy, eds.), pp. 117-124. Neurosciences Research Program Bulletin, Vol. 14. N.R.P., Boston.
Schou, M. (1976d). *Neuropsychobiology* **2**, 161-191.
Schou, M. (1978). *In* 'Lithium in Medical Practice' (F.N. Johnson, S. Johnson, eds.), pp. 21-39. M.T.P. Press, Lancaster.
Schou, M., Thomsen, K. (1975). *In* 'Lithium Research and Therapy' (F.N. Johnson, ed.), pp. 63-84. Academic Press, London.
Schou, M., Thomsen, K., Baastrup, P.C. (1970). *Int. Pharmacopsychiat.* **5**, 100-106.
Shaw, D.M. (1978). *In* 'Lithium in Medical Practice' (F.N. Johnson, S. Johnson, eds.), pp. 115-121. M.T.P. Press, Lancaster.
Shopsin, B., Stern, S., Gershon, S. (1972). *Arch. Gen. Psychiat.* **26**, 566-571.
Singer, I. (1976). *In* 'The Neurobiology of Lithium' (W.E. Bunney, D.L. Murphy, eds.), pp. 175-177. Neurosciences Research Program Bulletin, Vol. 14, N.R.P., Boston.
Smith, D.F. (1977). *Compr. Psychiat.* **18**, 449-452.
Squire, P.W. (1916). 'Companion to British Pharmacopoeia', 19th edn. Churchill, London.
Stern, K.H., Amis, E.S. (1959). *Chem. Rev.* **59**, 1-64.
Stern, S., Frazer, A., Mendels, J., Frustaci, C. (1977). *Life Sci.* **20**, 1669-1674.
Thilenius, G. (1882). 'Handbuch der Balneotherapie'. Hirschwald, Berlin.
Thellier, M., Steltz, T., Wissocq, J.C. (1976a). *J. Microscopie. Biol. Cell.* **27**, 157-168.
Thellier, M., Steltz, T., Wissocq, J.C. (1976b). *Biochim. Biophys. Acta.* **437**, 604-627.

Thomsen, K. (1976). *J. Pharmacol. Exp. Ther.* **199**, 483-489.
Thomsen, K., Olesen, O.V., Jensen, J., Schou, M. (1976). *In* 'Current Developments in Psychopharmacology' (Valzelli, L., Essman, W.B., eds.), Vol. 3, pp. 157-177. Spectrum Pub., Inc., New York.
Thornhill, D.P. (1978). *Brit. J. Clin. Pharmacol.* **5**, 352P.
Tyrer, S., Hullin, R.P., Birch, N.J., Goodwin, J.C. (1976). *Psychol. Med.* **6**, 51-58.
Wheat, J.A. (1971). *Applied Spectrosc.* **25**, 328-330.
Williams, R.J.P. (1970). *Quart. Rev.* **24**, 331-365.
Williams, R.J.P. (1973). *In* 'Lithium, Its Role in Psychiatric Research and Treatment' (S. Gershon, B. Shopsin, eds.), pp. 15-31. Plenum Press, New York and London.
Winkler, R. (1976). *In* 'The Neurobiology of Lithium' (W.E. Bunney, D.L. Murphy, eds.), pp. 139-142. Neurosciences Research Program Bulletin, Vol. 14, N.R.P., Boston.
Wissocq, J.C., Steltz, T., Thellier, M. (1976). *Newsletter Applic. Nucl. Meth. Biol. Agric.* **6**, 21-23.
Wolff, J. (1976). *In* 'The Neurobiology of Lithium' (Bunney, W.E., Murphy, D.L., eds.), pp. 178-180. Neurosciences Research Program Bulletin, Vol. 14, N.R.P., Boston.
Worrall, E.P. (1978). *In* 'Lithium in Medical Practice' (F.N. Johnson, S. Johnson, eds.), pp. 69-77. M.T.P. Press, Lancaster.
Van der Velde, C.D., Gordon, M.W. (1969). *Arch. Gen. Psychiat.* **21**, 478-485.
Vendsborg, P.B., Prytz, S. (1976). *Acta Psychiat. Scand.* **53**, 64-69.
Vizi, E.S. (1975). *In* 'Lithium Research and Therapy' (F.N. Johnson, ed.), pp. 391-409. Academic Press, London.

12   ELEMENT DETERMINATION IN BIOLOGICAL MATERIALS USING ELECTRON PROBE MICROANALYSIS

C. Lechene

Harvard Medical School, Boston, Massachusetts

Introduction

Discoveries and progress in experimental science rely
upon the development of tools and preparative methods for
observing what is not seen with the naked eyes, for
measuring and analysing what could not be manipulated
with the bare fingers.

In biological sciences, electron microscopy has revealed the morphological structure of cells and tissues
in the nanometer range, molecular biology is discovering
the structure and function of the genome. However, physiology has not yet described adequately cellular function.
This is due partly to our ignorance of the chemical element composition of the various cellular and extra or
intercellular compartments being studied. X-ray microanalysis with electron probe excitation is becoming a
tool of major importance in physiology by providing re-
searchers with the possibility to measure the chemical
element content within cells and their microenvironment.

Historics

The use of X-ray microanalysis was devised and imagined
by G. von Hevesey. It is by the study of characteristic
x-ray lines that he discovered the new element, Hafnium.
He applied a method based on fluorescent radiations

excited by x-rays to calculate the relative abundance of
different chemical elements. In late 1940, in France,
A. Guinier imagined a technique for elemental microanalysis based on the analysis of the characteristic
x-ray lines emitted when a sample is excited by a beam
of accelerated electrons. R. Castaing developed the tool
and laid the foundation of a quantitative x-ray microanalysis in his doctorate thesis (Castaing, 1951). The
instrument was soon commercialized and the method of
electron probe x-ray microanalysis found immediate and
broad applications in mineralogy and in metallurgy.

Principles and Instrumentation

Electron probe microanalysis is well described in an
elementary book (Birks, 1971) and a review article
(Beaman and Isasi, 1971). Each chemical element emits
characteristic x-ray lines when it is excited by high
energy radiation. By determining the wavelength of these
characteristic x-ray lines, the element can be recognized. By measuring the intensity of the characteristic
x-ray light emitted, one can measure how much of the
element is contained in the sample. In the electron probe,
the exciting radiation is an electron beam. Under the
interaction of the accelerated electrons of the beam with
the elements constituting the matter of the sample, there
is ionization of the electrons of the K, L or M shells of
some of the atoms in the sample. This phenomenon is at
the origin of the emission of the characteristic x-ray
lines. The quantum energy is precisely defined and is a
function of the nuclear charge. Other electrons of the
incident beam will be decelerated by the electrical field
produced by the nucleus. The energy lost can appear as an
x-ray photon with a quantum energy from zero up to the
energy of the incident electron. It will produce a
background x-ray spectrum called continuum, white radiation or bremsstrahlung above which characteristic x-ray

lines have to be recognized.

In an electron probe the sample, in a vacuum, is excited by an electron beam accelerated between a few Kilovolts up to 100 Kilovolts under a current of nanoamperes or picoamperes. The size of the incident beam and the current density are regulated by electromagnetic lenses. The exciting part of the instrument is very similar to an electron microscope. The incident electron beam can be kept static on the sample or scan the sample in a raster pattern. The sample can be of different thicknesses, from bulk, opaque to electrons, to ultrathin, transparent to the incident electrons. The characteristic x-ray lines emitted are selected with diffractive crystal spectrometers and their intensity is measured with proportional counter and adapted electronics. Another kind of x-ray detector can analyse simultaneously all the energy of the x-ray lines received. These solid state detectors, however, have a spectral resolution poorer than crystal diffractometers, receive proportionally more continuum than characteristic lines and should be followed by complex deconvolution techniques to sort out the characteristic peaks from the background and to recognize interfering peaks.

Samples can be viewed in the instrument using a photonic microscope or the electron optic. Bulk sample can be viewed by scanning microscopy, thick sample by scanning transmission microscopy and ultra thin sample by true transmission electron microscopy when the instrument is equipped with magnetic lenses below the sample.

The advantages of electron probe microanalysis make it the most powerful ultramicroanalytical tool for the biologist. Very small samples can be analysed, volume down to 6 to 12 orders of magnitude smaller than with usual ultra-microanalytical methods. As the sample is not destroyed during the analysis, any number of elements of the period table above boron can be analysed in the sample

providing their local concentration is greater than 10^{-4}M.
The smaller volume which can be analysed with the existing
equipment are less than one attoliter and the minimum
detectability of an element is of the order of $1 - 10^{-18}$g
(Shuman and Somlyo, 1976). There are no theoretical limitations for these limits to be scaled down.

Use of Electron Probe Analysis in Biology

The use of electron probe microanalysis in biology has
been reviewed recently (Coleman and Terepka, 1974; Gupta
et al., 1977; Lechene and Warner, 1977; Lechene, 1977).
Electron probe microanalysis had been applied to biology
very early after its development. It was used to study
the composition of hard tissues: bone (Brooks et al.,
1962), teeth or pathological inclusions (Boyde et al.,
1961). In these cases, the preparation of the sample was
very similar to the preparation of the geological hard
sample. However, the main problems for electron probe
analysis to give its full potential in biology was to
develop methods for sampling and preparing for analysis
samples which in the living state are characterized by
the presence of diffusible ions in compartments of very
different compositions and by the existence of very important ionic gradients over very short distances across
biological membranes.

Several kinds of biological samples can be analysed,
liquid droplet, isolated cells and biological tissues.
To each category corresponds a different sample size,
from the picoliter $(1 - 10^{-12}$L) to the attoliter
$(1 - 10^{-18}$L) range, and a general method of sample preparation.

The methods for preparing biological samples must
preserve the ionic gradients which exist between different biological compartments and must avoid adding to the
preparation over the elements present in vivo.

Liquid Droplet

Preparation

Liquid samples are collected by micropuncture with capillary glass micropipettes in canaliculi of living organisms using the techniques of microphysiology. Using calibrated volumetric micropipettes aliquot volumes in the picoliter range of samples and of standard solutions are freeze dried on an adequate support. On the same support, several hundreds of such spots can be prepared. Each spot is then excited in the microanalyser with the electron beam and the characteristic x-ray intensities are recorded and compared to those of the standards to obtain quantitation (Lechene, 1974). Such a preparation can be done routinely and for example seven elements of biological interests, Na, K, Ca, Mg, Cl, P, S can be analysed in picoliter volumes. The analytical times will be less than 3 hours for one hundred samples. Variations of this method have been proposed for using energy dispersive spectrometers attached to a scanning electron microscope (Quinton, 1975; Rick *et al.*, 1977).

Applications

Applications of electron probe analysis of liquid droplets are increasing in cell physiology. Indeed, one can analyse many more elements in much smaller samples than with any other method. Electron probe analysis has provided new data in many areas of physiology. In renal physiology, there are being studied: the sequential analysis along the nephron of the transports of major ions and their variations (Le Grimellec *et al.*, 1974); the localization and mechanism of action of parathyroid hormone on phosphate transport (Greger *et al.*, 1977); the mechanism of isotonic transport in the proximal tubule (Warner and Lechene, 1976). In reproductive biology: the microenvironment of the early embryo (Borland *et al.*,

1977); the mechanism of fluid accumulation in the blasto-
cyst (Biggers *et al.*, in press); the microenvironment
of the male gamete in the rete testis, the semineferous
tubules and along the epididymis. In auditory physiology:
the composition of the perilymph and the endolymph in
several species (Peterson *et al.*, in press). Other ongoing
projects study the fluid formation and composition in the
sweat gland and the bile formation in the bile canaliculi.
Liquid droplet analysis can also be applied to recognize
chemical elemental inhibitions in biochemical preparation (Cantley *et al.*, 1977).

Special Aspects

Electron probe microanalysis does not differentiate
between the chemical state of a given element. One analyses the sulfur, phosphorus etc. content of a sample and
not the sulfate, phosphate etc. concentration. In order
to differentiate between the different chemical forms of
a given element, methods of analytical chemistry have
been scaled down at the level of liquid droplet. One can
use precipitation techniques on picoliter volumes. After
specific precipitation, one can analyse the supernatant.
Or, by using freeze substitution on droplets, one can
eliminate the supernatant and analyse the precipitate
(Bonventre and Lechene, 1974). Such methods can be used
for example to differentiate between total sulfur and
organic sulfur, measured after having precipitated the
sulfate by adding Ba Cl_2 to the droplets. The sulfate
will precipitate under Ba SO_4. Sulfate will be deduced
by substracting organic sulfur from the measurement of
total sulfur. Sulfate can also be directly measured on
the Ba SO_4 precipitate.

Such precipitation methods could be applied to measure
with the electron probe organic compounds in picoliter
volume samples by precipitating quantitatively the organic
compound with a reagent containing an element foreign to

the sample which will be measured by x-ray analysis. In
special cases where the sample contains only one organic
compound, it can be measured directly by measuring the
carbon and/or nitrogen signal. For example, urea content
in renal micropuncture samples or raffinose content in
renal standing droplet experiments (Warner and Lechene,
1976) can be directly measured and correlated with the
ionic content measured simultaneously in the same samples.
A precipitation technique has been developed for urea
where, elegantly, the reagent is added simultaneously to
all droplets of samples after being dissolved in the oil
covering the droplets (Beeuwkes *et al.*, 1978).

Isolated Cell Analysis

Preparation of isolated cells use techniques which are
intermediate between the preparation of liquid droplets
and that of whole tissue analysis. A satisfactory preparation should provide a large number of cells, well separated from each other and with a minimum of contamination
by the medium in which the cells were suspended. The
best technique appears to spray the cells on a support
with low background (Lechene *et al.*, 1977). It is applied
to the study of the elemental distribution in human red
blood cells. Absolute quantitation can be obtained by
using standards made with cells which have been preloaded
with known elemental composition. However, such preparations do not eliminate entirely the possibility of contamination by the suspending medium, the risk of ionic
leak from the cells, or redistribution during drying and
the superposition of signals emitted from several compartments (for example nucleus and cytoplasm) when studying
systems more complex than human red blood cells. Analysis of the cells maintained frozen hydrated would avoid
such pitfalls. Suspending medium and cells could be
shock frozen and they could be cut and analysed either on
a bulk or on a thin preparation. Cutting could be made

by using the same techniques which are being developed
for whole tissue analysis.

Tissue Analysis

When analysing biological tissue, the aim is to measure
the chemical element content of contiguous compartments
in the biological tissue samples. One wants to know the
composition immediately outside a transporting cell, between the cells of an epithelia, in the cell nuclei, in
the different cell organelles, etc. The volumes to be
analysed are in the range of femtoliters [1.10^{-15} liter or
$(1\mu)^3$] to attoliters [1.10^{-18} liter or $(100 \text{ nm})^3$]. The
method used to prepare the tissue sample should avoid any
gain, loss, or redistribution of the elements to be analysed. Preservation of the chemical makeup is the primary
aim of the preparation even if it is at the expense of
the morphological appearance. Preparation for analysis of
elements which are free in the extracellular or intracellular water, which are submitted to important gradients
of concentration over short distances should avoid or
minimize any step favouring translocation by diffusion.
The preparations used in histology or electron microscopy, aimed at preserving morphological appearance and
at increasing contrast, are unsuitable for electron probe
microanalysis of diffusible elements.

The best technique presently available for preserving
the elemental distribution of free elements is to shock
freeze the samples. In order to avoid the formation of
large ice crystals with inhomogeneous distribution of the
solute content the eutectic temperature should be reached
with extreme rapidity. Ideally the cooling rate should be
faster than $1.10^4 °\text{K/s}$. After freezing, any physical change
to the sample should be avoided. The tissue should be
maintained under liquid nitrogen during the preparation
for analysis. At liquid nitrogen temperature, biological
tissue samples can be prepared as if they were a piece of

rock by exposing one side of a bulk sample following techniques which have been developed by mineralogists: cutting with a diamond blade or with a wire loop saw and grinding with diamond coated tools (Lechene *et al.*, 1975; Lechene and Bonventre, 1976). Analysis of intracellular organelles could make it necessary to cut thin (Coleman, 1976) or ultrathin sections (Somlyo *et al.*, 1977); cutting should be made at the lowest temperature possible ideally below -130°C. Once prepared, the sample should be transferred from the laboratory at atmospheric pressure to the column under a vacuum avoiding any rewarming and avoiding trapping any frost. Towards this aim, development of special tools, interlock chamber attached to the probe column and specimen stage cooled at or close to liquid nitrogen temperature are necessary. Excellent conditions for transfer are realized when the sample is maintained under liquid nitrogen in the interlock chamber, until the nitrogen is evacuated. Best cooling of the specimen stage is provided by directly circulating liquid nitrogen in the specimen holder within the probe column. Results from several laboratories indicate that maintaining the sample frozen hydrated is the only general method for electron probe microanalysis of diffusible elements in biological tissues (Lechene, 1977). Analysis of bulk tissue maintained frozen hydrated has revealed the longitudinal elemental gradient profile in antidiuretic or water diuretic rat kidneys, giving new insight on the mechanism of water conservation by the mammalian kidney.

Analysis of thin sections maintained frozen hydrated has been pioneered by the group of T.A. Hall in Cambridge, England. The analysis of ion distribution in and around the cells of malpighian tubules and in salivary glands provide the first measurements of concentration profile in the inter-cellular spaces of transporting tissues and is likely to open the way for important applications in cell physiology (Gupta *et al.*, 1977).

In certain special situations perhaps one could avoid
to maintain the sample frozen hydrated during the analysis by using freeze drying. Drying could permit satisfactory preparation, provided that the tissue is an ultra
thin section prepared at very low temperature (Somlyo *et
al.*, 1977) or that relatively large volumes are analysed
in an homogeneous tissue (Merriam *et al.*, 1975).

More conventional techniques of tissue sample preparation could be used for analysing the composition of pathological inclusions. Usual techniques of preparation of
samples for electron microscopy can preserve composition
and localization of insoluble inclusions (Stuve and Galle,
1970).

A special application of electron probe microanalysis
could be in cytochemistry. Cytochemical methods can be
validated by x-ray analysis (Beeuwkes and Rosen, 1975).
However, electron probe analysis holds the potential to
develop a new cytochemistry in which reaction products
will not need to be visible by being either coloured or
electron dense. Reaction products could be recognized,
although invisible, if they bear a chemical element tag
that can be analysed with the electron probe. Three very
important possibilities have been mentioned: (1) specific
new histochemical techniques could be used; (2) enzymatic
amplification could be used for increasing the amount of
reaction product, and (3) kinetics of accumulation of a
reaction product could be studied in a tissue sample
(Hale, 1962; Engel *et al.*, 1968). Moreover, several tags
could be analysed in the same preparation, as in a nuero-
anatomical study in which different dendrites could be
recognized in the same section after iontophoretic injections of different tagged markers in different cell
bodies (Kirkham *et al.*, 1975).

In conclusion, electron probe microanalysis is the
most powerful spectrometric method that the biologist can
use to analyse the entire elemental content in volumes

scaled down to 9 orders of magnitude smaller than with
any usual ultra-microanalytical method. Analysis of
liquid samples in the picoliter volume range is routinely
performed. Analysis of smaller volumes in biological
tissue *in situ* has been developed enough to become a most
important tool in the field of cell physiology.

References

Beaman, D.R., Isasi, J.A. (1971). *Mater. Res. Stand.* **11**, 8-78.
Beeuwkes, R. and Rosen, S. (1975). *J. Histochem. Cytochem.* **23**, 828-839.
Beeuwkes, R. III, Amberg, J.A. and Essandoh, L. (1977). *Kid. Int.* **12**, 438-442.
Biggers, J.D., Borland, R.M. and Lechene, C.P. (1978). *J. Physiol.* (In press).
Birks, L.S. (1971). 'Electron Probe Microanalysis', 2nd edn., p. 190. Wiley-Interscience, New York.
Bonventre, J.V. and Lechene, C. (1974). *Proc. Microbeam Anal. Soc., 9th Ann. Conf., Ottawa, Canada* 8A-8D.
Borland, R.M., Hazra, S., Biggers, J.D. and Lechene, C. (1977). *Biol. of Reprod.* **16**, 147.
Boyde, A., Switsur, V.R., Fearnhead, R.W. (1961). *J. Ultrastruct. Res.* **5**, 201-207.
Brooks, E.J., Tousimis, A.J., Birks, L.S. (1962). *J. Ultrastruct. Res.* **7**, 56-60.
Cantley, L.C., Josephson, L., Warner, R.R., Yanagisawa, M., Lechene, C. and Guidotti, G. (1977). *J. Biol. Chem.* **252**, 7421-7423.
Castaing, R. (1951). Ph.D. thesis, Univ. Paris, 92 pp.
Coleman, J.R. and Terepka, A.R. (1974). 'Principles and Techniques of Electron Microscopy' (M.A. Hayat, ed.), Vol. IV, pp. 159-207. Van Nostrand, New York.
Coleman, J.R. (1976). *Proc. Microbeam Anal. Soc., 11th Ann. Conf. Miami*, 58A-58H.
Engel, W.K., Resnick, J.S. and Martin, E. (1968). *J. Histochem. Cytochem.* **16**, 273-275.
Greger, R.F., Lang, F.C., Knox, F.G. and Lechene, C.P. (1977). *Am. J. Physiol.* **232**(3), F235-238.
Gupta, B.L., Hall, T.A. and Moreton, R.B. (1977). *In* 'Transports of Ions and Water in Animals' (B.L. Gupta, R.B. Moreton, J.L. Oschman and B.J. Wall, eds.), pp. 83-143, Academic Press, London.
Hale, A.J. (1962). *J. Cell Biol.* **15**, 427-435.
Kirkham, J.B., Goodman, L.J. and Chappell, R.L. (1975). *Brain Res.* **85**, 33-37.
Lechene, C.P. (1974). *In* 'Microprobe Analysis as Applied to Cells and Tissues' (T. Hall, P. Echlin and R. Kaufmann, eds.), pp. 351-368, Academic Press, London.
Lechene, C., Strunk, T., Warner, R.R. and Conty, C. (1975). *Proc. Microbeam Anal. Soc., 10th Ann. Conf., Las Vegas, Nevada*, **40A-40G**.
Lechene, C. and Bonventre, J.V. (1976). *Proc. Microbeam Anal. Soc., 11th Ann. Conf. Miami, Florida*, **61A-61G**.

Lechene, C. (1977). *Am. J. Physiol.* **1** (5), F391-396.

Lechene, C., Bronner, C. and Kirk, R.G. (1977). *J. Cell. Physiol.* **90**, 117-126.

Lechene, C. and Warner, R.R. (1977). *Ann. Rev. Biophys. Bioeng.* **6**, 57-85.

Le Grimellec, C., Roinel, N., Morel, F. (1974). *Pfleugers Arch.* **346**, 189-204.

Merriam, G.R., Naftolin, F. and Lechene, C. (1975). *Proc. Microbeam Anal. Soc., 10th Ann. Conf., Las Vegas, Nevada,* **42A-42F**.

Peterson, S.K., Firshkopf, L.S., Lechene, C., Oman, C.M. and Weiss, T.F. (1978). *J. Comp. Physiol.* (In press.)

Quinton, P.M. (1975). *Proc. 10th Ann. Conf. Microbeam Analysis Society, Las Vegas, Nevada* **50A-50B**.

Rick, R., Horster, M., Dorge, A. and Thurau, K. (1977). *Pflugers Arch.* **369**, 95-98.

Shuman, H. and Somlyo, A.P. (1976). *Proc. Natl. Acad. Sci. U.S.A.* **73**, 1193-1195.

Somlyo, A.V., Shuman, H. and Somlyo, A.P. (1977). *J. Cell. Biol.* **74**, 828-857.

Stuve, J. and Galle, P. (1970). *J. Cell. Biol.* **44**, 667-676.

Warner, R.R. and Lechene, C. (1976). *Proc. 11th Ann. Conf. Microbeam Anal. Soc.* **59A-59E**.

13 INTERACTION OF THE CHEMICAL ELEMENTS WITH BIOLOGICAL SYSTEMS

J.J.R. Fraústo da Silva

*Centro de Estudos de Química Estrutural das
Universidades de Lisboa, Portugal*

Introduction

Analysis of over 300 plant species among the existing 0.5
million and of about 200 among the probable 3 million
animal species have enabled the specialists in the field
to establish the number and identity of the chemical elements present in biological systems and to suggest those
which are to be considered as 'essential' for vegetable
and animal life. The limited number of species examined,
the difficulties of the analytical work (particularly for
elements present at trace level) and the lack of detailed
knowledge concerning the role of each element, make it
necessary to confirm the postulated 'essentiality' by
means of delicate tests, in which the plants or animals
are allowed to develop while being fed diets deficient in
the particular element considered. In several cases these
studies have indeed enabled not only the classification
of a certain element as 'essential' or 'not essential'
but also the evaluation of the effects of its deficiency.
At present, however, some elements such as As, Br or Sn,
among others, are still considered as 'possibly essential'
whereas others like Al, Ba, etc. are considered to have
no definite biological role, perhaps because they have
not yet been properly tested in adequate conditions or
because the requirements are so low that the uptake is

virtually undetectable. Measurements of the distribution
of the chemical elements in several organs or tissues
have also been carried out using a variety of techniques.
However, here again, the difficulties associated with the
preparation of element-free diets, with the analytical
work, the sometimes extremely low requirements for certain elements (e.g. for cobalt and molybdenum), and the
well proven possibilities of synergism or of antagonism
between certain elements, etc. leave room for considerable
doubt about the conclusions which can be drawn.

Note too that the classification referred to above
assumes a certain definition of essentiality; although
opinions may vary in detail some criteria have been proposed (Arnon, 1950; Cotzias, 1967) and these are now
commonly accepted. The 'state-of-the-art' as far as
essentiality is concerned is presented in standard works
which are given as general references in the present
article.

Although many of the observations concerning the essentiality of the chemical elements are indeed quite definite
(and the same can be said about their toxicity, which is
not a different but an entirely related problem) relatively little is known on the role of most chemical elements in biology. The relevance of such knowledge for two
major problems in today's society — nutrition and pollution — gave a new impact and generated a new interest in
questions related to the uptake and utilization of the
chemical elements by biological systems. This new field
of interdisciplinary research is calling the attention of
the inorganic chemists among other more traditional
specialists.

Here we propose to summarize what is known on the possible interactions of the chemical elements in biological
systems and to discuss in somewhat greater detail:

1. The uptake of the chemical elements by biological
systems and the mechanisms used to achieve selectivity

of uptake.

2. The physico-chemical basis of the competition of elements for biological molecules or structures.

3. The results of natural or artificial substitution of the normal elements in living tissues by other elements and this possible mechanism of toxic action.

Chemical Elements in Biological Systems

General Aspects

The biology of the chemical elements is largely the biochemistry of the light elements, atomic number less than 30, with a few exceptions (those of Mo, Se and I, and perhaps, As, Br and Sn). Eleven of these elements are present in appreciable amounts (four of them — H, O, C and N — correspond to 99% of the total number of atoms in the human body and the remaining seven — Na, K, Ca, Mg, S, P, Cl — correspond to 0.7% of that total). Besides these, at least sixteen other elements have been considered as essential for both animal and vegetable life, but are found only at the trace level. The following have been confirmed:

V, Mn, Cr, Fe, Co, Ni, Cu, Zn, Mo, Si, Se, F, I, As (?), Br (?), Sn (?)

It is possible, from what has been said above, that this list will increase in the future. Some 30 other elements do not meet the requirements for essentiality as far as studies have gone, but occur irregularly in living tissues and can be accumulated by certain species of plants or animals.

Among these Al, Ba, Bi, Ge, Li, Pb, Rb, Sr, Ti are worth mentioning, but the prevalent view is that these elements have been acquired through contact with the environment, either by physical processes (e.g. particles

ingested through respiration, contamination of skin,
hair and nails by dust or smoke, etc.) or by chemical
processes, that is competition with the 'normal' elements
for the binding sites of the biological system. It is
possible that if the 'abnormal' conditions in which this
competition now takes place become permanent, the biological systems will adapt themselves to these conditions
and change the (in principle) adverse effect of an element
into a beneficial one. They will then have started to
'use' the element. Actually, this is one of the general
principles of evolution, as discussed in another paper in
the present volume for a major but not always realised
case, that of oxygen. Hence, among other questions, it is
of obvious interest to discuss in greater detail the
problem of the competition between different elements for
the same site. Two different situations must be considered:

   a. when there is an excess of ligands relative to
those elements;

   b. when there is an excess of the elements relative to
the available ligands.

   We will return to this problem after summarizing some
aspects of the selectivity achieved by biological systems
in the uptake of the chemical elements they need.

The Uptake of the Chemical Elements

   Since the plants are the primary source of chemical
elements for animal life, we will deal mainly with them.

   The uptake of a certain element M from the soil involves a series of systems which are related as shown
below:

$$M - soil \xrightarrow{1} M - soil\ solution\ (near\ the\ surface)$$

$$2 \downarrow Diffusion$$

$$M - soil\ solution\ (near\ the\ roots)$$

$$3 \downarrow$$

$$ML - Plant\ ligand\ (external\ or\ at\ the\ surface)$$

(with branch labeled $1'$ from $M - soil$ to $ML - Plant\ ligand$)

$$4 \downarrow \quad ML + L_1 = ML_1 + L$$

$$ML_1 - deeper\ into\ a\ cell$$

$$5 \downarrow \quad ML_1 + L_2 = ML_2 + L_1$$

$$ML_2 - final\ site$$

Not all these reactions will come to rapid equilibrium and we shall have to deal with both thermodynamic and kinetic control of both acid/base and redox type equilibria in the overall uptake. Step 1 is determined, essentially, by the solubility of the mineral containing the element M in a soil solution containing natural complexing agents. In some cases step 1' is operative, when the plant sends out a complexing agent which solubilizes and captures the necessary element. Step 2 is a diffusional process and depends on the nature of the chemical species containing M, the distance to the plant roots, and the nature (physical and chemical) of the soils. Steps 3, 4 and 5 are under thermodynamic and kinetic biological control; particularly, step 5 may be a kinetic device to ensure irreversible uptake of a certain element. As one might expect, in some of these steps oxidation-reduction reactions occur and this must be the case for Fe^{3+} and MoO_4^{2-}, as it is for NO_3^- and SO_4^{2-}.

One point to note in such a scheme is the possible reasons for the selection of less than thirty biological elements. They are the combined result of redundancy in chemical properties of the elements, abundance and availability of the chemical elements (Frausto da Silva and Williams, 1976). Most of the unrequired elements have

been excluded either because they were not as abundant
or because they were more difficult to extract from their
sources, mainly for reasons of solubility of the naturally
occurring minerals, than were the presently required
elements. The rejected elements either have no biological
role or can be deleterious to life. If they are made
relatively abundant and available to the plant (or animal),
e.g. because of accidental contamination or environmental
pollution, they may compete successfully with the essential elements and they may be preferentially captured.
Indeed, the uptake mechanisms are not truly specific,
at least in their primary phase, and given the possibility
of competition, selection can, to a marked extent, be
based on the concentration of the various elements near
the plant roots, as the affinity of elements for the
common binding centres (carboxylate, phosphate, sulphate,
etc.) does not vary much. The values of binding constants
for the complexes of the alkaline-earth and transition
metals with such ligands (Table I) shows little selectivity. Hence, many plants can concentrate elements which
are not considered 'biological' and, in some cases, these
can be toxic. However that toxicity depends on the concentration level and all the elements, even the essential
ones, can be toxic if the uptake is excessive. We will
come to this subject again, later.

The case of animals is not very different from that of
the plants in the basic aspects; the principles are analogous even if the conditions and detailed mechanisms vary
and a much finer degree of selectivity and control is
achieved in the higher species. Accordingly, in the following sections, we will indiscriminately deal with 'biological systems' in general, occasionally restricting the
considerations to one case or another.

TABLE I

Stability constants (log K_{ML}) of alkaline-earth and first series transition metal ions with several carboxylate, phosphate and sulphate ligands (T = 20 or 25°C)

Ligands	Log K_{ML}									
	Mg^{2+}	Ca^{2+}	Sr^{2+}	Ba^{2+}	Mn^{2+}	Fe^{2+}	Co^{2+}	Ni^{2+}	Cu^{2+}	Zn^{2+}
Acetate	1.25	1.24	1.19	1.15	1.40	-	1.56	1.43	2.23	1.57
Malonate	1.95	1.85	(1.25)	1.34	(2.3)	(2.5)	2.98	3.30	5.55	2.97
Phosphate	1.60	1.33	1.0	-	2.58	-	2.18	2.08	3.2	2.4
Methylphosphate	1.52	1.49	-	-	2.19	-	2.00	1.91	-	2.16
Sulphate	2.20	2.31	2.1	2.3	2.0	2.3	2.47	2.40	2.4	2.3
Thiosulphate	1.84	1.98	2.04	2.21	1.95	2.17	2.05	2.06	-	2.30

Source: Stability constants - The Chemical Society Special Publications no. 17 and 25, London, 1964, 1971.

Mechanisms of Uptake - Basis of Selectivity

The uptake of a chemical element is strongly influenced and conditioned by the nature and composition of the media from which it is extracted and into which it is extracted. Thus, pH, chemical composition, redox potential, ionic strength and solvent play important roles, determining the chemical form in which the element to be extracted is found and its affinity for any liganding agent. On the otherhand, this agent may be in different forms depending on the pH of the media and on its pK_a; at pH = 7, ligands with $pK_a > 7$ will be nonionized, and, thus, favour the uptake of anions, and ligands with $pK_a <$ 7 will be ionized, favouring the uptake of cations. Here 'anions' does not imply non-metals and 'cations' does not mean necessarily metal ions; the same chemical element can be present in the media as positive, negative or neutral species. In sea water, at pH = 8, for instance, chromium is found as $Cr(OH)^{2+}$ 85% and CrO_2^- 14% and cadmium as $CdCl_2$ 50%, $CdCl^+$ 40% and $CdCl_3^-$ 6%. The problem of speciation, i.e., of identifying the nature of the species present in a certain medium, is a vital one and efforts are being made at present to establish analytical methods which enable their identification, since the mathematical calculations based on previously determined formation constants do not seem to give very reliable results (Stumm and Brauner, 1975). It is very important to recognise the fact that biological systems can have chosen to take up an element in a particular chemical form because this allows a higher degree of selectivity.

In any case, biology seems to have devised a series of processes to select the element it needs from among those available in a set of determined conditions. The subject has been discussed in a previous paper (Frausto da Silva and Williams, 1976a), and here we will summarize only a few major aspects. The main mechanisms of selection appear to be:

a. Selection by charge-type;

b. Selection by ion-sizes;

c. Selection by liganding atom;

d. Selection by preferential coordination geometry;

e. Selection by spin-pairing stabilization;

f. Selection by control of concentration;

g. Selection by kinetic control;

h. Selection by chemical reaction.

Below, we give some examples of the possible occurrence of these various types of mechanisms:

a. Selection by charge type The elements of the first transition series are normally found in oxidation states $\geqslant$ +3 from Sc to Fe and +2 from Co to Zn. Hence these two groups can easily be differentiated; the appropriate ligands for the first group will be small first row anions such as $\diagdown N^-$, RO^-, F^-. RS^- ligands cannot be used externally since they are oxidised to RS-SR or other species at E^0 = 0.8V. Since F^- is unlikely to be involved in capture devices and the concentrations of $\diagdown N^-$ will be small at pH = 7, the main ligand used is RO^- which occurs in enolates, phenolates, etc. This is the uptake system for Fe^{3+} and the same ligands may well take up Mn^{3+} and Cr^{3+} complexes but in the latter case there is still no clear evidence for Cr(III) phenolates in biology.

b. Ion sizes The order of stability constants of different metal complexes should be that of the ionic potential of the metal ions if covalence, steric hindrance, and other specific factors did not intervene. Hence for the ionic complexes of the alkali and alkaline-earth metals one would expect the following orders to be followed

$$Li^+ > Na^+ > K^+ > Cs^+ > Rb^+$$

$$Be^{2+} > Mg^{2+} > Ca^{2+} > Sr^{2+} > Ba^{2+}$$

These orders are seldom found and many inversions in order between pairs of elements occur depending mainly on the type of ligand used and its basicity. The simplest reason for the inversion of an order is of a stereochemical nature, particularly with polydentate ligands and the smallest metal ions, e.g. the case of Mg^{2+} - EGTA compared with Ca^{2+} - EGTA. More generally, however, they are the result of energy balances involving hydration energies of the metal ions and the ligands to which we turn later.

For the anions, the position is quite similar. F^-, Cl^-, Br^- and I^- should follow a normal order of decreasing stability in their *ionic* complexes, but inversions are usually found, which can also be ascribed to balances involving hydration energies and stereochemical factors involving polydentate binding. Thus it has been possible to synthesise ligands which are so tailored that they strongly favour the complexation of one ion on the basis of its size — namely the 'crown-ethers' and the macrocyclic ligands. Unfortunately, there is no evidence for the occurrence of such ligands in biology in higher systems although they are well documented in bacterial antibiotics.

A curious recent example of differentiation between two similar ions of different size is that of the use of 1, 4, 7, 10-tetraazacyclotridecane- N,N',N'',N'''-tetracetic acid (as I), which gives complexes with Ca^{2+} and Sr^{2+} of stability: Sr^{2+} (log K_{ML} = 11.70) > Ca^{2+} (log K_{ML} = 8.06). Previously for about 15/20 years ligands capable of decontaminating ^{90}Sr by complexation, without affecting Ca^{2+}, were looked for without much success (see Stetter and Frank, 1976).

$$\text{HOOC CH}_2 \diagdown \underset{\text{N}}{\diagup} (\text{CH}_2)_m \diagdown \underset{\text{N}}{\diagup} \diagdown \text{CH}_2 \text{ COOH}$$

$$\text{HOOC CH}_2 \diagup \underset{\text{N}}{\diagdown} (\text{CH}_2)_m \diagup \underset{\text{N}}{\diagdown} \diagup \text{CH}_2 \text{ COOH}$$

I

1,4,7,10-tetraazacyclotridecane-
N,N',N'',N'''-tetracetic acid

In more generic terms, however, and although some
proteins are quite mobile and flexible molecules, such
'holes' can be simulated by an adequate positioning of
liganding groups when the protein is folded. In this way
the complexation of an ion with a specific size may be
favoured. As one would expect the 'holes' or 'sites' for
cations and for anions must be different; whereas for
cations they should be lined with groups such as R_2O,
ROH, R_2S, RCOOH, R_3N, etc., some of them ionised, for the
anions they must be lined with positive, neutral or
hydrogen-bond forming groups. They can also be formed by
hydrophobic groups but these must surround a buried
charge (such as a metal ion or an ammonium group) to be
able to pick up anions. Finally whereas almost all cations
are spherical monatonic ions many anions have a shape,
compare sulphate and carbonate, so that holes can be
designed to fit these shapes.

c. *Selection by liganding atom* This mechanism can be
used for metals and is based on the affinity of A-group
metal ions for ligands coordinating through oxygen atoms
whereas B-group metal ions prefer sulphur or nitrogen
ligands. This topic has been reviewed several times and
we will not discuss it in detail. Note only as an example
that Cu(II) has the remarkable ability to bind to ionized
peptide nitrogen atoms, contrarily to other metal ions,
e.g. Zn^{2+}. This effect alone can favour the uptake of

Cu^{2+} relatively to Zn^{2+} by a factor perhaps as great as 10^4.

The equivalent destination between anions can be based upon binding to different types of metal ion or on the basis of the different abilities to form hydrogen bonds.

d. Selection by preferential geometry   In certain cases uses can be made of the preference of certain metal ions for coordinating sites of a particular geometry. The cases of Cu^{2+} and Zn^{2+} can again be given as examples, as well as the cases of Mn^{3+} and Fe^{3+}. Thus, Cu^{2+} and Mn^{3+} prefer square planar (or distorted octahedral) ligand arrangements whereas Zn^{2+} and Fe^{3+} prefer tetrahedral sites. By combining these preferences with the different affinities of the metal ions for various types of ligands, such as phenolates, ionised peptide or thiolate groups, selective sites for each one of them can be designed which can operate effectively despite differences in availability. Superoxide dismutases offer examples for this kind of selection; the blue-green Cu^{2+}/Zn^{2+} superoxide dismutase has a zinc site with an approximately tetrahedral distribution of nitrogen ligands and a copper site with a distorted tetragonal hole with one ionised nitrogen of imidazole. For Mn^{3+} (and Fe^{3+}) the preferred ligands are the phenolates in tyrosines and thiolate groups; tetragonal sites will take up Mn^{3+} and tetrahedral or octahedral sites will preferably select Fe^{3+} — see Table II. Curiously all four metals are found in this one group of enzymes.

e. Selection by spin-pairing stabilization   For transition metal ions there is a further possibility to increase selectivity (which can be expressed by the energy change when a certain specific ion is taken up by the appropriate ligand relatively to another metal ion). We refer to the case when a spin-state change of the metal ion occurs on binding, giving a more perfect fit to the ligands. The spin-states will generally be as shown in Tables III & IV.

TABLE II

Selection by adequate combination of donor atom and geometry of site

Geometry of site	Donor atom	Phenolate RO⁻	Ionised Peptide N⁻	Thiolate RS⁻	Nitrogen -N
Tetrahedral		Fe^{3+}	-	Zn^{2+} Fe^{3+}	Zn^{2+}
Tetragonal		Mn^{3+}	Cu^{2+}	-	Cu^{2+}

TABLE III

Usual spin-state for some metal ions

	M^{2+}	M^{3+}
Mn	High-spin (d^5)	Usually high-spin (d^4)
Fe	Low-spin, high-spin (d^6)	Usually high-spin (d^5)
Co	Usually high-spin (d^7)	Low-spin (d^6)

TABLE IV

Cation	d-electrons	High-spin	Low-spin
Mn^{2+}	5	The vast majority	CN^-
Fe^{2+}	6	NH_3, H_2O, Cl^-	CN^-, NO_2^-, dipy, phen.
Co^{2+}	7	The vast majority	CN^-
Ni^{2+}	8	The vast majority	CN^-
Mn^{3+}	4	The vast majority, phen.	CN^-
Fe^{3+}	5	H_2O, F^-, Cl^-	CN^-, phen., dipy
Co^{3+}	6	F^-	All the other complexes

It is then to be expected that strong donor N or S ligands will bind Co^{3+} > Mn^{3+} > Fe^{3+} (and this may apply also with RO^-) while the order for divalent ions will be Fe^{2+} > Co^{2+} > Mn^{2+}. This may be the principle behind vitamin B_{12} selection for cobalt and, possibly, behind the selection of iron in preference to copper (which cannot give spin-changes) in cytochromes.

f. Selection by control of concentration There may be conditions in which sites which are not too specific will have to capture certain elements from among others which may be more abundant and available in order to generate a special chemical (biological) advantage. A possibility which is open and is certainly used by biological systems is that of 'sequestering' the undesirable elements using a secondary ligand, thus lowering the effective concentration of these elements. Disregarding acidity effects, i.e. assuming that the ligands do not lose or take up protons for the sake of simplicity, the situation can be represented by

$$M_1 + M_2 + L_1 + L_2 \rightarrow M_1L_1 + M_2L_2$$

$$M_1L_2 - \text{excluded by lack of } M_1$$
$$M_2L_1 - \text{excluded by lack of } M_2$$

The problem can be formulated in terms of 'conditional stability constants':

$$K'_{M_1L_1} = \frac{[M_1L_1]}{[M_1]^*[L_1]^*}$$

where $[M_1]^*$ is the concentration of M_1 *not bound* to L_1 and $[L_1]^*$ is the concentration of L_1 *not bound* to M_1.
    One can easily derive:

$$[M_1]^* = \alpha_{M_1} \cdot [M_1] = [M_1]\{1 + K_{M_1L_2}[L_2]\}$$

$$[L_1]^* = \alpha_{L_1} \cdot [L_1] = [L_1]\{1 + K_{M_2L_1}[M_2]\}$$

If one wants to avoid the binding of M_2 to L_1, we have to decrease $[M_2]$ to such an extent that the product $K_{M_2L_1} \cdot [M_2]$ becomes much smaller than say 0.01.
    In this case, since

$$[M_2] = \frac{[M_2L_2]}{[L_2] \cdot K_{M_2L_2}}$$

one obtains:

$$\frac{K_{M_2L_1}}{K_{M_2L_2}} \cdot \frac{[M_2L_2]}{[L_2]} < 10^{-2}$$

Hence, the regulating factors are the stability constants of the complexes of M_2 with both L_1 and L_2 and the relative proportions of M_2L_2 and free L_2. If L_2 is in excess relative to M_2, e.g. 10 times more L_2 is available than M_2, it is enough that $K_{M_2L_2} > 10\ K_{M_2L_1}$ to decrease M_2 to a level sufficient to decrease interference with M_1, down to 1%. To avoid binding of M_1 by L_2, the same reasoning can be applied and one obtains

$$\frac{K_{M_1L_2}}{K_{M_1L_2}} \cdot \frac{[M_1L_1]}{[L_1]} < 10^{-2}$$

and the problem is now related to the degree of 'saturation' of ligand L_1 by M_1.

For this type of control to be possible it is then necessary that there is an excess of ligands over metal ions, so that the concentration of each metal is controlled by its own binding site while the other metals are excluded.

We have made use of this argument (Frausto da Silva and Williams, 1976b) to advance a possible mechanism for the biological action of lithium. In the case of prolonged lithium therapy, the concentrations of the common alkaline and alkaline-earth metals in the cytoplasm of cells will be close to those given in Table V. Adenosine triphosphate (ATP) is also often present in a very comparable concentration range. Now, under these *concentration*

TABLE V

*Concentration of common alkali, alkaline-earth metals
and ATP in the cytoplasm of cells*

Species	Concentration (M)
Li^+	10^{-3}
Na^+	10^{-2}
K^+	10^{-1}
Mg^{2+}	$10^{-3} - 10^{-2.5}$
Ca^{2+}	10^{-7}
ATP	$10^{-3} - 10^{-2.5}$

conditions inside the cytoplasm of cells, lithium will be
selected preferentially to sodium and potassium by a
ligand if it forms stronger complexes with the ligand
by a factor of $10^2 - 10^3$, which is not too difficult to
imagine. It will be selected preferentially to calcium
by any complexing agent which gives with lithium a complex whose stability constant is not more than 10^3 smaller.
(Even if it is 10^5 smaller there will be appreciable competition by lithium for calcium sites since the calcium
concentration is so low.) Magnesium poses a different
problem, but since it gives rather strong complexes with
ATP, it may be sequestered to a point when lithium may
compete successfully for normal magnesium sites. These
sites could be (O^-, N, O^-, O^-) sites for magnesium which
could have evolved in proteins. Although the proposal is
rather speculative (for a more general discussion of the
role of lithium see Birch in the present volume) it is
probable that just such a situation arises in the internal
fluids of the adrenal chromaffin granule and in many other
vesicles, for example in the transmitter vesicles of nerve
cells, which have a high concentration of ATP. Thus it
could well happen that Li^+ is indeed challenging Mg^{2+} and
so affecting the metabolic rates of reactions requiring
the latter metal ion.

A more general example of control of concentration to achieve selectivity is that of the blood stream, where the metal ions are removed one after another by powerful complexing agents so as to avoid mutual interference. Thus, copper is removed by a special ligand containing an ionised peptide bond, iron is oxidised and stored as Fe.O.OH in ferrition, zinc is removed by zinc-albumin but Na^+, K^+, Mg^{2+} and, to some extent, Ca^{2+} are left free in the blood solution. If extra Ca^{2+} is required, it cannot travel as free Ca^{2+} in the blood since an increase in concentration would exceed the solubility products of its phosphate or carbonate. There are special calcium carrier proteins generated for the purpose. If it was not for the fact that Cu, Fe and Zn were removed, these would strongly bind the calcium sites and the process would be impossible.

g. *Selection by kinetic control* All the processes discussed so far are determined by thermodynamic factors, but, as we have mentioned in the initial considerations (see paragraph 2) several steps of uptake can be under kinetic control. Particularly, the insertion of a certain element into its final site may be a kinetic device to ensure irreversibility, see step 5 in paragraph 2.

We may have

$$M + L_1 \underset{}{\overset{K_{ML_1}}{\rightleftharpoons}} ML_1 \xrightarrow[+\ L_2]{k} ML_2 + L_1$$

Selectivity can then be based on *rates* of transfer 'k' from ML_1 to ML_2, which can be very different from the stability orders of K_{ML_1} or K_{ML_2}. We believe that the insertion of magnesium (of chlorophyll) into the chlorin ring for example, may be governed by such a device which prevents the formation of zinc-chlorophyll, which is more stable in thermodynamic terms. To achieve this end one needs a carrier protein L_1 so that MgL_1 can be recognised by L_2 but ZnL_1 is not. Then if MgL_2 is formed in an irreversible step this reaction will block the entrance of zinc and the insertion will be specific. It seems likely that the correct insertion of iron in protoporphyrin, cobalt in corrin and copper in uroporphyrin can only be brought about by similar controlling kinetic devices to achieve the purity observed in the different biological systems.

Although the processes involving chemical reactions can also be considered kinetic traps or devices, we will examine a few cases in a separate section.

  h. Selection by chemical reaction For non-metals one possible process of uptake involves change in oxidation state followed by covalent binding to an organic moiety. Common examples are:

$$Br^-, \ I^- \ \rightarrow \ Br^o, \ 1^o \ \rightarrow \ \text{bromo and iodo compounds;}$$
$$NO_3^- \ \rightarrow \ NH_3 \ + \ R\text{-}CHO \ \rightarrow \ RCH = NH \ \rightarrow \ RCH_2 - NH_2$$
$$SO_4^{2-} \ \rightarrow \ \text{intermediates} \ \rightarrow \ \text{thiols}$$

Fluorine and chlorine are usually handled as F^- and Cl^-; phosphate, silicate and borate can condense with hydroxy-groups forming semi-labile species, e.g.:

$$HPO_4^{2-} \ + \ HOR \ \rightarrow \ RPO_4^{2-} \ + \ H_2O$$

$$B(OH)_3 \ + \ \begin{matrix} HO\text{-}C \\ | \\ HO\text{-}C \end{matrix} \ \rightarrow \ HO\text{-}B \begin{matrix} O\text{-}C \\ | \\ O\text{-}C \end{matrix}$$

Similar reactions can occur for silicate. Examples of all cases are known, boron is found in boromycin and silicon in mucopolysacharides in states similar to those above.

Although many cases and certainly some other possibilities have been left out, we believe that most mechanisms of uptake are based on the series of processes already discussed. We will still devote a few paragraphs to the problem of hydration, which is of a major although not always realised importance. This is the subject of the next section.

Hydration and Binding of Cations and Anions in Biology

Leaving structural or other specific effects aside, the binding of cations and anions by biological molecules depends on the radii and charge of the ions and on the nature of the binding site, which conditions the energy of the corresponding interactions. However these bindings are measured against a background since both cations and anions in biological media are hydrated and the dielectric constant of the region of the binding site can vary considerably. Hence, the free energy change on binding of cations and anions to their respective sites from aqueous media must be a balance between the various types of site interactions and that of hydration and it is not surprising that this balance can generate orders which differ from those simply based on ionic sizes or potentials. To give an example, for the five alkali metal ions there have been found 11 different sequences of activity in various biological mechanisms among the possible 120 permutations. Similar different sequences have also been found for the halides F^-, Cl^-, Br^- and I^-, from the normal order of increasing size to the completely reversed order.

These effects can be explained in qualitative or semi-quantitative terms (in a few cases) by considering the effects of hydration of all species involved. It must be stressed that there is little biological specificity in the phenomenon; the problem is a general chemical one and the basic factors are the same as those which regulate processes such as the preferential precipitation of large cations by large anions (e.g. Cs^+ by ClO_4^- and Li^+ by F^-) or the inversions in stability orders for the complexes of the same groups of metals with a series of analogous ligands but of variable basicity. A full treatment of the problem requires the consideration of both enthalpic and entropic changes on binding of cations or anions to their respective sites. As the case of cations has been fully developed elsewhere we will concern ourselves mainly with the case of anions interacting with proteins, but the case of cations is analogous, except for the additional effects to be expected when dealing with transition metals, i.e. ligand field effects.

The net free energy change in binding to a site from aqueous solution can be written as

$$\Delta G_{binding} = \Delta G_{elect.} \text{ (+ any } \Delta G_{specific}) - \Delta G_{hyd}$$

Here $\Delta G_{binding}$ is the net free energy balance between the found (hydrated) state and the free hydrated forms, $\Delta G_{elect.}$ is the free energy change corresponding to the electrostatic interactions, $\Delta G_{specific}$ is the contribution to the free energy change derived from other interactions, e.g. covalent binding, hydrogen-bonding, stereochemical or other effects and ΔG_{Hyd} is the hydration energy of the free and bound species. All terms on the right hand side are calculated from gaseous free ion states. In the absence of specific effects we can write, using the Born equation for coulombic attraction between species of opposite charge,

$$\Delta G_{binding} = Z\left(\frac{A}{\varepsilon_1 d_1} - \frac{B}{\varepsilon_2 d_2}\right) + \frac{Bz'}{\varepsilon_2 d_3}$$

where Z is the charge of the anion, ε_1 is the dielectric constant of the protein (organic) solvent, ε_2 is the dielectric constant of water, d_1 is the equilibrium distance between the bound anions and the positive site in

the protein, d_2 is the distance between the anions and
the centre of the water dipole and d_3 is the distance
between the bound complex and the centre of the water
dipoles. A and B are constants — related to the charges
at the protein site and the dipole moment of water; z'
is the residual charge of the bound species. Hence, the
first term represents the difference in free energy
between the bound and free states of the interacting
species and the second term is the energy of residual
hydration of the complex.

Although the different terms are difficult to evaluate,
it can be seen, from the form of the equation, that different orders of $\Delta G_{binding}$ can be generated depending
on the relative values of the separated terms. Generally
speaking, hydration opposes binding if the sites are
hydrophobic and the less hydrated anions will be bound
preferentially. One obtains then the so called 'lyotropic'
or Hofmeister series (Voet, 1937), which is just the
reverse of the hydration free energy, i.e.

$$I^- > Br^- > Cl^- > F^-$$

$$ClO_3^- > BrO_3^- > IO_3^- > CO_3^{2-} > HPO_4^{2-}$$

These non-polar sites often contain histidine residues.
However, in other cases, the opposite order is found,
i.e., the more hydrated anions, like SO_4^{2-} or $P_2O_7^{4-}$, are
exactly those which are bound preferentially. (They are
often the substrates of corresponding enzymes.) This
occurs when the binding sites are polar and heavily hydrated. X-ray studies have shown that, quite often, these
regions contain arginine and lysine amino groups, which
also undergo extensive hydrogen-bonding to the anion.

To go deeper into the analysis of these observations,
one must consider the contributions of the enthalpy and
entropy changes to the overall free energy change. Small

and highly charged anions have high heats of hydration
$\Delta H_{hyd.}$ but very unfavourable entropies of hydration
$\Delta S_{hyd.}$; large and lowly charged anions have lower heats
of hydration but less unfavourable entropies of hydration.
When an ion combines with a protein both are partly (or
completely) dehydrated and the process can be favoured by
an entropic gain due to the freed water molecules, but
opposed by the breaking of the ion-water bonds, i.e. by
$\Delta H_{hyd.}$. However, since $\Delta Hyd. > T\Delta S_{hyd}$, the process is
dominated by the heat term; this means that for binding
highly hydrated anions (or cations) *overall favourable* ΔH
of binding is necessary, requiring one or more interacting charges in the protein (or covalent bonding, or extensive hydrogen bonding); single binding sites may not
be able to bind these anions at all. Hence the binding of
anions like SO_4^{2-} or $P_2O_7^{4-}$, as in kinases and dehydrogenases, requires several *positive* charges in the protein,
usually with a hydrogen bonding capability such as the
$-NH_3^+$ groups of arginine and lysine sometimes suitably
arranged for specific matching. Again hydrophylic sites,
where the species do not undergo extensive dehydration
and the bound species is solvated will then favour the
binding of these anions and the order of binding will
tend to be that of the decreasing ionic potentials (and
decreasing hydration energy); the higher the ionic potential the stronger the binding.

For large and lowly charged anions binding is not so
strongly opposed by the heat of hydration and they combine with hydrophobic sites following the lyotropic
series, i.e., the reverse of the hydration energies. One
such site can be generated by a 'buried' charge, e.g. a
metal ion, as in carbonic anhydrase, where Zn^{2+} is bound
to neutral ligands only, but lies in a deep narrow pocket
to which water has only limited access. We stress that
in this case, it is not the nature of the metal which
determines the order of binding (Pocker and Stone, 1967);

$$I^- > Br^- > Cl^- > F^-$$

$$CNO^- > SCN^- > N_3^- \gg ClO_4^- > NO_3^- > \text{Acetate}$$

since this is what one would expect for a (b) class metal which Zn^{2+} is not. It is the protein pocket which has changed the balance of energy changes involved, that is it is the degree of hydrophobicity of the site which regulates the order of binding.

For the alkali and alkaline-earth metal ions, changes in stability orders can also be explained on the basis of the interplay of the several coulombic interactions involving the ions, the ligands and water molecules. Ignoring the hydration of the final complex and adapting currently adapted expressions to calculate the hydration energies of these types of metals we can derive the following expression for the overall free energy change on binding to a biological molecule from aqueous solution, Phillips and Williams (1966)

$$\Delta G = - \frac{nB}{r_{eff} + r^-} + \frac{167}{r_{eff}}$$

where n is the coordination number of the metal, B is a constant determined by the charge of the site, r^- is the site radius of the ion plus a correcting additive term equal to 0.85 Å. This expression gives calculated values of ΔG as a function of n_{eff}, using nB and r^- as parameters and it can be shown that it generates the several orders found in the study of biological mechanisms, e.g. the selectivity for transfer across excitable membranes for the monovalent alkali cation series (Nobel, 1975). The first inversions, as the binding strength nB is increased, are generated in the selectivity order for the larger cations, as expected from their smaller energy of hydration. Another example where hydration effects play a role is that of the competition between calcium and magnesium,

but the situation here is complicated by a certain degree
of covalency in magnesium nitrogen bonds and by the demand
of this ion for sites with a well defined octahedral
symmetry. This problem has been extensively discussed
and reference can be made to early reviews (Williams, 1976,
1977).

Toxicity of the Chemical Elements

General Aspects

As has been mentioned above, all elements, including
the 'essential' ones can be toxic if the uptake is excessive. This situation can be represented by the graphs in
Fig. 1, where the two patterns of behaviour are self-
explanatory. In some cases, a decrease in concentration
of a certain essential element results in some beneficial
effect (e.g. growth in plants); this is known as the
Steenbjerg effect (Steenbjerg, 1951); see the right hand

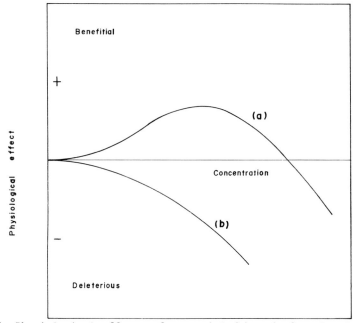

Fig. 1. Physiological effects of essential (a) and of toxic elements
(b) according to dosage.

side of Fig. 1(a). The same can be said for the common
toxic elements, some of which, when adequately used and
administered at *low* concentrations, can act as drugs and
have beneficial effects. Well known cases are those of
gold salts used against arthritis and platinum complexes
used against cancer. Arsenicals, antimonials, bismuth and
mercury salts have also been extensively used, but their
therapeutic action is based on their toxicity towards
some infectious agents.

The curves of figure 1 correspond to an idealised situation. For each element in one specific chemical form and
for each species of biological organism a different curve
will be obtained. To be even more precise, different
curves may be obtained for different individuals of a
given species and they can still vary for each individual
with the age and even the time and form of administration.
Cases are known of drugs tested for toxicity where results (normally expressed as LD_{50}, the dose which kills
50% of the population tested) vary enormously depending
on the hour at which the test was carried out. As a
consequence there cannot be a precise general classification of toxic elements nor of degrees of toxicity. Some
elements are very toxic for some species e.g. fungi and
quite harmless for others, e.g. angiosperms. One well
known example is sulphur.

The problem is, however, of vital importance today,
when the activity of man is bound to change, sometimes
drastically, the conditions in which he himself and
other species will live, and many studies are being
carried out to try to understand and control the effects
of the toxic elements, particularly in animals. It may
perhaps be said that under *normal* circumstances 'natural'
deficiencies constitute for plants a more serious problem
than 'natural' toxicity. For animals the reverse is more
generally true, since they are exposed to a much greater
variety of environmental conditions, feed on elements

already extracted and concentrated by plants or other animals and are consequently more susceptible to poisons.

Well known examples of the latter case are those of infant intoxication by high NO_3^- and NO_2^- content in vegetables (particularly spinach and lettuce) (Emerich, 1974) and of the poisoning of grazing cattle by plants with a high content of selenium or fluorine, Velu (1938).

Some elements are rapidly excreted, but some others can accumulate in the organisms and reach toxic levels after some time; cases of current interest are those of mercury, lead and cadmium, but this is very much a matter of 'fashion' determined by some particular circumstance. Other cases which were much studied earlier were those of plutonium, arsenicals and antimonials, etc. in connection with atomic energy in the case of Pu, wide use of compounds of As and Sb against syphilis and bacterial infections.

The problem is therefore an open one and even some elements, considered relatively harmless until recently, e.g. aluminium, are nowadays the subject of intensive observation (Burrows, 1977; Crapper *et al.*, 1976). For an inorganic chemist major interest centres on binding of particular metals to organic ligands and the competition between metal ions for different sites.

Competition between toxic and 'biological' elements

Due to their great abundance and availability there is not, for the main essential metals in normal conditions, a problem of competition with other elements. For *trace* elements the possibility of competition is obvious. Let us suppose that a biological system is faced with a choice of two or more elements M_1, M_2, etc. which are made available to it either due to 'natural' reasons or to fortuitous (or continuous) contamination. Two cases can then occur: a) The elements are supplied to the biological system in such conditions that the normal selection

controls are operative; this can happen if the 'ligands'
used by biological system are in excess over the elements
reaching them. b) The elements are supplied to the biological system in such conditions that they exceed the
amount of ligands available for their uptake.

We have already commented on the first case when discussing the mechanisms of uptake; in the latter case one
may write assuming just one ligand L:

$$\frac{[M_2L]}{[M_1L]} = \frac{K_{M_2L}[M_2]}{K_{M_1L}[M_1]}$$

and since the metals are in excess relative to the
ligands

$$[M_2] \sim C_{M_2}$$

$$[M_1] \sim C_{M_1}$$

Hence

$$\frac{[M_2L]}{[M_1L]} = \frac{K_{M_2L} \cdot C_{M_2}}{K_{M_1L} \cdot C_{M_1}}$$

i.e. the interference of M_2 with M_1 depends on the relative values of the products $K_{ML} \cdot C_M$ where the K_{ML} are the
stability constants of the M_1L and M_2L complexes and
C_{M_1} and C_{M_2} the total concentration of the two metals.

Assuming various tolerance limits for interference
without causing a toxic effect, one may represent the
results in graphical form - see Fig. 2. In this figure
one sees that if $K_{M_2L} \cdot C_{M_2}$ is to be made smaller than
$K_{M_1L} \cdot C_{M_1} \times 10^{-3}$, the concentration of M_2 has to be
smaller than 10^{-3} times that of M_1 if the affinities for
the ligand are of the same order of magnitude. Only with
more selective sites can this requirement be lowered.
Examples of both cases occur and the parallel between

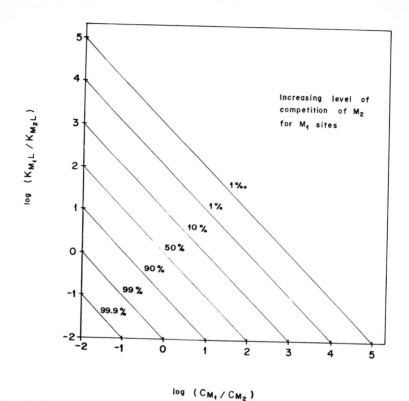

Fig. 2. Relationship between log (K_{M_1L}/K_{M_2L}) and log (C_{M_1}/C_{M_2}) to decrease the interference of M_2 for M_1 sites to the x% level.

toxicity and orders of stability of metal complexes, see below, can be explained on this basis.

Fig. 3 shows the effect of the donor atoms in bidentate ligands on the relative stability of their complexes with the first transition series metal ions. Whereas for (0,0) oxygen ligands the relative differences are small (and toxic effects caused by mutual interference will be determined mainly by concentration differences), for (N,N) nitrogen and sulphur ligands the relative differences in stability are more pronounced and this factor becomes more important. In biological systems when the various binding centres are usually more complex, this

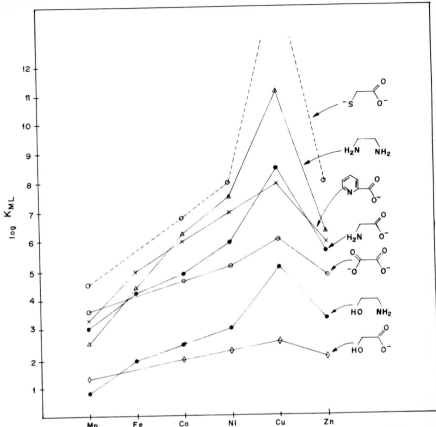

Fig. 3. Stability constants ($\log K_{ML}$) of the 1:1 complexes of first series transition metals with several bidentate ligands with different donor atoms.

effect may become even more pronounced but only by studying the relative affinities of those centres for the several metal ions can the results be predicted.

Mechanisms of toxic action

We must enquire now as to the possible sources of disturbance or imbalance when the toxic elements compete successfully with the normal biological elements.

Theoretically, one can forsee several possibilities:

a) Binding at centres not normally binding a metal.

Some of the toxic elements have a very high affinity for
ligands, e.g. mercury, and can bind under conditions where
none of the biologically active elements would bind.
Probably this is the most important case of toxicity. Its
details are straightforward.

b) Substitution of some essential elements, particularly metals in active sties of enzymes, by other elements, thereby hindering their functions or modifying
their activity. Probably this is the most important mechanism of the toxic action of many metals and it is interesting to note that the most deleterious of these metals
belong to Chatt's class (b) of acceptors; i.e., they are
heavy elements which form their most stable complexes
with the most polarisable donors, such as the thiolate
group. Consequently, elements like mercury and lead can
replace other metals, like iron, in enzymes containing
that donor in its active site, thereby blocking or altering their action. The same can be said for other common
toxic elements, such as silver, tin or arsenic. This possibility is not restricted to thiolate groups; metals like
silver, cadmium, copper, cobalt and nickel also have a
pronounced affinity for amino and imino groups, which are
also found in many active sties of enzymes and can compete with their 'natural' metals. Their effect will be so
much more pronounced and permanent the higher the affinity
for such groups and the more inert the species formed,
when cumulative effects are to be expected - see Table VI.
In many of these cases, the order of toxicity of the
elements follows the order of stability of their complexes, the order of the solubility products of their
sulphides and even the order of electronegativities (the
most toxic are those with highest electronegativity).
All those observations support the idea of a kind of competitive inhibition which is the essence of this type of
behaviour; the analytical treatment in Section 2 can be
applied as a rough guide to predict the effects to be

TABLE VI

Activities of Metallocarboxy-Peptidases

| Metal | Activity (v/v_{zinc} × 100) | |
	Peptidase*	Esterase**
Apoenzyme	0	0
Zinc	100	100
Cobalt	200	114
Nickel	47	43
Manganese	27	156
Cadmium	0	143
Mercury	0	86
Rhodium	0	71
Lead	0	57
Copper	0	0

*pH 7.5, 0 C (0.02 h benzyloxycarbonylshycyl-L-phenyladenine)
**pH 7.5, 25 C (0.01 M benzolyglycyl-DL-phenyllacetate).
Data from Coleman, J. and Vallee, B., *J. Biol. Chem.* **236**, 2244 (1961).
v is the velocity of the enzyme under specified conditions.

expected in simple cases.

   c) Binding of certain elements to radicals different from those in the active site of enzymes but resulting in changes of their conformation and reactivity. The biological molecules and, particularly, the enzymes contain many potential donors besides those of the active site which tend to complex certain elements, especially metal ions. Being very flexible molecules this can result in changes of conformation and reactivity and one may speak in this case of a kind of non-competitive inhibition, but in some cases there can be a stimulation of the enzyme function, see Table VII and Fig. 4 for various examples studied *in vitro*. It can be seen that both Al^{3+} and Zn^{2+} (at lower concentrations) stimulate the enzyme δ-aminolaevulinate hydrolase whereas Pb^{2+} depresses its activity; the effects of Al^{3+} and Zn^{2+} are additive suggesting that these metals interact with the enzyme in different sites (Meredith

TABLE VII

In vitro *effect of Al, Zn and Pb, separately and together,*
on erythrocyte δ-aminolaevulinate hydrolyase
(average of five experiments)

Metal in assay	Concentration	Control Activity %
Aluminium	1.85 mmol/1 Al^{5+}	242 ± 9.8
Zinc	0.20 mmol/1 Zn^{2+}	259 ± 18.4
Lead	1.86 mmol/1 Pb^{2+}	41 ± 8.6
Aluminium + Zinc	1.85 mmol/1 Al^{3+} + 0.20 mmol/1 Zn^{2+}	391 ± 20.4
Aluminium + Lead	1.85 mmol/1 Al^{3+} + 1.86 mmol/1 Pb^{2+}	173 ± 7.2
Zinc + Lead	0.20 mmol/1 Zn^{2+} + 1.86 mmol/1 Pb^{2+}	193 ± 7.2

From: P.A. Meredith, M.R. Moore and A. Goldberg (1977) *Enzyme* **22,**
22-27.

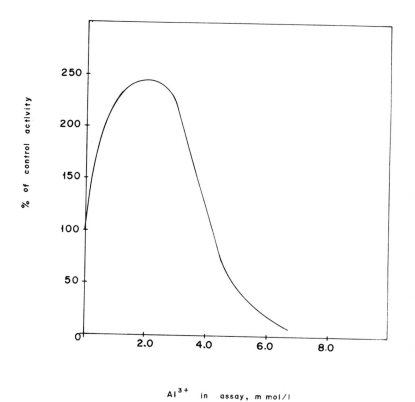

Fig. 4. Effect of Al^{3+} in the activity of erythrocyte δ-aminolevu-
linate hydrolyase.

et al., 1977). The figure illustrates the effect of concentration of the metal, Al^{3+} in this case (for Zn^{2+}, a similar behaviour is found); the curve is similar to that of Fig. 1, showing that there is an optimum concentration for maximum activity, but at higher concentrations the metal can block the enzyme completely.

d) Replacement of certain groups in biological molecules: certain polyatomic ions, like arsenate, silicate, selenate, tungstate etc. have dimensions, charge and shape which are similar to those of other groups, particularly phosphate. If this group is replaced by any of the previous ones anomalies will result which can be included among the general definition of toxicity - see Table VIII.

TABLE VIII

Effects of vanadate, arsenate, chromate and selenate on the uptake of phosphate by respiring mitochondria

Treatment	Relative Uptake
Control	15,295
1 mM VO_4	5,289
40 mM AsO_4	9,680
1 mM CrO_4	14,267
40 mM SeO_4	15,921

Each solution contained 4mM Pi.
C. Hill and G. Matrone (1970). *Fed. Proc.* **29**, 1434.

e) Complexation or precipitation of the metals of metallo-enzymes or other important elements or groups. The ingestion of certain complexing or precipitating agents can alter the metabolic reactions and cause toxic effects; this is the case of cyanide and carbon monoxide, which can bind to and block the iron atom in heme enzymes and proteins, and also the case of sulphide, which is a precipitating agent for many biological elements, thereby hindering their proper metabolic role. The toxic effects of elements like aluminium, beryllium, titanium, germanium, zirconium and others may also be due to the fact

that they can precipitate the phosphate group; the same
can be said for barium, which precipitates sulphate, and
for iron, which strongly complexes ATP. The toxicity of
these metals can be roughly correlated with the solubility
products or the stability constants of the corresponding
salts or complexes (Shaw, 1954, 1961).

f) Alterations in membrane permeability. Biological
membranes have important regulatory and control functions;
muscle action and nervous message transmission are, among
others, phenomena regulated by the membrane permeability
to ions. Hence, elements capable of changing this permeability can be toxic. Free radicals like $OH^{\cdot}$ (obtained
on reduction of O_2) or $Cl^{\cdot}$ (which may be formed from
chlorine covalent compounds) can even disrupt the membranes completely and originate serious diseases. Other
elements, e.g. Ag, Au, Cu, Hg, Pb, U, etc. can combine
with various groups in proteins and alter the permeability
of the membranes in which they are enclosed, thus dis-
turbing the normal metabolic functions based on such
mechanism.

g) Replacement of elements with electrochemical roles.
A final possibility, which is sometimes used for therapeutic purposes, is the replacement of certain ions which
fulfil osmotic or electrochemical roles in the cells.
We have already commented on the use of relatively high
doses of lithium salts (carbonate) in the therapy of
manic-depressive states, when this ion probably competes
with sodium and magnesium. Another well-known example
is that of bromide which can compete with chloride and
finds use as a mild sedative.

It should be stressed that the chemical form under
which the element is administered is vitally important;
to give an example, CN^{-} is an extremely toxic agent,
whereas HCO_3^{-} is quite harmless, although in both cases
we have the same element *carbon*. Metallo-organic compounds
also behave differently and are generally much more toxic

than simple inorganic or other organic compounds given their usually higher lipo-solubility. Well-known examples are those of alkyl-mercury compounds which are able to cross the brain-blood barrier and cause mental disturbances. On the contrary, certain metal compounds and complexes can be less toxic than the free metal ion, e.g. copper salicylaldoxime, the reason being the fact that the ligand competes with the biological molecules and effectively sequesters the toxic element.

The physical form of the toxic substance and the process of ingestion must also be considered; it is not indifferent whether the substance is given orally or by intravenous injection or whether it is in solution or as solid particles absorbed through the lungs.

All these possible variables make the study of toxicity an extremely complicated subject and many contradictory findings are reported in the literature. Here we have restricted ourselves to a few general considerations to stress the fact that both the uptake and utilisation of both the biological and the toxic chemical elements by the living beings are regulated by analogous physico-chemical principles, quite familiar to the inorganic chemists, who are therefore able to make useful contributions to this fascinating 'new' field of contemporary research.

References

General References

Bowen, H.J.M. (1966). 'Trace elements in biochemistry', Academic Press, London and New York.
Underwood, E.J. (1971). 'Trace elements in human and animal nutrition, 3rd edn. Academic Press, London and New York.
Hewitt, E.J. and Smith, T.A. (1975). 'Plant mineral nutrition. The English Universities Press, London.
Mills, C.F. and Livingstone, E.R.S. (eds.) (1970). 'Trace element metabolism in animals'. Edinburgh and London.
Hoekstra, W., Suttie, J.W., Ganther, H.E., Merz, W. (eds.) (1974). 'Trace element metabolism in animals - 2'. University Park Press, Baltimore, Maryland.
Mortvedt, J.J., Giordano, P.M., Lindsay, W.L. (eds.) (1972). 'Micronutrients in agriculture'. Soil Science Society of U.S.A.,

Madison, Wisconsin.
Carson, E.W. (1974). 'The plant root and its environment'. University
    Press of Virginia, Charlottesville.
Phipps, D.A. (1976). 'Metals and metabolism'. Clarendon Press, Oxford.

Specific References

Arnon, D.I. (1950). *Lotsya* **3**, 31.
Burrows, W.D. (1977). *CRC Crit. Rev. Envir. Control* **7**, 167.
Cotzias, G.C. (1967). 'Proc. First ann. Conf. Trace Substances
    Environ. Health' (D.D. Hemphill, ed.), Columbia, Missouri, p. 5.
Crapper, D.R., Krishnan, S.S. and Quittkat (1976). *Brain* **99**, 67.
Emerick, R.J. (1974). *Federation Proc.* **33**, 1183.
Fraústo da Silva, J.J.R. and Williams, R.J.P. (1976). *Structure and
    Bonding*, no. 29, pp. 67-121.
Fraústo da Silva, J.J.R. and Williams, R.J.P. (1976). *Nature* **263**, 237.
Meredith, P.A., Moore, M.R. and Goldberg, A. (1977). *Enzyme* **22**, 22.
Nobel, D. (1975). *In* 'Biological Membranes' (D.S. Parsons, ed.).
    Oxford University Press, Oxford, p. 133.
Phillips, C. and Williams, R.J.P. (1966). 'Inorganic Chemistry'.
    Oxford University Press, Oxford, Vol. I, p. 159.
Pocker, Y. and Stone, J.F. (1968). *Biochemistry* **6**, 686 (1967); **1**,
    2936.
Shaw, W.H.R. (1954). *Science, N.Y.* **120**, 361
Shaw, W.H.R. (1961). *Nature, London*, **192**, 754.
Steenbjerg, F. (1951). *Plant Soil* **3**, 97.
Stetter, H. and Frank, W. (1976). *Angew. Chem. Int. Ed. Engl.* **15**,
    686.
Stumm, W. and Brauner, P.A. (1975). *In* 'Chemical Oceanography'.
    (J.P. Riley and G.B. Kirrow, eds.), Academic Press, London.
Velu, H. (1938). *C.R. Soc. Biol.* **108**, 854.
Voet, A. (1937). *Chem. Rev.* **20**, 167.
Williams, R.J.P. (1976). *In* 'Calcium in Biological System Symposium
    XXX, Soc. Exp. Biol.', Cambridge University Press, Cambridge, p. 1.
Williams, R.J.P. (1977). *In* 'Calcium Binding Proteins and Calcium
    Function (R.H. Wasserman *et al.*, eds.). Elsevier North-Holland,
    Inc. Amsterdam.

SUBJECT INDEX

abundance of
  calcium, 266
  elements, 210, 449-
  phosphorus, 266
acrodermatitis enteropathica, 52
adenylate cyclase and lithium,
  421-6
affective disorders, 358, 389,
  392
ageing, 202, 235
alcohol dehydrogenase, 19, 28-32,
  50
alkali metals, 457
alkaline
  earths, 457
  phosphatase, 19, 25-33, 50, 295
aluminium, 479
alumino-silicate, 216
Alzeimer's disease, 237
amines, 427
amine oxidases, 147
analysis,
  biological tissues, 440
  bone, 280
  by electron microscopy, 437
  frozen hydrated cells, 443
  isolated cells, 440
  liquid droplet, 440
  lithium, 413
  microchemical methods, 14
  organic compounds, 443
  picoliter volumes, 440
  precipitation methods, 442
  preparation of samples, 441
  zinc, 13-20, 22
anions, uptake, 456
antioxidants, 233, 253
apatite, 287
arsenic, 259, 360, 449
arthritis, 201
association constants: conditional, 429
artherosclerosis, 243

atomic
  absorption, 14, 19, 413
  emission, 14, 438
  fluorescence, 14
ATP-ase and lithium, 429
autoxidation, 186

baleen, 318
barium salts, 356, 359
blood clotting, 196
bloodstream, 465
blue oxidases, 60-75
body
  fluids, 430
  weight, 416-19
bone, 261
  analysis, 280
  cells, calcification, 301
  composition, 277-80
  crystals, 281-5
  electron microscopy, 291
  epitaxy, 297
  evolution, 346
  formation, 293, 329-335
  function, 342
  metabolism, 290
  resorption, 303
  salts, nature of, 310
  silicon in, 227-31
  structure, 277-80
  ultrastructure, 288
  vesicles, 296, 312
  zinc, 25, 45
boron, 7, 466
bremstrahlung, 438

cadmium, 44
calcification, 229, 301
calcium, 5, 261
  abundance of, 266
  carbonate, 261
  in man, 25, 272
  phosphate,

calcium (*cont.*)
  in cells, 313
    clusters, 315
    exchange, 304
  resorption, 335
    in water, 267
    and zinc, 45-50
cancer, 40, 176, 370
carbon
  redox potentials, 130
  -selenium bonds, 255
carbonic anhydrase, 25-61
carboxypeptidase, 28-32, 479
catalase, 162
cations uptake, 456
cells,
  analysis, 441
  division, 34
ceruloplasmin, 60-7
chelating agents, 33
chromium, 450
chrysotherapy, 363-5
cirrhosis, 50
cobalt, redox potentials, 169
collagen, 231-40, 262
copper, 142, 180
  ascorbate oxidases, 60
  azurin, 60, 61, 64
  blue oxidases, 60, 62, 65, 67,
    69, 70, 71, 74, 75
  carbonic anhydrase, 61
  ceruloplasmin, 60, 61, 67
  enzymes, 60, 142, 167
  proteins, 60
  redox potentials, 62, 75, 169
  type 1, 62-75
  type 2, 65-75
  type 3, 67-75
cupreins, 188
cytochemistry, 446
cytochromes, 60, 75, 142, 151,
  179, 196

dentine, 280, 357
depression and lithium, 394
dermatan sulphate, 231
detoxification, 137-, 256
diamine oxidase, 60
diatoms, 223
dioxygen, 173-
dioxygenase, 148
disproportionation, 181
DNA
  and platinum, 370-

  and zinc, 38
dose/effect response, 3

elastase, 236, 244
electron
  affinity of oxygen, 176
  microscopy, 291, 437-
  probe microanalysis, 238, 438-
  transfer mechanisms, 75, 111
elements,
  biological systems, 449
  distribution, 450
  essential, 449
  selectivity, 450
  toxicity, 472
  uptake, 450, 452
emission spectra, 17
enamel of teeth, 280
entatic state, 27, 59, 168
enzymes,
  evolution, 122
  redox, 136-
EPR spectroscopy, 64-66, 86, 180
equisetum, 212-221
erythrocuprein, 61

fluoride in bone, 7, 269
formate dehydrogenase, 83, 102,
  258
fossil records, bone, 261
frozen cell analysis, 443

galactose oxidase, 60
gamma-carboxyglutamate, 197
genetics and zinc, 27
germanium, 223
glutathione, 258
  peroxidase, 257
glycine reductase, 258
gold salts, 363
  and arthritis, 363-
growth and zinc, 34-

Haber-Weiss reaction, 187
halogens, 7, 170, 458
hemocyanin, 60
heparan sulphate, 231
hepatocuprein, 60
hexosamines, 230
Hofmeister series, 469
hormones and zinc, 43, 48
hyaluronic acid, 231
hydrated radii, 400, 467
hydride, selenium, 255

hydrogen-bonding, 469
hydrophobic sites, 470
hydroxide complexes, 154
hydroxyapatite, 269
hydroxyl radicals, 71, 176, 187
hydroxylation, 144

indoleamine-2,3-dioxygenase, 196
inorganic medicine, 4
insulin and zinc, 43
ion size, selectivity, 457
ionophores, 404
iron, 26-48, 79-119, 137-70, 190

keratin, 318
kinetic control of selectivity, 465

laccase, 60-74
lead, 479
leucocytes and zinc, 25-
leukemia, 41, 49
lichens, 216
lithium, 389, 463
  absorption and distribution, 403
  chemical aspects, 398, 400
  clinical trials, 391
  complexes of, 401
  distribution
    in brain, 404
    in tissue, 407
  effects on endocrine systems, 420
  isotopes, 408-10, 412-14
  metabolic effects, 416
  mode of action, 429
  pharmacology, 405
  physical properties, 398-400
  poisoning, 391
  renal effects, 415
  side effects and toxicity, 397
  solution chemistry, 400, 463
  therapy, practical aspects, 395
lyotropic series, 469

macrophages, 239
magnesium, 5
  complexes, 401, 463
  enzymes, 425
  and lithium, 399, 424-9
  and zinc, 32, 50
manganese, 38, 51, 155, 164, 186
manic depressive, 392

mercury, 357, 360
metals in biology, survey, 6
metal substitution, 27-33
micro-wave plasmas, 18
mineralisation, 230, 242
mitochondria, 124
molybdenum
  enzymes, 79
  E.P.R., 86
  hydroxylases, 85
  -protein, *D. gigas*, 83, 86, 89, 111
monoamine oxidases, 60
monoxygenases, 179
mucopolysaccharides, 227-
myeloperoxidases, 197

neoplastic processes and zinc, 40
neutron activation analysis, 408-
nickel complexes, 368
nitrate reductase, 83-111, 152
nitrogen couples, 94, 156
NMR probes, 6
non-metals, survey, 7, 128
nutrition and zinc, 44-9

osteoblasts, 231
oxidation
  reactions, 140-
  -reduction potentials, 130-80
    copper, 64-75
    molybdenum, 105
    selenium, 257
  state diagrams, 130-180
oxocations, 143
oxogenases, 148
oxygen, 123
  and superoxide, 173
  binding, 138
  evolution, 155
  and metals, 137
  oxidation states, 127

pancreas and zinc, 25, 44
peroxidase, 162
peroxide, 124, 162, 182, 253
phagocytosis, 201, 239
pharmacology of lithium, 411
phenol oxidase, 60, 146
phosphate, 9
  in bone, 277
  metabolic pathways, 273
phosphorite, 350
phosphorus, 261

phosphorus (*cont.*)
  abundance of, 266
  in biology, 271
  in bone, 275, 336-41
  geology, 266
  in man, 272
  resorption, 335
  in rocks, 265-71
  in water, 267
photosynthesis, 155
phytoplankton, 212
plasma lithium, 396
plastocyanin, 60-7
platinum drugs, 370-
polysaccharides, 221-
porcine parakeratosis, 46
porphyrins, 43
potassium, 5-9
prostate and zinc, 24, 44
probes, NMR, 6
prothrombin, 196

quinones, 182

radicals, 73, 140-70
radiolysis of water, 187
redox reactions, *see* oxidation
reduction reactions, *see* oxidation
response curves, dose/effect, 3
reverse transcriptose, 19, 40
rheumatism, 243
rhodium drugs, 382-
RNA polymerases, 34, 45
  amoebae, 39
  Escherichi coli, 39
  Euglena gracilis, 19, 34, 36-40
  Helianthus tuberosus, 39
  liver, 39
  purification procedure, 36
  sea urchin, 39
  yeast, 19, 38
  ruthenium drugs, 368

schizoaffective disorders, 393
sections for EM, 445
selectivity of ion reactions, 465-
selenium, 102, 156, 253
  antioxidants, 233
  arsenic, 259
  and cancer, 253
  carbon-selenium bonds, 255
  detoxification, 256
  enzymes, 258-
  ethers, 255

  formate dehydrogenase, 258
  glutathione, 253
    peroxidase, 257
  glycinate reductase, 258
  hydride stability, 255
  oxidation reduction potentials:
      selenium, 257
  oxide stability, 255
  peroxides, 253
  reactions, 254
selenols, 255
selenoenzymes, 156
sequestration, 462
serum lithium, 414
  zinc, 25, 46
shells, 262
silica, 209-50
silicates, 209-16
silicon, 210-40
sodium, 5
  deprivation, 415
  intake, 415
  potassium
    A.T.P.-ases, 428
    pump, 428
  reabsorption, 415
  transport, 428
soil analysis, 209, 453
spirestomum ambiguum, 315
stability constants, 454-
stellacyanin, 60-65
steric hindrance, 460
superoxide, 173
  and carbon dioxide, 188
  dismutase, 53, 60
  disproportionation of, 181
  enzymes, 190
  and γ-carboxyglutamate, 197
  and quinones, 182
  and radicals, 187
  and vitamin K, 182
sulphite oxidase, 86, 152
sulphur-gold complexes, 363

teeth and silicon, 213
thermoneutron activation analy-
    sis, 88
tin, 449
tissue
  analysis, 444
  structure, 286
toxicity, 239, 450, 472-
  of zinc, 22, 53
transition metals: survey, 6